BAYESIAN ANALYSIS
OF LINEAR MODELS

STATISTICS: Textbooks and Monographs

A SERIES EDITED BY

D. B. OWEN, Coordinating Editor

*Department of Statistics
Southern Methodist University
Dallas, Texas*

Vol. 1: The Generalized Jacknife Statistic, *H. L. Gray and W. R. Schucany*
Vol. 2: Multivariate Analysis, *Anant M. Kshirsagar*
Vol. 3: Statistics and Society, *Walter T. Federer*
Vol. 4: Multivariate Analysis: A Selected and Abstracted Bibliography, 1957-1972, *Kocherlakota Subrahmaniam and Kathleen Subrahmaniam* (out of print)
Vol. 5: Design of Experiments: A Realistic Approach, *Virgil L. Anderson and Robert A. McLean*
Vol. 6: Statistical and Mathematical Aspects of Pollution Problems, *John W. Pratt*
Vol. 7: Introduction to Probability and Statistics (in two parts), Part I: Probability; Part II: Statistics, *Narayan C. Giri*
Vol. 8: Statistical Theory of the Analysis of Experimental Designs, *J. Ogawa*
Vol. 9: Statistical Techniques in Simulation (in two parts), *Jack P. C. Kleijnen*
Vol. 10: Data Quality Control and Editing, *Joseph I. Naus* (out of print)
Vol. 11: Cost of Living Index Numbers: Practice, Precision, and Theory, *Kali S. Banerjee*
Vol. 12: Weighing Designs: For Chemistry, Medicine, Economics, Operations Research, Statistics, *Kali S. Banerjee*
Vol. 13: The Search for Oil: Some Statistical Methods and Techniques, *edited by D. B. Owen*
Vol. 14: Sample Size Choice: Charts for Experiments with Linear Models, *Robert E. Odeh and Martin Fox*
Vol. 15: Statistical Methods for Engineers and Scientists, *Robert M. Bethea, Benjamin S. Duran, and Thomas L. Boullion*
Vol. 16: Statistical Quality Control Methods, *Irving W. Burr*
Vol. 17: On the History of Statistics and Probability, *edited by D. B. Owen*
Vol. 18: Econometrics, *Peter Schmidt*
Vol. 19: Sufficient Statistics: Selected Contributions, *Vasant S. Huzurbazar (edited by Anant M. Kshirsagar)*
Vol. 20: Handbook of Statistical Distributions, *Jagdish K. Patel, C. H. Kapadia, and D. B. Owen*
Vol. 21: Case Studies in Sample Design, *A. C. Rosander*
Vol. 22: Pocket Book of Statistical Tables, *compiled by R. E. Odeh, D. B. Owen, Z. W. Birnbaum, and L. Fisher*
Vol. 23: The Information in Contingency Tables, *D. V. Gokhale and Solomon Kullback*
Vol. 24: Statistical Analysis of Reliability and Life-Testing Models: Theory and Methods, *Lee J. Bain*
Vol. 25: Elementary Statistical Quality Control, *Irving W. Burr*
Vol. 26: An Introduction to Probability and Statistics Using BASIC, *Richard A. Groeneveld*
Vol. 27: Basic Applied Statistics, *B. L. Raktoe and J. J. Hubert*
Vol. 28: A Primer in Probability, *Kathleen Subrahmaniam*
Vol. 29: Random Processes: A First Look, *R. Syski*

Vol. 30: Regression Methods: A Tool for Data Analysis, *Rudolf J. Freund and Paul D. Minton*
Vol. 31: Randomization Tests, *Eugene S. Edgington*
Vol. 32: Tables for Normal Tolerance Limits, Sampling Plans, and Screening, *Robert E. Odeh and D. B. Owen*
Vol. 33: Statistical Computing, *William J. Kennedy, Jr. and James E. Gentle*
Vol. 34: Regression Analysis and Its Application: A Data-Oriented Approach, *Richard F. Gunst and Robert L. Mason*
Vol. 35: Scientific Strategies to Save Your Life, *I. D. J. Bross*
Vol. 36: Statistics in the Pharmaceutical Industry, *edited by C. Ralph Buncher and Jia-Yeong Tsay*
Vol. 37: Sampling from a Finite Population, *J. Hajek*
Vol. 38: Statistical Modeling Techniques, *S. S. Shapiro*
Vol. 39: Statistical Theory and Inference in Research, *T. A. Bancroft and C.-P. Han*
Vol. 40: Handbook of the Normal Distribution, *Jagdish K. Patel and Campbell B. Read*
Vol. 41: Recent Advances in Regression Methods, *Hrishikesh D. Vinod and Aman Ullah*
Vol. 42: Acceptance Sampling in Quality Control, *Edward G. Schilling*
Vol. 43: The Randomized Clinical Trial and Therapeutic Decisions, *edited by Niels Tygstrup, John M. Lachin, and Erik Juhl*
Vol. 44: Regression Analysis of Survival Data in Cancer Chemotherapy, *Walter H. Carter, Jr., Galen L. Wampler, and Donald M. Stablein*
Vol. 45: A Course in Linear Models, *Anant M. Kshirsagar*
Vol. 46: Clinical Trials: Issues and Approaches, *edited by Stanley H. Shapiro and Thomas H. Louis*
Vol. 47: Statistical Analysis of DNA Sequence Data, *edited by B. S. Weir*
Vol. 48: Nonlinear Regression Modeling: A Unified Practical Approach, *David A. Ratkowsky*
Vol. 49: Attribute Sampling Plans, Tables of Tests and Confidence Limits for Proportions, *Robert E. Odeh and D. B. Owen*
Vol. 50: Experimental Design, Statistical Models, and Genetic Statistics, *edited by Klaus Hinkelmann*
Vol. 51: Statistical Methods for Cancer Studies, *edited by Richard G. Cornell*
Vol. 52: Practical Statistical Sampling for Auditors, *Arthur J. Wilburn*
Vol. 53: Statistical Signal Processing, *edited by Edward J. Wegman and James G. Smith*
Vol. 54: Self-Organizing Methods in Modeling: GMDH Type Algorithms, *edited by Stanley J. Farlow*
Vol. 55: Applied Factorial and Fractional Designs, *Robert A. McLean and Virgil L. Anderson*
Vol. 56: Design of Experiments: Ranking and Selection, *edited by Thomas J. Santner and Ajit C. Tamhane*
Vol. 57: Statistical Methods for Engineers and Scientists. Second Edition, Revised and Expanded, *Robert M. Bethea, Benjamin S. Duran, and Thomas L. Boullion*
Vol. 58: Ensemble Modeling: Inference from Small-Scale Properties to Large-Scale Systems, *Alan E. Gelfand and Crayton C. Walker*
Vol. 59: Computer Modeling for Business and Industry, *Bruce L. Bowerman and Richard T. O'Connell*
Vol. 60: Bayesian Analysis of Linear Models, *Lyle D. Broemeling*

OTHER VOLUMES IN PREPARATION

BAYESIAN ANALYSIS OF LINEAR MODELS

Lyle D. Broemeling
Oklahoma State University
Stillwater, Oklahoma

MARCEL DEKKER, INC.   New York and Basel

Library of Congress Cataloging in Publication Data

Broemeling, Lyle D., [date]
  Bayesian analysis of linear models.

  (Statistics, textbooks and monographs ; v. 60)
  Includes bibliographical references and index.
  1. Linear models (Statistics) 2. Bayesian statistical
decision theory. I. Title. II. Series.
QA276.B695 1985   519.5   84-19902
ISBN 0-8247-7230-X

COPYRIGHT © 1985 by MARCEL DEKKER, INC. ALL RIGHTS RESERVED

Neither this book nor any part may be reproduced or transmitted in
any form or by any means, electronic or mechanical, including photo-
copying, microfilming, and recording, or by any information storage
and retrieval system, without permission in writing from the publisher.

MARCEL DEKKER, INC.
270 Madison Avenue, New York, New York 10016

Current printing (last digit):
10 9 8 7 6 5 4 3 2 1

PRINTED IN THE UNITED STATES OF AMERICA

PREFACE

This book presents the basic theory of linear models from a Bayesian viewpoint. A course in linear model theory is usually required for graduate degrees in statistics and is the foundation for many other courses in the curriculum. In such a course the student is exposed to regression models, models for designed experiments, including fixed, mixed and random models, and perhaps some econometric and time series models.

The approach taken here is to introduce the reader to a wide variety of linear models, not only those most frequently used by statisticians but those often used by engineers and economists. This book is unique in that time series models such as autoregressive moving average processes are treated as linear models in the same way the general linear model is examined. Discrete-time linear dynamic systems are another class of linear models which are not often studied by statisticians; however, these systems are quite popular with communication and control engineers and are easily analyzed with Bayesian procedures.

Why the Bayesian approach? Because one may solve the inferential problems in the analysis of linear models with one tool, namely Bayes theorem. Instead of learning the large number of sampling theory techniques of estimation, hypothesis testing, and forecasting, one need only remember how to apply Bayes theorem.

Bayes statistics is now rapidly becoming accepted as a way to solve applied statistical problems and has several special features which combine to make it appealing for solving applied problems: First, it can be applied to a wide range of statistical and other problems; second, it allows the formal use of prior information, and thereby it offers a well-defined and straightforward way of analyzing any problem. Also, it assists the understanding and formulation of applied problems in statistical terms, and, finally, it faces squarely the problem of optimal action in the face of uncertainty.

This book is intended for second-year statistics graduate students who have completed introductory courses in probability and mathematical statistics, regression analysis, and the design of experiments. Graduate students of econometrics and communication engineering should also benefit, and the book should serve well as a reference for statisticians, econometricians, and engineers.

Many people have helped me to write this book, and I am indebted to former students Samir Shaarawy, Joaquin Diaz, Peyton Cook, David Moen, Diego Salazar, Muthyia Rajagopalan, Mohamed Al Mahmeed, Mohamed Gharraf, Margaret Land, Juanita Chin Choy, Albert Chi, Yusoff Abdullah, and Don Holbert.

Secretarial and clerical assistance was provided by Leroy Folks of Oklahoma State University and William B. Smith of Texas A & M University. Some of the topics in this book are a result of the research sponsored by the Office of Naval Research, Contract No. N000-82-K-0292.

Lyle D. Broemeling

CONTENTS

Preface iii
Introduction ix

1 BAYESIAN INFERENCE FOR THE GENERAL LINEAR MODEL 1

Introduction 1
The Parametric Inference Problem 1
Bayes Theorem 2
Prior Information 3
Posterior Analysis 5
An Example 7
Inferences for the General Linear Model 8
Predictive Analysis 14
Summary and Guide to the Literature 18
Exercises 21
References 22

2 LINEAR STATISTICAL MODELS AND BAYESIAN INFERENCE 25

Introduction 25
Linear Statistical Models 26
Bayesian Statistical Inference 39
Summary 58
Exercises 58
References 60

v

3 THE TRADITIONAL LINEAR MODELS — 65

Introduction — 65
Prior Information — 66
Normal Populations — 70
Linear Regression Models — 84
Multiple Linear Regression — 94
Nonlinear Regression Analysis — 104
Some Models for Designed Experiments — 117
The Analysis of Two-Factor Experiments — 123
The Analysis of Covariance — 134
Comments and Conclusions — 137
Exercises — 138
References — 140

4 THE MIXED MODEL — 143

Introduction — 143
The Prior Analysis — 145
The Posterior Analysis — 147
Approximations — 150
Approximation to the Posterior Distribution of σ^2 — 153
Approximation to the Posterior Distribution
 of the Variance Components — 154
Posterior Inferences — 155
Summary and Conclusions — 177
Exercises — 177
References — 178

5 TIME SERIES MODELS — 181

Introduction — 181
Autoregressive Processes — 183
Moving Average Models — 187
Autoregressive Moving Average Models — 196
The Identification Problem — 203
Other Time Series Models — 207
A Numerical Study of Autoregressive Processes
 Which Are Almost Stationary — 216
Exercises — 232
References — 234

6 LINEAR DYNAMIC SYSTEMS — 237

Introduction — 237
Discrete Time Linear Dynamic Systems — 238

CONTENTS vii

Estimation	239
Control Strategies	244
An Example	251
Nonlinear Dynamic Systems	267
Adaptive Estimation	274
An Example of Adaptive Estimation	294
Summary	302
Exercises	303
References	305

7 STRUCTURAL CHANGE IN LINEAR MODELS 307

Introduction	307
Shifting Normal Sequences	308
Structural Change in Linear Models	315
Detection of Structural Change	346
Structural Stability in Other Models	351
Summary and Comments	369
Exercises	370
References	371

8 MULTIVARIATE LINEAR MODELS 375

Introduction	375
Multivariate Regression Models	376
Multivariate Design Models	384
The Vector Autoregressive Process	389
Other Time Series Models	400
Multivariate Models with Structural Change	410
Comments and Conclusions	425
Exercises	426
References	428

9 LOOKING AHEAD 431

Introduction	431
A Review of the Bayesian Analysis of Linear Models	431
Future Research	434
Conclusions	438
References	438

APPENDIX

Introduction	441
The Univariate Normal	441
The Gamma	442

The Normal-Gamma	442
The Univariate t Distribution	443
The Multivariate Normal	443
The Wishart Distribution	444
The Normal-Wishart	444
The Multivariate t Distribution	445
The Poly-t Distribution	447
The Matrix t Distribution	448
Comments	449
References	449
Index	451

INTRODUCTION

There are many who do not accept Bayesian techniques because of the controversy surrounding the use of prior distributions and the rejection, by Bayesian doctrine, that the sampling distribution of statistics is relevant for making inferences about model parameters. See Tiao and Box (1973) for an account of this aspect of the controversy. They essentially say the sampling properties of Bayes estimators are irrelevant, since they refer to hypothetical repetitions of an experiment which in fact will not occur. Lindley (1971) rejects many sampling-theory techniques because they violate the likelihood principle and thus violate the axioms of utility and probability. For more about the advantages and disadvantages of Bayesian inference and other inferential theories, Barnett (1973) gives an unbiased view. Chapter 2 also discusses some of the more controversial topics of Bayesian inference.

Bayes theorem and its use in inference and prediction are introduced with a study of the general linear model. The general linear model is the first model to be encountered in the first linear model course, and from a non-Bayesian viewpoint, Graybill (1961) and Searle (1968) present the sampling theory viewpoint. The treatment given here is quite different, since one theorem on the posterior analysis provides one with a way to make all inferences (estimation and hypothesis testing) about the parameters of the model. The techniques of the prior, posterior, and predictive analyses presented in Chapter 1 will be repeated as each class of models is analyzed.

Chapter 2 gives a brief history of Bayesian inference and explains the subjective interpretation of probability and the implementation of prior knowledge. It is argued that the subjective interpretation of the probability distribution of a parameter is no more subjective than the frequency interpretation of the distribution of an observable random variable. Prior information about the parameters of a model is implemented by fitting the hyperparameters of a proper density either to past data or to prior values of the observations. This chapter also gives a brief discussion of how each of the various models is used in other areas of interest.

The general linear model includes as special cases the regression models and the models of designed experiments, and since they are so important, Chapter 3 gives a detailed account of how a Bayesian would use them.

The mixed linear model is studied in Chapter 4, in which the marginal posterior distribution of the variance components is derived. Using a normal approximation to the posterior distribution of the random effects, one is able to devise Bayes estimators of the primary parameters of the model.

The subject of time series analysis is often given separate treatment; however, with the Bayesian approach to statistical inference, one is able to study ARMA (autoregressive moving average) processes in much the same way one studies the general linear model. First autoregressive processes are considered, in which one is able to find the posterior distribution of the parameters and the predictive distribution of future observations. The marginal posterior distribution of the autoregression coefficients is a multivariate t, and the error precision has a gamma posterior distribution. The predictive distribution of future observations is shown to be a t distribution. The moving average class of models is more difficult to analyze, although the exact analysis of an MA(1) and ARMA(1,1) processes is reported. Other time series models which are studied are the regression model with autocorrelated errors and the so-called lagged variable model in econometrics. This chapter shows that the Bayesian approach to the study of time series models is in its infancy but that the approach will produce some interesting results.

Very few statisticians learn about the linear dynamic model and its use by communication engineers with monitoring and navigation problems. The model is called dynamic because the parameters of the model are always changing, from one observation to the next; thus it is natural to study such processes from a Bayesian approach.

There are several important problems in connection with a dynamic linear model. The first is estimation of the parameters of the system, and estimation consists of filtering, smoothing, and prediction. Filtering is estimating the current state (parameter) of the system, while smoothing is estimating the past states of the system. Prediction, unlike time series analysis, is estimating future states of the system. Chapter 6 develops the Bayesian way of estimating the parameters by finding the posterior distribution of the states of the system, and the Kalman filter is shown to be the mean of the posterior distribution of the current state.

Chapter 6 also treats other aspects of the linear dynamic model, including adaptive estimation and control.

INTRODUCTION

Chapter 7 is a complete study of structural change in linear models, in which structural change means some of the parameters of the model will change during the time the observations will be taken. In this type of system the parameters change only occasionally, whereas in linear dynamic systems they change with each additional observation. Thus models with structural change stand between the static class, such as the traditional linear models (regression models, for example), and linear dynamic models.

There are two ways to model structural change: One is with a shift point, and the other is with a transition function. Both are used in Chapter 7, with regression models and time series processes.

Most of the material in Chapter 8 on multivariate processes is given by Box and Tiao (1973) and Zellner (1971). Multivariate linear regression models, models for designed experiments, and multivariate autoregressive processes are introduced. The only new material is a section on structural change of a multivariate regression model.

Many examples illustrate the prior, predictive, and posterior analyses, which are the three main ingredients of a Bayesian inference.

The necessary statistical background is given in the Appendix, in which the following distributions are defined: the univariate normal, the gamma, the normal-gamma, the multivariate normal, the Wishart, the normal-Wishart, the multivariate t, the poly t, and the matrix t. If one learns the material in the Appendix, one will be able to understand the way a Bayesian makes statistical inferences about linear models.

The books by Box and Tiao (1973) and Zellner (1971) have given us the foundation for the Bayesian analysis of linear models; thus this book can best be thought of, first, as a review of the material in those books and, second, as a presentation of new material. This Preface and Chapters 1, 2, and 3 are mostly a review of the Bayesian techniques of estimation, hypothesis, testing, and forecasting as applied to the traditional populations, while the remaining chapters present material which is either new or an extension of existing work.

REFERENCES

Barnett, V. (1973). *Comparative Statistical Inference.* John Wiley and Sons, Inc., New York.

Box, G. E. P. and G. C. Tiao (1973). *Bayesian Inference in Statistical Analysis.* Addison-Wesley, Reading, Ma.

Graybill, F. A. (1961). *An Introduction to Linear Statistical Models.* McGraw-Hill, New York.

Lindley, D. V. (1971). *Bayesian Statistics, A Review,* Reg. Conf. Ser. Applied Math., 2. SIAM, Philadelphia.

Searle, S. R. (1968). *Linear Models*. John Wiley and Sons, Inc., New York.

Tiao, G. C. and G. E. P. Box (1973). "Some comments on Bayes estimators," The American Statistician, Vol. 27, No. 1, pp. 12-14.

Zellner, Arnold (1971). *An Introduction to Bayesian Inference in Econometrics*. John Wiley and Sons, Inc., New York.

BAYESIAN ANALYSIS
OF LINEAR MODELS

1
BAYESIAN INFERENCE FOR THE GENERAL LINEAR MODEL

INTRODUCTION

Many books begin with the general linear model and this is no exception. The general linear model includes many useful and interesting special cases. For example, independent normal sequences of random variables, simple and multiple regression models, models for designed experiments, and analysis of covariance models are all special cases of the general linear model which will be explained in more detail in the following chapters.

This chapter introduces the prior, posterior, and predictive analysis of the linear model and thereby lays the foundation for the remainder of the book where the analyses are reported with other models.

THE PARAMETRIC INFERENCE PROBLEM

Let θ be a $p \times 1$ vector of real parameters, $Y = (y_1, y_2, \ldots, y_n)'$ a $n \times 1$ vector of observations, X a $n \times p$ known design matrix. Then the general linear model is

$$Y = X\theta + e, \quad (1.1)$$

where $e \sim N(0, \tau^{-1} I_n)$, and τI_n is the precision matrix of e, which has covariance matrix $\sigma^2 I_n$, and $\sigma^2 = \tau^{-1} > 0$ is unknown.

This is the general linear model and our objective is to provide inferences for θ and τ when observing $s = (y_1, y_2, \ldots, y_n)$, where y_i is the i-th observation. The word inference is somewhat of a vague word, but it usually implies a procedure which extracts information about θ from the sample s.

Introductory statistics books explain inferential techniques in terms of point and interval estimation, tests of hypotheses, and forecasts.

For the Bayesian, all inferences are based on the posterior distribution of θ, which is given by Bayes theorem.

BAYES THEOREM

Suppose one's prior information about θ is represented by a probability density function $\xi(\theta,\tau)$, $\theta \in R^p$, $\tau > 0$, then Bayes theorem combines this information with the information contained in the sample. The likelihood function for θ and τ is

$$L(\theta,\tau|s) \propto \tau^{n/2} \exp -\frac{\tau}{2}(Y - X\theta)'(Y - X\theta), \qquad (1.2)$$

where $\theta \in R^p$ and $\tau > 0$, where $\propto$ means proportional to (as a function of τ and θ). The likelihood function is one's sample information about the parameters and is the conditional density function of the sample random variables given θ and τ.

Bayes theorem gives the conditional density of θ given s

$$\xi(\theta,\tau|s) \propto L(\theta,\tau/s)\xi(\theta,\tau), \quad \theta \in R^p, \quad \tau > 0. \qquad (1.3)$$

The posterior density of θ is $\xi(\theta,\tau|s)$ and represents one's knowledge of θ and τ after observing the sample s. On the other hand our information about θ and τ before s is observed is contained in the prior density.

Note the posterior density (1.3) is written with a proportional symbol and ξ is used to denote both the prior and posterior densities. If one uses an equality sign, the posterior density is

$$\xi(\theta,\tau|s) = K \cdot L(\theta,\tau|s)\xi(\theta,\tau), \quad \theta \in R^p, \quad \tau > 0 \qquad (1.4)$$

where K is the normalizing constant and is given by

PRIOR INFORMATION

$$K^{-1} = \int_0^\infty \int_{R^p} L(\theta, \tau | s) \xi(\theta, \tau) \, d\theta \, d\tau,$$

which is the marginal probability density of Y.

Obviously it is easier to omit the normalizing constant and use the proportional symbol and this convention will be adhered to in this book.

Before continuing the analysis, one must find the posterior density of θ and τ.

PRIOR INFORMATION

The prior information about the parameters θ and τ is given in two ways. The first is when $\xi(\theta, \tau)$ is a normal-gamma prior density, namely,

$$\xi(\theta, \tau) = \xi_1(\theta|\tau) \xi_2(\tau), \quad \theta \in R^p, \quad \tau > 0, \tag{1.5}$$

where

$$\xi_1(\theta|\tau) \propto \tau^{p/2} \exp -\frac{\tau}{2}(\theta - \mu)' P(\theta - \mu), \quad \theta \in R^p \tag{1.6}$$

and μ is a p × 1 given vector and P is a known p × p positive definite matrix. Thus ξ_1 is the conditional density of θ given τ and is normal with mean vector μ and precision matrix τP.

The marginal prior density of τ is gamma with parameters $\alpha > 0$ and $\beta > 0$.

$$\xi_2(\tau) \propto \tau^{\alpha-1} \exp -\tau\beta, \quad \tau > 0. \tag{1.7}$$

What does this imply about the marginal prior density of θ? Since (1.5) is the joint prior density of θ and τ, the marginal density of θ is

$$\xi_1(\theta) \propto \int_0^\infty \xi(\theta, \tau) \, d\tau$$

$$\propto \int_0^\infty \tau^{((p+2\alpha)/2)-1} \exp -\frac{\tau}{2}[2\beta + (\theta - \mu)' P(\theta - \mu)] \, d\tau$$

$$\propto [2\beta + (\theta - \mu)' P(\theta - \mu)]^{-(p+2\alpha)/2}, \quad \theta \in R^p, \tag{1.8}$$

which is a t density with 2α degrees of freedom, location vector μ and precision matrix $(2\alpha)(2\beta)^{-1}P$.

By using the normal-gamma density as a prior for the parameters, one cannot stipulate one's prior information about θ separately from that of τ. The parameters of the marginal distribution of θ involve α and β, which are parameters of the prior distribution of τ, but the marginal prior density of τ does not involve parameters of the marginal of θ.

The parameter μ is one's prior mean for θ, while one's opinion of the correlation between the components of θ is given by $P^{-1}(2\beta)(2\alpha - 2)^{-1}$, which is the marginal prior dispersion matrix of θ. Since this involves α and β, the marginal prior information about τ depends on the choice of the dispersion or precision matrix of θ.

With regard to the information about τ, it is convenient to think of τ as the inverse of the residual variance σ^2 and that

$$E(\tau^{-1}) = \beta(\alpha - 1)^{-1}, \quad \alpha > 1 \tag{1.9}$$

$$\text{var}(\tau^{-1}) = \beta^2(\alpha - 1)^{-2}(\alpha - 2)^{-1}, \quad \alpha > 2 \tag{1.10}$$

or that

$$E(\tau) = \alpha/\beta \quad \text{and} \quad \text{var}(\tau) = \alpha/\beta^2.$$

These two equations together with

$$E(\theta) = \mu \tag{1.11}$$

and

$$D(\theta) = P^{-1}(2\beta)(n + 2\alpha - 2)^{-1}, \tag{1.12}$$

which are the mean vector and dispersion matrix of θ, will assist one in choosing the four hyperparameters for the prior distribution of θ and τ.

As will be seen, the normal-gamma prior density is a member of a conjugate class of distributions, that is the posterior density $\xi(\theta,\tau|s)$ of (2.3) is also a normal-gamma density. Conjugate families have the advantage that one has a scale by which to judge the amount of information added by the sample, beyond the amount given a priori.

Early writers such as Jeffreys (1959) and Lindley (1965) use improper densities for prior information. For example,

POSTERIOR ANALYSIS 5

$$\xi(\theta,\tau) \propto 1/\tau, \quad \theta \in R^p, \quad \tau > 0 \tag{1.13}$$

is called a vague non-informative prior density for θ and was developed by Jeffreys, because it satisfied certain rules of invariance and conveys "very little" information about the parameters. This density, although improper, produces a normal-gamma posterior density for θ and τ, which of course is proper. Thus, improper prior densities may result in proper posterior densities, but this does not hold in general. The Jeffreys prior implies that, a priori, θ and τ are independent and that θ has a constant density over R^p and that the marginal prior density of τ is $\xi_2(\tau) \propto 1/\tau$, $\tau > 0$.

Both forms (1.5) and (1.13) of prior information will be employed in the posterior analysis which follows.

POSTERIOR ANALYSIS

Using Bayes theorem given by (1.3), then using the normal-gamma prior density (1.6), the posterior density of θ and τ is

$$\xi(\theta,\tau|s) \propto \tau^{((n+2\alpha+p)/2)-1} \exp -\frac{\tau}{2} [2\beta + (\theta - \mu)'P(\theta - \mu)$$
$$+ (Y - X\theta)'(Y - X\theta)] \tag{1.14}$$

for $\theta \in R^p$ and $\tau > 0$.

Now, completing the square on θ gives

$$\xi(\theta,\tau|s) \propto \tau^{((n+2\alpha)/2)-1} \exp -\tau$$

$$\times \left[\beta + \frac{Y'Y - (X'Y + P\mu)'(X'X + P)^{-1}(X'Y + P\mu)}{2} \right]$$

$$\times \tau^{p/2} \exp -\frac{\tau}{2} [\theta - (X'X + P)^{-1}(X'Y + P\mu)]'(X'X + P)$$

$$\times [\theta - (X'X + P)^{-1}(X'Y + P\mu)]$$

which is a normal-gamma density, hence the marginal posterior density of τ is gamma with parameters

$$(n + 2\alpha)/2 \text{ and } \beta + \frac{Y'Y - (X'Y + P\mu)'(X'X + P)^{-1}(X'Y + P\mu)}{2}.$$

The marginal posterior density of θ is found by integrating (1.14) with respect to τ and yields

$$\xi_1(\theta|s) \propto \{2\beta + Y'Y - (X'Y + P\mu)'(X'X + P)^{-1}(X'Y + P\mu)$$
$$+ [\theta - (X'X + P)^{-1}(X'Y + P\mu)]'(X'X + P)$$
$$\times [\theta - (X'X + P)^{-1}(X'Y + P\mu)]\}^{-(n+2\alpha+p)/2}, \quad \theta \in R^p,$$
(1.15)

which is a p-dimensional t density with $n + 2\alpha$ degrees of freedom, location vector

$$\mu^* = (X'X + P)^{-1}(X'Y + P\mu) \quad (1.16)$$

and precision matrix

$$D^*(\theta|s) = (X'X + P)(n + 2\alpha)[2\beta + Y'Y - (X'Y + P\mu)'(X'X + P)^{-1}$$
$$\times (X'Y + P\mu)]^{-1}. \quad (1.17)$$

The reader is referred to the Appendix for properties of the normal-gamma, multivariate t and gamma distributions.

The marginal posterior moments of τ^{-1} are

$$E(\tau^{-1}|s) = [2\beta + Y'Y - (X'Y + P\mu)'(X'X + P)^{-1}$$
$$\times (X'Y + P\mu)](n + 2\alpha - 2)^{-1} \quad (1.18)$$

and

$$Var(\tau^{-1}|s) = E(\tau^{-1}|s)^2(n + 2\alpha - 4)^{-1}. \quad (1.19)$$

The posterior analysis of the general linear model reveals the joint distribution of θ and τ is a normal-gamma distribution, the marginal distribution of θ is a multivariate t, and the marginal of τ a gamma if the prior of the parameters is a normal-gamma.

What is the posterior analysis if the improper density $\xi(\theta,\tau) \propto 1/\tau$, $\theta \in R^p$, $\tau > 0$ is used as prior information? The following theorem should be verified.

Theorem 1.1. If θ and τ are the parameters of the general linear model (1.1) where X'X is of full rank and if the prior density of θ and τ is Jeffreys' improper density

AN EXAMPLE

$$\xi(\theta,\tau) \propto 1/\tau, \quad \tau > 0, \quad \theta \in R^p, \quad (1.20)$$

then the posterior density of θ and τ is normal-gamma where the marginal posterior density of τ is gamma with parameters $(n - p)/2$ and $[Y'Y - Y'X(X'X)^{-1}X'Y]/2$, and the conditional posterior density of θ given τ is normal with mean vector $(X'X)^{-1}X'Y$ and precision matrix $\tau X'X$. Also, the marginal posterior density of θ is a p-dimensional t distribution with $n - p$ degrees of freedom, location vector

$$E(\theta|s) = (X'X)^{-1}X'Y \quad (1.21)$$

and precision matrix

$$D^*(\theta|s) = (n - p)(X'X)[Y'Y - Y'X(X'X)^{-1}X'Y]^{-1}. \quad (1.22)$$

These results can be obtained by assuming the prior distribution of θ and τ is a normal-gamma with hyperparameters α, β, P, and μ, then letting $\beta \to 0$, $\alpha \to -p/2$, and $P \to 0(p \times p)$ in the joint posterior distribution of θ and τ. Note, this is not equivalent to letting $\beta \to 0$, $\alpha \to -p/2$, and $P \to 0(p \times p)$ in the joint prior distribution of θ and τ because the prior density would not be Jeffreys improper prior, (1.20).

If one uses the normal gamma density as a prior for θ and τ, $X'X$ may be singular, but if one uses Jeffreys improper prior, $X'X$ must be nonsingular, otherwise the posterior density of θ and τ is improper.

AN EXAMPLE

Consider a special case of the general linear model, where θ is a scalar ($p = 1$) and the design matrix is a $n \times 1$ column vector of ones, then

$$y_i = \theta + e_i \quad (1.23)$$

where the e_i are n.i.d. $(0, \tau^{-1})$, $\tau > 0$, $\theta \in R$.

The y_i, $i = 1, 2, \ldots, n$ represent a random sample of size n from a normal population with mean θ and precision τ. Suppose the prior information about θ and τ is a normal-gamma with parameters α, β, P, and μ, where α, β, and P are positive scalars and μ is any real number, then the marginal posterior density of θ is a univariate t with $n + 2\alpha$ degrees of freedom, mean $(n + P)^{-1}(\Sigma Y_i + P\mu)$ and precision $(n + P)(n + 2\alpha)$

$\times [2\beta + \Sigma Y_i^2 - (\Sigma Y_i + P\mu)^2(n+P)^{-1}]^{-1}$. The marginal posterior density of τ^{-1} is gamma with mean

$$E(\tau^{-1}|s) = [2\beta + \Sigma Y_i^2 - (\Sigma Y_i + P\mu)^2/(n+p)](n + 2\alpha - 2)^{-1},$$

and variance

$$Var(\tau^{-1}|s) = [E(\tau^{-1}|s)]^2 (n + 2\alpha - 4)^{-1}.$$

Using the improper Jeffreys prior density, the marginal posterior density of θ is a t with $n - 1$ degrees of freedom, mean $\bar{y}$ and precision $n(n-1)/\Sigma(Y_i - \bar{y})^2$, and the marginal posterior distribution of τ is gamma with parameters $(n-1)/2$ and $\Sigma(Y_i - \bar{y})^2/2$. The results are obvious from Theorem 1.1.

INFERENCES FOR THE GENERAL LINEAR MODEL

From the Bayesian viewpoint, all inferences are based on the joint posterior distribution of θ and τ, but what is meant by inference? Within the Bayesian approach, there are two ways to study the parameters. The first is called inference and consists of plotting the posterior distribution of the parameters and computing certain characteristics, such as the mean, variance, mode, and median of the distribution. Usually, one wants to estimate the parameters with point and interval estimates or test hypotheses about the parameters. These methods are to be explained and will be illustrated in the book.

The second way to study the parameters of the model is called Bayesian decision theory approach which utilizes a loss function $L(d,\phi)$, which measures the loss when d is the decision and ϕ is the value of the parameter. The loss d* which minimizes the average loss $EL(d,\phi)$ with respect to the posterior distribution of ϕ is called the Bayes decision.

This more formal method of studying the parameters will not be used, however the reader is referred to the end of the chapter, where the current literature is discussed.

Point Estimation

This section will present point and interval estimates of the parameters θ and τ of the general linear model. Joint estimates of θ and τ will be

INFERENCES

based on the joint posterior distribution of θ and τ. If τ is a nuisance parameter and θ is the parameter of interest, estimates of θ will be based on the marginal posterior distribution of θ. In the same way, if θ is the nuisance parameter and τ is of interest, estimates of the latter will be based on the marginal posterior distribution of τ.

Consider the joint estimation of θ and τ and the joint marginal distribution of θ and τ, assuming a normal-gamma prior density with hyperparameters α, β, P, and μ. Recall from the section on posterior analysis (pp. 5-7) that the joint posterior distribution of θ and τ is also normal-gamma with parameters

$$\alpha^* = (n + 2\alpha)/2$$

$$\beta^* = \frac{2\beta + Y'Y - (X'Y + P\mu)'(X'X + P)^{-1}(X'Y + P\mu)}{2}$$

$$P^* = (X'X + P)\tau \qquad (1.24)$$

and

$$\mu^* = (X'X + P)^{-1}(X'Y + P\mu).$$

If one estimates θ and τ jointly, what characteristic of the joint distribution should one use as an estimate? There is no definitive answer, but since θ is a normal random variable with mean μ^* and μ^* does not depend on τ, and since α^* and β^* do not depend on θ, then $[\mu^*, \beta^*(\alpha^* - 1)^{-1}]$ appears to be a reasonable choice with which to estimate (θ, τ^{-1}).

The marginal distribution of τ is $G[\alpha^*, \beta^*]$, hence the mean of τ^{-1} is

$$E(\tau^{-1}|s) = \beta^*(\alpha^* - 1)^{-1}$$

and its mode is

$$M(\tau^{-1}|s) = \beta^*(\alpha^* + 1)^{-1}.$$

Since the gamma distribution is asymmetric, whether one takes the mean or mode to estimate τ^{-1} is a matter of personal choice.

Suppose τ is a nuisance parameter and θ is to be estimated. The marginal mean vector of θ is μ^* and the dispersion matrix of θ is, from (1.17),

$$D(\theta|s) = (X'X + P)^{-1}(n + 2\alpha - 2)^{-1}[2\beta + Y'Y - (X'Y + P\mu)'$$
$$\times (X'X + P)^{-1}(X'Y + P\mu)]. \quad (1.25)$$

Since the posterior density of θ is symmetric, the mean, median, and mode coincide and μ^* is the natural estimate of θ.

Suppose θ is partitioned into two subvectors, θ_1 which is $p_1 \times 1$ and θ_2 which is $(p - p_1) \times 1$, where $\theta = (\theta_1', \theta_2')'$ and we want to estimate θ_1, ignoring θ_2 and τ. Using the properties of the t distribution (see the Appendix), the marginal posterior distribution of θ_1 is a p_1 dimensional t distribution with mean vector μ_1^* and precision matrix $D_1^*(\theta_1|s)$, where $\mu^* = (\mu_1^{*'}, \mu_2^{*'})'$, where μ_1^* is $p_1 \times 1$, and

$$D_1^*(\theta|s) = D_{11}^*(\theta|s) - D_{12}^*(\theta|s) D_{22}^{*-1}(\theta|s) D_{21}^*(\theta|s) \quad (1.26)$$

and

$$D^*(\theta|s) = \begin{bmatrix} D_{11}^*(\theta|s) & D_{12}^*(\theta|s) \\ D_{21}^*(\theta|s) & D_{22}^*(\theta|s) \end{bmatrix}, \quad (1.27)$$

where $D^*(\theta|s)$ is the precision matrix of θ, given by (1.17), and $D_{11}^*(\theta|s)$ is a $p_1 \times p_1$ matrix.

The natural estimator of θ_1 is the mean of the marginal posterior distribution of θ_1, which is μ_1^*.

If one uses a point estimate to estimate θ and τ, the mean of θ, μ^*, and either the mean or mode of τ will suffice, however these estimates are special characteristics of the joint posterior distribution of the parameters, hence they convey only partial information about these parameters. It is important to remember that the whole of the posterior distribution should be used to make inferences and that one should not be restricted to a few special properties of the distribution.

A distinctive feature of the Bayesian approach to inference is that one is not interested in the sampling properties of estimators since the posterior distribution of the parameters is conditional on the observed values s of the sample random variables y_i, $i = 1, 2, \ldots, n$. Future values of the estimators in hypothetical repetitions of the experiment are not relevant, and neither are the sampling distributions of the

INFERENCES

estimators, therefore the sampling properties of the estimators such as variance, mean square error, bias, correlation, etc., will not be pursued.

Nevertheless, there are interesting connections between those estimators which are employed by non-Bayesians, and certain moments of the posterior distribution of θ and τ, when one assumes a Jeffreys prior density.

For example, the "usual" estimate of θ is the least squares estimate $\hat{\theta} = (X'X)^{-1}X'Y$, which has a $N[\theta, \tau(X'X)]$ sampling distribution, but the marginal posterior distribution of θ is a t density with mean $\hat{\theta}$ and dispersion matrix $(X'X)^{-1}(n - p - 2)^{-1}(Y - X\hat{\theta})'(Y - X\hat{\theta})$. In other words, the least squares estimate is the mean vector of the marginal posterior distribution of θ and the dispersion matrix of θ is the estimated dispersion matrix (except for a constant) of the sampling distribution of $\hat{\theta}$. Also, the dispersion matrix of the posterior distribution of θ is a subjective measure of how the posterior probability of θ is concentrated about the mean $\hat{\theta}$, whereas the estimated dispersion matrix of the sampling distribution of $\hat{\theta}$ is a subjective measure of the variability of future values of $\hat{\theta}$ in hypothetical repetitions of the experiment. There is a close connection between the two ways of estimating θ, because on the one hand, probability is interpreted with regard to hypothetical repetitions, and on the other, posterior probability is interpreted in the subjective sense of confidence, that is in one's personal confidence, but either way the interpretation is subjective.

Using a proper prior for the parameters such as the normal-gamma will not produce estimators which are "close" (except in a limiting sense) to their sampling-theory counterparts, because the Bayesian estimators are functions of the hyperparameters α, β, P, and μ.

The early workers in Bayesian inference employed Jeffreys' prior distribution and the corresponding Bayes estimators were closely related in their sampling properties to the sampling theory counterparts, thus Bayes estimators were more or less tolerated. With the introduction of proper prior distributions, Bayes estimators were viewed with more suspicion since they differed more and more from the traditional estimators.

Interval and Region Estimation

In 1965, Box and Tiao defined regions of highest posterior density or HPD regions as they are called and used them to estimate the parameters θ and τ of the linear model. Their formulation refers to a posterior probability density $\xi(\phi|s)$, $\phi \in \Phi$ of some parameter ϕ, where $s = (y_1, y_2, \ldots, y_n)$ are the observed values of the sample random variables y_i, $i = 1, 2, \ldots, n$. A subset R of Φ is called a $1 - \gamma$, $0 < \gamma < 1$ HPD region for ϕ if

(i) $P[\phi \in R | s] = 1 - \gamma$
(ii) If $\phi_1 \in R$ and $\phi_2 \notin R$, then $\xi(\phi_1 | s) \geq \xi(\phi_2 | s)$.

From the definition, one may prove that (a) among all regions of Φ which have posterior probability $(1 - \gamma)$, R has minimum volume, (b) R is a unique $(1 - \gamma)$ HPD region if the posterior density $\xi(\phi | s)$ is not uniform over every region of the parameter space, and (c) if $\xi(\phi_1 | s) = \xi(\phi_2 | s)$, then either both ϕ_1 and ϕ_2 belong to R or both are not contained in R; also, the converse holds.

We now give the $(1 - \gamma)$ HPD region for θ, the parameter vector of the linear model.

Since the marginal distribution of θ is $t_p[\theta, n + 2\alpha, \mu^*, D^*(\theta | s)]$, i.e., a p-dimensional t distribution with $n + 2\alpha$ degrees of freedom, location vector μ^* and precision matrix $D^*(\theta | x)$, the random variable

$$F(\theta) = p^{-1}(\theta - \mu^*)'D(\theta|s)(\theta - \mu^*), \quad \theta \in R^p \quad (1.28)$$

has an F distribution with p and $n + 2\alpha$ degrees of freedom and a $1 - \gamma$ HPD region for θ is

$$C_{1-\gamma}(\theta) = \{\theta : F(\theta) \leq F_{\gamma; p, n+2\alpha}\} \quad (1.29)$$

To prove this the reader should refer to the properties of the t distribution and the definition of an HPD region.

Letting $\alpha \to -p/2$, $\beta \to 0$, and $P \to 0(p \times p)$ in the posterior distribution of θ is equivalent to using the Jeffreys prior density, thus under these conditions the $1 - \gamma$ HPD region for θ is given by (1.29) with the necessary adjustments.

An HPD region for τ, based on $\xi_2(\tau | s)$, which is gamma with parameters α^* and β^*, must be done numerically, because the gamma density is asymmetric. An approximate HPD region is easily found by using the chi-square tables. Note, since $(\tau | s) \sim G(\alpha^*, \beta^*)$, then $2\beta^*\tau \sim \chi^2(2\alpha^*)$, and from this one may find an interval estimate for τ and τ^{-1}.

Testing Hypotheses

The Neyman-Pearson approach is based on sampling theory and is the prevailing way to test hypotheses. Within the Bayesian framework, one

INFERENCES 13

has several techniques with which to test hypotheses about the parameters θ and τ of the linear model.

Consider a null hypothesis of the form H_0: $A\theta = b$ versus the alternative hypothesis H_1: $A\theta \neq b$, where θ is the vector of parameters in the linear model, A is a known m × k matrix, and b a known m × 1 vector. If the null hypothesis is true, then m linear functions of θ are determined to be given constants. Most hypotheses in regression analysis and those arising in the design of experiments are of this type and will be explained in later chapters.

The approach taken here is based on the HPD region for θ and was initiated by Lindley (1965) and Box and Tiao (1965). This approach is chosen because it produces the usual confidence region for θ when a Jeffreys' prior density is appropriate.

Suppose prior information for θ and τ is given by a normal-gamma density $\xi(\theta,\tau)$, (1.5), then the HPD region for θ of content $1 - \gamma$ was derived in the previous section. Since the null hypothesis H_0 is given in terms of $U(\theta) = A\theta$ (=b), we now derive a $1 - \gamma$ HPD region for $U(\theta)$ and reject the null hypothesis whenever b is not a member of the region. What is the HPD region for $U(\theta)$?

Now $U(\theta)$ is a linear function of θ and since θ has a t distribution with $n + 2\alpha$ degrees of freedom, mean μ^*, and precision matrix D^*, $U(\theta)$ also has a t distribution and the parameters are $n + 2\alpha$ degrees of freedom, mean $A\mu^*$, and m × m precision matrix $(AD^*A')^{-1}$, where A is chosen so that AD^*A' is nonsingular. If AD^*A' is singular the posterior distribution of $U(\theta)$ is not proper. The distribution of $U(\theta)$ is denoted by

$$U(\theta) \sim t_m[u(\theta); n + 2\alpha, A\mu^*, (AD^*A')^{-1}], \qquad (1.30)$$

where the m indicates an m-dimensional distribution.

Since $U(\theta)$ has a t distribution, the random variable

$$G[U(\theta)] = m^{-1}[U(\theta) - A\mu^*]'(AD^*A')^{-1}[U(\theta) - A\mu^*] \qquad (1.31)$$

has an F distribution with m and $n + 2\alpha$ degrees of freedom and a $1 - \gamma$ HPD region for $U(\theta) = U$ is given by

$$C_{1-\gamma}(u) = \{u: G(u) \leq F_{\gamma;m,n+2\alpha}\}, \qquad (1.32)$$

where $F_{\gamma;m,n+2\alpha}$ is the upper $100\gamma\%$ point of the F distribution with m and $n + 2\alpha$ degrees of freedom. The null hypothesis is rejected if $b \notin C_{1-\gamma}(u)$ or when $G(b) > F_{\gamma;m,n+2\alpha}$.

If the prior density is improper $\xi(\theta,\tau) \propto \tau^{-1}$ for $\theta \in R^p$ and $\tau > 0$, the $1 - \gamma$ HPD region for θ is the set of all θ, where $Q(\theta) = (\theta - \hat{\theta})'X'X(\theta - \hat{\theta}) \leq ps^2 F_{\gamma;p,n-p}$, where $\hat{\theta} = (X'X)^{-1}X'Y$ is the least squares estimate of θ and the estimate of σ^2 is $s^2 = (Y - X\hat{\theta})' \times (Y - X\hat{\theta})(n - p)^{-1}$. This is the standard confidence region of content $1 - \gamma$ for θ, and if the null hypothesis is $H_0: \theta = \theta_0$ and the alternative is $H_1: \theta \neq \theta_0$, then H_0 is rejected if $Q(\theta_0) > ps^2 F_{\gamma;p,n-p}$.

This method of testing hypotheses is not the only way and other tests are available. For example, one may use posterior odds ratios or a decision theory approach, when a loss function is available to describe the costs of making a wrong decision. DeGroot (1970) presents various Bayesian tests of hypotheses while Zellner (1971) has a good introduction to posterior odds ratios.

With the HPD approach, one must specify the hyperparameters μ, P, α, and β of the normal gamma prior density and this information must reflect the statistician's opinion about the parameters. One's opinion about θ and τ, when the null hypothesis is true, is perhaps different than when the alternative is true, so one must be careful in choosing the prior density.

PREDICTIVE ANALYSIS

Predictive analysis is the methodology that is developed in order to forecast future observations. Of course, forecasting is a very important activity in business and economics, where sophisticated time series techniques are used.

The Bayesian uses the so-called Bayesian predictive density to forecast future observations, so this will be defined in this section. There are many ways to predict observations, however the Bayesian approach is natural in that one bases one's prediction on the conditional distribution of the future given the past. In order to do this, one must treat the parameters of the model as random, that is, as will be seen, the joint distribution of the future observations and the parameters, given the past observations, is averaged over the posterior distribution of the parameters.

In this section the Bayesian predictive density is defined and illustrated with linear models, then in the later parts of the book the same approach will be employed with time series models.

Suppose $y_1, y_2, \ldots, y_n$ is a random sample from a normal population with mean θ and precision τ and that $w_1, w_2, \ldots, w_k$ are future observations, how does one forecast the values of the future observations?

PREDICTIVE ANALYSIS

The Bayesian predictive density is the conditional density of $w = (w_1, w_2, \ldots, w_k)'$ given the sample $s = (y_1, y_2, \ldots, y_n)'$. The density of w given θ, τ, and s is

$$f(w|\theta,\tau,s) \propto \tau^{k/2} \exp - \frac{\tau}{2} \sum_{i=1}^{k} (w_i - \theta)^2 \qquad (1.33)$$

where $w \in R^k$ and does not depend on s, since given θ and τ, w and s are independent.

Now suppose θ and τ have a prior normal-gamma density, say

$$\xi(\theta,\tau) \propto \tau^{\alpha-1} e^{-\tau\beta} \tau^{1/2} e^{-(\tau/2)q(\theta-\mu)^2}, \qquad \theta \in R, \quad \tau > 0, \quad (1.34)$$

where $\alpha > 0$, $\beta > 0$, $\mu \in R$, and $q > 0$ are the hyperparameters.

The product of the posterior density of θ and τ and (1.33) is the joint conditional density of w, θ, and τ given s, which when averaged with respect to θ and τ produces the Bayesian predictive density of w.

The posterior density of θ and τ is

$$\xi(\theta,\tau|s) \propto \tau^{(n+2\alpha)/2-1} \exp - \frac{\tau}{2}[2\beta + \Sigma(y_i - \bar{y})^2 + n(\theta - \bar{y})^2] \qquad (1.35)$$

for $\theta \in R$ and $\tau > 0$, and when this is multiplied by (1.33), gives

$$g(w,\theta,\tau|s) \propto \tau^{(n+2\alpha+k)/2-1} \exp - \frac{\tau}{2}\left[2\beta + \sum_{1}^{n}(y_i - \bar{y})^2 + n(\theta - \bar{y})^2 + \sum_{1}^{k}(w_i - \theta)^2\right] \qquad (1.36)$$

for $w \in R^k$, $\theta \in R$, and $\tau > 0$, which is to be integrated with respect to θ and τ over $R \times (0,\infty)$.

Completing the square on θ and integrating with respect to θ over R, then integrating (1.36) with regard to τ over $(0,\infty)$, then completing the square on w, results in

$$h(w|s) \propto \{[w - A^{-1}B]'A[w - A^{-1}B] + C - B'A^{-1}\}^{(n+2\alpha+k)/2},$$

$$w \in R^k \qquad (1.37)$$

as the predictive density of w, where A is a k × k matrix with diagonal elements $(n + k - 1)(n + k)^{-1}$ and off-diagonal elements $-(n + k)^{-1}$. The column vector B has elements $n\bar{y}(n + k)^{-1}$, thus we see one would forecast w with a k-dimensional t distribution with $n + 2\alpha$ degrees of freedom, location vector $A^{-1}B$, and precision matrix $(n + 2\alpha)A(C - B'A^{-1}B)^{-1}$, where $C = 2\beta + \Sigma(y_i - \bar{y})^2 + n\bar{y}^2 - n^2\bar{y}^2(n + k)^{-1}$.

It is important to remember one's prediction of w is to be based on the whole of the predictive distribution, that is to say one has available an entire distribution with which to forecast w. A point forecast of w is given by $A^{-1}B$, the mean of the distribution and the confidence or the forecast error of this prediction is given by the dispersion matrix of w namely $A^{-1}(C - B'A^{-1}B)(n + 2\alpha - 2)^{-1}$. Note, the degrees of freedom do not depend on the number of observations to be predicted but on the number of observations in the sample and the prior parameter α.

An ad hoc way to make a forecast about w_1, one future observation, is to treat $w_1 \sim n(\theta, \tau)$ and estimate θ by $\bar{y}$ and τ^{-1} by $(n - 1)^{-1} \Sigma_1^n (y_i - \bar{y})^2$, then predict w_1 with $\bar{y}$, since $\bar{y}$ is the mean of the conditional distribution of w_1, given θ and τ.

The same prediction arises from the Bayesian method if one uses a Jeffreys' vague prior density for θ and τ, and uses the mean $\bar{y}$ of the predictive density of w_1 given s; however the ad hoc method is based on a $n(\theta, \tau)$ distribution, while the Bayesian forecast is based on a t distribution; but the latter is equivalent to the former, when the number of observations is large, in which case the t distribution approaches the normal.

In the more general case of a general linear model, how does one forecast future observations w, where

$$w = Z\theta + \varepsilon, \tag{1.38}$$

w is a k × 1 vector of future observations, Z a known k × p matrix, θ the p × 1 parameter vector of the linear model

$$Y = X\theta + e \tag{1.1}$$

and ε a k × 1 vector of future errors?

The forecasting principle is the same here as in the previous example, one must find the conditional distribution of w given y, thus assuming ε and e are independent normal errors, where $\varepsilon \sim N(0, \tau^{-1}I_k)$

PREDICTIVE ANALYSIS

and using a normal-gamma prior density for θ and τ, namely (1.5), it is easy to derive the Bayesian predictive density of w. The previous example is a special case of forecasting w in the general linear model.

Suppose the prior density for θ and τ is the normal-gamma density (1.5) with hyperparameters $\alpha > 0$, $\beta > 0$, $\mu \in R^p$, and P, a positive definite $p \times p$ matrix, then the reader may verify that the predictive density of w is

$$g(w|y) \propto \{[w - A^{-1}B]'A[w - A^{-1}B] + C - B'A^{-1}B\}^{(n+k+2\alpha)/2},$$

$$w \in R^k, \quad (1.39)$$

where

$$A = I - Z(Z'Z + X'X + P)^{-1}Z',$$

$$B = Z(Z'Z + X'X + P)^{-1}(X'Y + P\mu),$$

and

$$C = Y'Y + \mu'P\mu - (X'Y + P\mu)'(Z'Z + X'X + P)^{-1}(X'Y + P\mu) + 2\beta.$$

Thus w has a k-dimensional t distribution with $n + 2\alpha$ degrees of freedom, location vector $A^{-1}B$, and precision matrix $(n + 2\alpha)A(C - B'A^{-1}B)^{-1}$, and it may be confirmed that this density reduces to the special case (1.37).

We have seen that if one has the t density with which to forecast w that one may use the mean $A^{-1}B$ as a point forecast of the k future values, but one may also want an interval or region prediction and some idea of the accuracy of the forecast.

Since w has a t distribution each component of w has a scalar t distribution (see the Appendix), hence to predict w_1 for example, one could construct a $1 - \gamma$ confidence interval for w_1 or one can perform simultaneous predictions for the k future observations by constructing a $1 - \gamma$ prediction region of w, with highest predictive density.

Let

$$Q(w) = (n + 2\alpha)(w - A^{-1}B)'A(w - A^{-1}B)(C - B'A^{-1}B)^{-1}/k,$$

$$(1.40)$$

then it may be shown that $Q(w)$ has a predictive distribution which is F with k and $n + 2\alpha$ degrees of freedom, and that $\{w: Q(w) \leq F(\gamma;k,n + 2\alpha)\}$ is a $1 - \gamma$ region for w with highest predictive density. Thus, regions of highest predictive density for w are constructed in the same way as regions of highest posterior density for θ. A future value of w will be contained in $Q(w)$ with a probability of $1 - \gamma$, giving one an idea of the error of the forecast.

The Bayesian predictive density will be used in a variety of ways, not only for forecasting future values, although this is its most important function. Another use of the predictive density is that it can be used to determine the parameters α, β, μ, and P of the prior density of the parameters θ and τ of the linear model. By employing the so-called prior predictive density, one may predict hypothetical future values w of the experimental values y, where $Y = X\theta + e$ is the model of the experiment, then fit the w values to the hyperparameters of the normal-gamma conjugate prior. These techniques will be introduced in the chapter on regression models and experimental design models.

Another use of the Bayesian predictive density is the control problem, which is encountered in regression and time series analysis and linear dynamic systems.

Consider a general linear model

$$Y = X_1 \theta_1 + X_2 \theta_2 + e \qquad (1.41)$$

where X and θ have been partitioned into $X = (X_1, X_2)$ and $\theta = (\theta_1', \theta_2')'$, and suppose X_2 is a $n \times p_2$ matrix of p_2 control variables; that is variables that can be set at one's discretion. The control problem is to choose X_2 in such a way so that the future values of $w = T$, where T is some target value and $w = z_1 \theta_1 + x_2 \theta_2 + \varepsilon$.

The predictive density of w will depend on x_2, hence one may choose x_2 so that w in some sense is close to the target value. These ideas will be explored in the other chapters.

One may use a loss function in conjunction with the Bayesian predictive density. Suppose $L(w,w^*)$ is the loss in predicting w by w^*, then w^* can be chosen to minimize the expected loss with respect to the predictive distribution of w. This decision-theory way of forecasting will not be used in this book.

SUMMARY AND GUIDE TO THE LITERATURE

We have seen that this chapter introduces the Bayesian analysis of the general linear model. An analysis consists of identifying the model

SUMMARY AND GUIDE TO THE LITERATURE 19

$Y = X\theta + e$, specifying a prior distribution for the parameters θ and τ, either by Jeffreys' vague prior or by a normal-gamma conjugate prior, deriving the posterior densities of the parameters so one may estimate or test hypotheses about the parameters, and then, if one wishes, forecast future values by the Bayesian predictive density. The parameters are estimated with point estimates, which are particular characteristics of the posterior distribution or with interval estimates constructed by regions of highest posterior density, which also provide a way to test hypotheses about the parameters. In the same way, regions of highest predictive density give simultaneous forecasts of future values.

It has been shown that only five distributions occur in the study of linear models, namely the normal, gamma, normal-gamma, the multivariate t, and the F distributions. The normal distribution specifies the errors of the model and the conditional prior and posterior distributions of θ given τ, while the gamma is the marginal prior and posterior distribution of τ. The joint marginal prior and posterior distribution of θ and τ is normal-gamma, which is also the conjugate family of distributions to the normal linear model. Lastly, the F distribution occurs as a transformation of the t distribution and gives one a method to construct regions of highest posterior and predictive density for θ and w respectively, where w is a vector of future values. The t distribution is the marginal prior and posterior distribution of θ and also that of the predictive density of w.

These five distributions will suffice until the posterior analysis of the mixed model is developed. At that time, it will be necessary to introduce a generalization of the t distribution, called the poly-t distribution (see the Appendix).

An advantage of the Bayesian approach is that Bayes theorem provides a way to solve all problems in the analysis of the normal linear model. We have seen that the inferences for the parameters of the model are all based on the posterior distribution of the pertinent parameters and that these posterior distributions are a consequence of Bayes theorem. Forecasting is accomplished by the predictive distribution which is derived on the basis of Bayes theorem, hence if one knows how to apply Bayes theorem, one is well on the way to learning the theory and methodology of analyzing data which can be modeled in terms of linear models.

The reader should be familiar with some of the literature which deals with the Bayesian analysis of linear models.

First, there are several books which introduce the reader to general Bayesian inferential procedures as well as those methods which are peculiar to the linear model. The earliest book is by Jeffreys (1939) who introduced the vague prior distribution, which we use in this chapter. Later Lindley (1965) wrote a book which contains very useful information about the linear model, but prior information is restricted to improper vague prior distributions. Box and Tiao's (1973) book actually deals exclusively with all types of linear models including mixed and random models. Zellner's (1971) book is not only an excellent introduction to linear model theory, but also provides the Bayesian analysis of time series and econometric models, including simultaneous equation models. He also discusses hypothesis testing techniques and control theory problems. The above four books are not oriented to the Bayesian decision theory approach, but instead, focus on informal inferential procedures, which is the way this book treats the subject.

For a more formal decision-theory approach, one should refer to Savage (1972), Raiffa and Schlaifer (1961), DeGroot (1970), and Berger (1980). The Savage book is one of the earliest formulations to present the axiomatic development of decision theory, where one postulates axioms of utility and probability, and as a consequence, proves that Bayesian decisions are the only optimal ones. This tradition is continued by Ferguson (1967), DeGroot (1970), and Berger (1980), and the former is a good introduction to Bayesian decision theory as well as other theories of decision. DeGroot (1970) has a short section on the theory of linear models and good accounts of conjugate family distributions and the distribution theory which is involved in the posterior and predictive analysis of linear models. This book is highly recommended to the reader.

In order to learn more about the implementation of prior information, the reader should consult Bernardo (1979), Jeffreys (1939), Box and Tiao (1973), Winkler (1977), and Zellner (1971, 1980). The Winkler and Zellner (1980) papers are about choosing the hyperparameters of the normal-gamma prior distribution, while the remaining references deal with improper prior distributions, which convey little or no information.

With regard to the posterior analysis of linear models, Box and Tiao's (1965) paper on regions of highest posterior density shows how to test hypotheses about the means and variances of normal theory models and their book (1973) is by far the best account of Bayesian methodology as applied to the linear model. Also, this book is the only one which studies random and mixed models and it gives many numerical examples of the analysis of all types of models. Needless to say, this book is highly recommended to those who want to know the detailed analysis of the traditional statistical models, that is regression models and models (fixed and random) employed to analyze designed experiments.

Both Box and Tiao (1973) and Zellner (1971) use regions of highest posterior density to estimate parameters and test hypotheses, but the latter specializes in posterior odds ratios to test hypotheses. For forecasting problems in economics, Zellner's (1971) book is the only one to give a detailed account of the Bayesian treatment of time series and econometric models.

Other Bayesian references on predictive analysis are Aitchison and Dunsmore (1975), Harrison and Stevens (1979) and Geisser (1971).

The controversy surrounding the Bayesian approach has somewhat subsided and it is interesting to read some of the earlier discussions on the subject. For example, Lindley and Smith's (1972) paper on exchangeable prior distributions provoked a heated debate about Bayesian methodology, but for a more balanced approach to the comparison of Bayesian and non-Bayesian theories of inference, read Barnett (1973) and Lindley's (1971) monograph.

Lindley's monograph is a review of Bayesian theory and methodology written up to 1971, and although much material has appeared since then, the serious student cannot afford not to read it. This reference contains criticism of some sampling theory methodologies, such as the Neyman-Pearson theory of testing hypotheses and the confidence interval approach to estimation, and Barnett presents an unbiased comparison of Bayesian and other theories of inference.

EXERCISES

1. Prove Theorem 1.1.
2. Referring to Theorem 1.1, show that if X'X is singular and if the prior density of the parameters is vague (1.20), then the joint posterior density of the parameters is improper.
3. Verify equation (1.28) and show (1.29) yields a $1 - \gamma$ HPD region for θ.
4. Show the region given by (1.32) is a $1 - \gamma$ HPD region for $A\theta$.
5. Show the random variable $Q(w)$, equation (1.40), has an F distribution with k and $n + 2\alpha$ degrees of freedom.
6. Assume $k = 2$ in problem 5 above and find an interval of highest predictive density for the second future observation.
7. The general linear model is given by (1.1). How would you represent a normal-theory simple linear regression model, that is, explain what Y, X, θ, and e are? Now, using the notation of the linear model (1.1), define a normal theory multiple regression model with $p - 1$ regressors.
8. Compare the classical and non-Bayesian approaches to the analysis of a linear model. For example, compare the way a Bayesian would estimate θ to the way a non-Bayesian would do it. What are the similarities and differences in the two approaches? How would a non-Bayesian predict future observations assuming the future observations are generated with the general linear model?
9. Let

$$Y = X\theta + e$$

be a linear model with the usual assumptions, except $e \sim N(0, P^{-1})$, where P is a n × n unknown precision matrix. Assume θ is a p × 1 unknown parameter vector and that θ and P have a normal-Wishart distribution. (See the Appendix.) Find the marginal posterior distribution of θ.

10. Let

$$Y = X\theta + e$$

be a linear model with the usual assumptions where $e \sim N(0, P^{-1})$ and P is a n × n *known* precision matrix. Using a conjugate prior density for θ, find the marginal posterior distribution of θ.

11. Let

$$Y = X\theta + e$$

with the usual assumptions except that θ is known and $e \sim N(0, \tau^{-1}I_n)$, and $\tau > 0$ is an unknown precision with a prior gamma density. Find the posterior density of τ.

REFERENCES

Aitchison, J. S. and J. R. Dunsmore (1975). *Statistical Prediction Analysis*, Cambridge University Press, London.

Barnett, V. (1973). *Comparative Statistical Inference*, John Wiley and Sons, Inc., New York.

Berger, James O. (1980). *Statistical Decision Theory, Foundations, Concepts, Methods*, Springer-Verlag, New York.

Bernardo, J. M. (1979). "Reference posterior distributions for Bayesian inference" (with discussion), Journal of the Royal Statistical Society, Series B, Vol. 41, No. 2, pp. 113-147.

Box, G. E. P. and G. C. Tiao (1965). "Multiparameter problems from a Bayesian point of view," Annals of Mathematical Statistics, Vol. 36, pp. 1468-1482.

Box, G. E. P., and G. C. Tiao (1973). *Bayesian Inference in Statistical Analysis*, Addison Wesley, Reading, Mass.

DeGroot, M. H. (1970). *Optimal Statistical Decisions*, McGraw-Hill Book Company, New York.

Ferguson, Thomas S. (1967). *Mathematical Statistics, A Decision Theoretic Approach*, Academic Press, New York and London.

Geisser, S. (1971). "The inferential use of predictive distributions," in *Foundations of Statistical Inference*, edited by V. P. Godambe and A. D. Sprott, Holt, Rinehart and Winston, Toronto.

Harrison, P. J. and C. F. Stevens (1976). "Bayesian forecasting" (with discussion), Journal of the Royal Statistical Society, Series B, Vol. 38, pp. 205-247.

REFERENCES

Jeffreys, H. (1939/1967). *Theory of Probability*, Third Edition, Clarendon Press, Oxford.

Lindley, D. V. (1965). *Introduction to Probability and Statistics from a Bayesian Viewpoint*, University Press, Cambridge.

Lindley, D. V. (1971). *Bayesian Statistics, A Review*, Reg. Conf. Ser. Appl. Math., 2. SIAM, Philadelphia.

Lindley, D. V. and A. F. M. Smith (1972). "Bayes estimates for the linear model," Journal of the Royal Statistical Society, Series B, Vol. 34, pp. 1-42.

Raiffa, Howard and Robert Schlaifer (1961). *Applied Statistical Decision Theory*, Division of Research, Harvard Business School, Boston.

Savage, L. J. (1972). *The Foundation of Statistics*, Dover, Inc., New York.

Winkler, Robert, L. (1977). "Prior distributions and model building in regression analysis," in *New Developments in the Applications of Bayesian Methods*, edited by Ahmet Aykac and Carlo Brumat, North-Holland, Amsterdam.

Zellner, Arnold (1971). *An Introduction to Bayesian Inference in Econometrics*, John Wiley and Sons, Inc., New York.

Zellner, Arnold (1980). "On Bayesian regression analysis with g-prior distributions." Technical Reprint, H. G. B. Alexander Research Foundation, Graduate School of Business, University of Chicago.

2
LINEAR STATISTICAL MODELS AND BAYESIAN INFERENCE

INTRODUCTION

This chapter will introduce the various linear models which are to be examined in the remaining chapters of this book. One class of models was introduced in Chapter 1, namely the so-called general linear model, which includes, as special cases, the fixed models, which are used for regression analysis and the analysis of designed experiments.

The regression and design models are the traditional ones in that they are the most frequently used in statistical practice. Many of the software packages such as SAS and BMD allow one to routinely analyze data with one of the traditional models.

A first cousin to the design model is the mixed model, which is often employed when some of the experimental factors are random, that is the levels of the factors are selected at random from a population of levels and one is interested in making inferences about the factor populations.

The time series models to be analyzed in this book are special Gaussian linear models, namely the regression model with autocorrelated errors, the ARMA class (autoregressive moving average models), and the distributed lag models of econometrics. During the past decade the ARMA models have been thoroughly examined from the classical analysis of Box and Jenkins (1970), while the other time series models, in addition to being analyzed from a classical perspective, have been approached, principally by Zellner (1971), from the Bayesian viewpoint.

Linear dynamic models are now being studied by statisticians, but in the past have received the most attention of engineers who are interested in communication theory, navigation systems, and tracking of satellites. This class of models should prove to be very useful in the statistical analysis of time series.

Linear models which have unstable parameters are studied in Chapter 7 under the title structural change, which is a term from the field of economics. If one is confident, a priori, that a change in model (population) parameters will occur, special techniques are necessary. We will see that the Bayesian analysis of such models has been a major contribution to the analysis of data in this area.

Most of this book deals with univariate linear models and it is only in Chapter 8 that multivariate models are first encountered. Multivariate regression, design, time series models, and some econometric models are examined in detail.

The linear logistic model is also mentioned in this chapter but will not be studied.

In what is to follow, each of the above models will be defined. An example of each is given, and a preview of the Bayesian analysis of the particular model is described.

LINEAR STATISTICAL MODELS

Regression Models

The regression model is a special case of the general linear model

$$Y = X\theta + e \tag{1.1}$$

of Chapter 1, where Y is a n × 1 vector of observations, X is a n × p known matrix, θ a p × 1 unknown real parameter vector, and e a n × 1 vector of observation errors. Regression models are employed to examine the relationship between a dependent variable y and q independent variables $x_1, x_2, \ldots, x_q$, thus one observes $(Y, x_1, x_2, \ldots, x_q)$ and the n observations are denoted by $(y_i, x_{1i}, x_{2i}, \ldots, x_{qi})$, $i = 1, 2, \ldots, n$, where y_i is the value of Y when one observes $x_{1i}, x_{2i}, \ldots, x_{qi}$. Presumably Y is observed with an error, but the independent variables are observed without error.

In terms of the general model (1.1)

$$y_i = \beta_0 + \sum_{j=1}^{q} x_{ji}\beta_j + e_i, \quad i = 1, 2, \ldots, n, \tag{2.1}$$

and the i-th component of Y is y_i, of e, e_i, the first column of X is j, and the second column is $(x_{11}, x_{12}, \ldots, x_{1n})'$, etc., where $p = q + 1$.

Thus, one is assuming the average value of the dependent variable is a linear function of q independent variables if the n errors each

LINEAR STATISTICAL MODELS

have a zero mean. The regression model is examined in detail in Chapter 3, where a complete Bayesian analysis is performed. It is shown how to estimate and test hypotheses about the parameters of the model and how to forecast future observations, using either a vague density or a conjugate density to express prior information. Chapter 1 explained how one is to do a Bayesian analysis and these ideas are applied to the regression model and design models of Chapter 3.

The regression and design models are quite similar and each is expressed in terms of the general linear model, however, with the design model the design matrix X of (1.1) is used to indicate the presence or absence of the levels of several factors of the experiment, and the components of X are either zero or one.

The Design Models

If the experiment has only one factor with, say, k levels, the model is

$$y_{ij} = \theta_i + e_{ij}, \qquad (2.2)$$

where y_{ij} is the j-th observation on the i-th level, where $i = 1, 2, \ldots, k$, and $j = 1, 2, \ldots, n_i$, and $n = \sum_{i=1}^{k} n_i$ is the number of observations and e_{ij} is the error associated with the ij-th observation. Of course, this example can be expressed in terms of the general linear model, where $y = (y_{11}, y_{12}, \ldots, y_{1n_1}; \ldots; y_{k1}, y_{k2}, \ldots, y_{nk_k})$,

$$X = \begin{pmatrix} 1 & 0 & \cdots & 0 \\ 1 & 0 & \cdots & 0 \\ \cdot & & & \\ \cdot & & & \\ \cdot & & & \\ 1 & 0 & \cdots & 0 \\ \cdot & & & \\ \cdot & & & \\ \cdot & & & \\ 0 & 0 & \cdots & 1 \\ 0 & 0 & \cdots & 1 \\ \cdot & & & \\ \cdot & & & \\ \cdot & & & \\ 0 & 0 & \cdots & 1 \end{pmatrix}$$

is a n × k matrix, where the first n_1 components of the first column are ones and the remaining elements are zero. The last n_k components of the k-th column are ones, thus the elements of X are either zeros or ones with a one in each row. The elements indicate the presence or absence of the k levels of the factor for each of the n observations, and the theory of Chapter 1 is applied in Chapter 3 in order to analyze one- and two-factor experiments, such as the completely randomized design and completely randomized block design.

If the k levels of the experiment are the only ones of interest to the experimenter, that is inferences are confined to the k means θ_i, the model is called fixed, otherwise if the k levels are selected at random from a population of levels, the model is said to be mixed or random.

In the fixed version of the one-way model (2.2), the e_{ij}'s are independent normal variables with zero mean and unknown precision τ, and the k parameters θ_i represent the effects of the k levels of the factor. In the classical approach to this problem, the θ_i's are regarded as fixed unknown constants, if the model is fixed, otherwise they are thought of as unobservable random normal variables with zero mean and constant variance (the variance component) if the k levels are selected at random from a population of levels, hence if the variance component is zero the population is degenerate and there is no treatment effect.

From the Bayesian viewpoint in the fixed case, the θ_i's are treated as random variables with some known prior distribution, but in the random case, a prior density is put on the variance component of the model. Thus if one adopts a Bayesian approach the distinction between fixed and random is not as meaningful as it is when one adopts the classical approach.

Consider an example by Davies (1967, page 105), which Box and Tiao (1973, page 216) analyze from a Bayesian approach.

The experiment consists of learning what effect the batch to batch variation of a raw material has on the yield of the product produced. Five samples from each of six batches were examined for product yield and the experiment is thought of as random, because the six batches were selected at random from a population of batches. Clearly, one's inferences should not be confined only to the six batches of the experiment, but to the population of batches, which is characterized by a variance component which explains the batch-to-batch variation.

The Bayesian analysis of random and mixed models is given in Chapter 4, but the approach differs from the method of Box and Tiao (1973) who use numerical integration to isolate the marginal posterior distribution of each of the variance components. The results of Chapter 4 are based on the dissertation of Rajagopalan (1980) who found a

LINEAR STATISTICAL MODELS   29

way to isolate the posterior density of each variance component via analytical approximation.

The mixed model is given by

$$y = x\theta + \sum_{i=1}^{c} u_i b_i + e \qquad (2.3)$$

where y is a n × 1 vector of observations, x is a n × p known matrix, θ a p × 1 unknown parameter vector, u_i is a n × m_i known matrix, the b_i's are independent normal unobservable vectors each with a zero mean vector and b_i has precision matrix $\tau_i J_{m_i}$, and e is independent of the b_i's and is a normal random vector with zero mean and precision matrix τI_n. Also, $\theta \in R^p$, $\tau_i > 0$, $\tau > 0$, and the c variance components are $\sigma_i^2 = \tau_i^{-1}$, while $\sigma^2 = \tau^{-1}$ is called the error variance.

The mixed model accommodates both fixed and random factors of a designed experiment and the θ vector has the levels of the fixed factors, while the levels of the c random factors are represented by $b_1, b_2, \ldots, b_c$. Box and Tiao (1973, page 341) introduce the additive mixed model

$$y_{ij} = \theta_i + c_j + e_{ij} \qquad (2.4)$$

to analyze a car-driver experiment with eight drivers and six cars. The main response was the mileage per gallon of gasoline and each driver drove each of the six cars, where i = 1, 2, ..., 8 and j = 1, 2, ..., 6. Thus y_{ij} is the gasoline mileage when driver i drives car j, the θ_i's are unknown constants, the c_j's are normal independent random variables with zero mean and variance component σ_c^2, and the e_{ij}'s are independent normal variables with zero means and error variance σ^2. Thus model (2.4) is seen to be a special case of the mixed model (2.3), where the cars were selected from a population of cars with variance σ_c^2, but inferences about the drivers are confined to the eight drivers of the experiment.

Chapter 4 is concerned with such experiments and the posterior analysis consists of determining the posterior distribution of each variance component and the vector of fixed effects.

Time Series Models

One of the most useful models to analyze time series data is the p-th order autoregressive model

$$y(t) = \sum_{i=1}^{p} \theta_i y(t-i) + e_t, \qquad (2.5)$$

where $y(t)$ is the observation at time t, θ_i is an unknown parameter vector, the e_t's, $t = 1, 2, \ldots, n$, are independent $n(0, \tau^{-1})$ random variables, and $y(0), y(-1), \ldots, y(1-p)$ are known constants.

Time series data are often correlated through time and models such as (2.5) allow one to introduce correlation into the model. For example, if $p = 1$, the correlation between observations s units apart is

$$\rho[y(t), y(t+s)] = \theta_1^s, \quad |\theta_1| < 1, \qquad (2.6)$$

and if $p \geq 2$, more complex correlations can be studied.

Still another useful model to analyze time series data is the q-th order moving average class

$$y_t = e_t - \sum_{i=1}^{q} \phi_i e_{t-1}, \qquad t = 1, 2, \ldots, n, \qquad (2.7)$$

where y_t is the observation at time t, the ϕ_i's are unknown real parameters (the moving average coefficients), and the e_t's are independent $n(0, \tau^{-1})$ random variables (white noise). If $q = 1$, the correlation, see Box and Jenkins (1970), is given by

$$\rho(y_t, y_{t+s}) = \begin{cases} -\dfrac{\phi_1}{1 + \phi_1^2}, & s = 1 \\ 0, & s \geq 2 \end{cases} \qquad (2.8)$$

and the moving average class introduces another form of correlation for the observations.

The autoregressive model has been used for many years for analysis of time series data and has received attention from both the Bayesians, see Zellner (1971), and from classical procedures, see for example Box and Jenkins (1970). On the other hand the moving average model has not been explored from a Bayesian viewpoint but has received a lot of attention from other perspectives.

LINEAR STATISTICAL MODELS

In Chapter 5, the Bayesian analysis of Zellner will be expanded for the autoregressive model, and the first and second order moving average models will be studied, where the posterior and predictive analysis will be derived.

The autoregressive and moving average models are combined into the ARMA class, thus an ARMA(1,1) model is given by

$$y_t - \theta y_{t-1} = \varepsilon_t - \phi \varepsilon_{t-1}, \tag{2.9}$$

where y_t is the observation at time t, ϕ is the moving average parameter, θ the autoregressive parameter, the ε_t's are white noise, and y_0 is a known constant. Thus there are three parameters to the model. The ARMA model is parsimonious, that is, it is able to represent time series data with a few parameters, thus for a given set of observations $\{y(t), t = 1, 2, \ldots, n\}$ an ARMA(1,1) model and an AR(8) (8th order autoregressive model) perhaps will give the same "fit" to the data. Parsimony was emphasized by Box and Jenkins (1970) in their development of the statistical analysis of time series which are generated by ARMA models. Indeed they say in practice one is not likely to encounter a case where either (or both) the moving average order q or (and) the autoregressive order p is in excess of two. The Box and Jenkins analysis of time series is a major contribution to statistical methodology and is today very popular. Compared to Box-Jenkins methodology, the Bayesian approach is in its infancy.

Nevertheless, the Bayesian posterior analysis and forecasting techniques of ARMA models are attempted in Chapter 5, where a complete treatment of autoregressive processes is given. The ARMA(1,1), MA(1) and MA(2) processes are also examined, but only partial success can be reported here for the MA(2) case. Monahan (1983) has recently developed a Bayesian analysis of ARMA models for low order processes.

Models for Structural Change

Two important references on structural change in linear models are Poirier (1976) and more recently the special issue of the Journal of Econometrics, which was edited by Broemeling (1982). The book by Poirier is a review of the statistical and econometric literature on structural change and presents new ideas about how one should model structural change with the aid of spline functions. The special issue contains ten articles on the subject and introduces the reader to a large variety of theory and methodology. This issue contains some material which was introduced before 1976, but for the most part includes ideas which appeared after Poirier's book.

The term "structural change" doesn't have a definite definition and is used by econometricians to denote a change in the parameters of a model which explains a relationship between economic variables. Also, the term is often employed to refer to a model which has been misspecified. Terms or phrases such as shift point, change point, transition function, separate regressions, switching regression, and two-phase regression, although not identical in meaning to structural change, are involved in some way with structural change.

When data are collected in an ordered sequence, often one is interested in the probabilistic structure of the measured variables from one subset of the time domain to the other. Suppose for example a major economic policy change is put into effect at some time, then the researcher is perhaps interested in assessing the effects of the change on the variables under study.

Two fundamental questions now occur:

(i) Has a change occurred among the variables under investigation during the observation period? This is called the detection problem because one wants to detect a change in the relationship.
(ii) Assuming a change has taken place, can one estimate the parameters of the model? For example, one may want to estimate the time where the change occurred as well as the other parameters of the model, namely those which explain the "before" and "after" relationship between the variables.

Problems of structural change are found in a wide variety of disciplines. Often the variables under study are economic where the policy change could be a new tax law (such as the windfall profits tax), a new government program (such as price supports), or a major disturbance to the economy (an oil embargo). In biology, one might focus on the life of an organism and the time when a change in growth pattern will occur. These examples and others are given by Holbert (1982), who reviews Bayesian developments in structural change from 1968 to the present.

An interesting example reported by Holbert is a two-phase regression problem illustrated with stock market sales volume.

On January 3, 1970, *Business Week* reported regional stock exchanges were hurt by abolition of give-ups (commission splitting) on December 5, 1968. McGee and Carleton (1970) analyzed the relationship between dollar volume sales of the Boston Stock Exchange and the combined New York and American Stock Exchanges. They did this to investigate the hypothesis that a structural change occurred (between dollar volume sales of the regional and national exchanges) after December 5, 1968, and their analysis was based on a piecewise regression methodology. Holbert (1982) reexamined the data, but used the

LINEAR STATISTICAL MODELS

Bayesian approach, and came to the same conclusion as McGee and Carleton, namely that abolition of split-ups did indeed hurt the regional exchanges.

The Bayesian approach was based on the two-phase regression model

$$y_i = \begin{cases} \alpha_1 + \beta_1 x_i + e_i, & i = 1, 2, \ldots, m \\ & 2 \leq m \leq n - 2 \\ \alpha_2 + \beta_2 x_i + e_i, & i = m + 1, \ldots, n, \end{cases} \quad (2.10)$$

where the e_i's are n.i.d. $(0, \tau^{-1})$, y_i is the sales volume on the combined exchanges at time i, x_i the corresponding total dollar volume on the Boston Stock Exchange at time i, m the unknown shift point, and α_i and β_i (i = 1,2) are unknown regression parameters. Thus the first m pairs (x_i, y_i) follow one relationship and the last n − m pairs (x_i, y_i) a regression relation with parameters α_2 and β_2. The main question here is what is m? Is m "close to" December 5, 1968? Does m have a large posterior probability corresponding to support the hypothesis the regional exchanges were damaged because of the abolition of give-ups?

Holbert's analysis adopted a vague improper prior for the parameters

$$\xi(\alpha_1, \beta_1, \alpha_2, \beta_2) \propto \text{constant}, \quad \alpha_i \in R, \; \beta_i \in R$$

$$\xi(m) = (n - 3)^{-1}, \quad 2 \leq m \leq n - 2 \quad (2.11)$$

$$\xi(\tau) \propto 1/\tau, \quad \tau > 0$$

which when combined with the likelihood function gave

$$\xi(m | \text{data}) \propto \left[m(n - m) \sum_{i=1}^{m} (x_i - \bar{x}_{1,m})^2 \sum_{i=m+1}^{n} (x_i - \bar{x}_{2,m})^2 \right]^{-1/2}$$

$$\times \left[\sum_{i=1}^{m} (y_i - \hat{y}_{i,m})^2 + \sum_{i=m+1}^{n} (y_i - \hat{y}_{i,m+1})^2 \right]^{-(n-4)/2},$$

(2.12)

where $2 \leq m \leq n - 2$, as the marginal posterior mass function of m. In this formula $\bar{x}_{1,m}$ is the mean of the first m x-values, $\bar{x}_{2,m}$ the mean of the last n − m values, $\hat{y}_{i,m}$ is the predicted y_i value based on the first m (x,y) pairs, and $\hat{y}_{i,m+1}$ the predicted value of y_i based on the last n − m (x,y) pairs. Holbert computes the function (normalized) (2.12) for the months from February, 1967, to November, 1969, and obtains a "large" posterior probability of .2615 for November, 1968, the month just before the change went into effect.

The two-phase regression model is one of the many models of structural change which appear in Chapter 7. Other models which are examined are normal sequences, time series processes such as the autoregressive and regression models with autocorrelated errors. The multiple linear regression model with one change is investigated in detail and a thorough sensitivity study is done. In most of the models either a vague improper prior density or a quasiconjugate prior density is used.

Models for structural change usually incorporate only one change (or at most two changes). On the other hand, models which incorporate parameters which always change (from one time point to the next) are also very important and as we will see have been quite valuable to engineers and scientists in the communication and space industries.

Dynamic Linear Models

The Kalman filter is one of the major achievements of mathematical (statistical) modeling, however, it is not well-known to statisticians. The Kalman filter is an algorithm which recursively computes the mean of the posterior distribution of a parameter of a dynamic model. Models are called dynamic when the parameters of the model are actually always changing, thus the traditional models of regression and the design of experiments are static, since the parameters of these models are not thought to be always changing their values.

The dynamic linear model is for t = 1,2, ...

$$y(t) = F(t)\theta(t) + U(t),$$
$$\theta(t) = G\theta(t - 1) + V(t),$$

(2.13)

where $y(t)$ is a m × 1 observation vector, $F(t)$ a m × p matrix, $\theta(t)$ a p × 1 state vector, $U(t)$ a m × 1 unobservable observation error vector, G a p × p systems transition matrix, and $V(t)$ a p × p unobservable systems error vector.

The primary parameters are the state vectors $\theta(0), \theta(1), \ldots$, and consequently one must estimate them as the observations become avail-

LINEAR STATISTICAL MODELS

able, that is, given y(1), what is θ(1), given y(1) and y(2), what is θ(2), etc. The first equation of the model is the observation equation which relates the observation in a linear way to the state vector and by itself is the usual linear model of regression and the design of experiments. The second equation gives the model its dynamic flavor, and is called the systems equation and tells us how the parameters of the model change. The second equation is a first-order vector autoregressive model (examined in Chapter 8), which tells us the parameter (state vector) is always changing in such a way that the next value is a linear function of the present value apart from a random error.

One immediately notices that the dynamic model contains a large number of parameters when compared to their static counterparts, thus one usually assumes that F(t), t = 1, 2, ... and the covariance matrices of the error vectors U(t) and V(t) are known. Assuming these quantities are known as well as making other assumptions such as normality of the error vectors, Kalman (1960) found the mean of the marginal posterior normal distribution of θ(t) given y(1), y(2), ..., y(t) for all t = 1, 2, ..., but more importantly, developed a recursive formula for its computation. It was very important to have a recursive formula since often the time between successive observations is very short, only a few seconds or less.

The Kalman filter is often used to compute the position of a satellite or a missile and tracking these craft requires the computation to be extremely fast, thus Kalman's filter was a major contribution.

Consider again the dynamic linear model, then θ(t) might represent the actual position of a satellite at time t, y(t) the measured position by radar or telemetry, and U(t) the error in the observed position. Knowing the relationship between the actual and observed positions allows one to know F(t) and knowing the noise characteristics of the radar or telemetry signals allows one to know the covariance matrix of the observation error vector U(t). Celestial mechanics assists one to find a law which relates the successive positions of the satellite from one time to the next, that is, one knows G, the transition matrix of the system equation. See Jazwinski (1970, page 324) for an example of orbit mechanics.

Tracking problems are easily handled by dynamic models (2.13) as are navigation problems where it is necessary to control a missile, aircraft, or submarine. A linear dynamic model for navigation is given by the observation equation of (2.13) but the systems equation is amended to give

$$\theta(t) = G\theta(t-1) + Hx(t-1) + V(t), \qquad (2.14)$$

where t = 1, 2, ..., H is a p × s known matrix, x(t − 1) a s × 1 control vector, G a p × p matrix, and V(t) a p × 1 error vector.

With navigation problems, one must accurately predict the position $\theta(t+1)$ of the vehicle one time unit ahead, then make the necessary adjustments so that the actual position of the vehicle is at a predetermined location. This problem of navigation is called the control problem in the literature.

At time t, one has t observations $y(1), y(2), \ldots, y(t)$ and one must control the vehicle so that its position $\theta(t+1)$ one time unit in the future is to be say $T(t+1)$, thus, how should one choose $x(t)$ so that $\theta(t+1)$ is "close" to $T(t+1)$? A plan to do this is called a control strategy and is well-developed in the literature.

Chapter 6 introduces the dynamic linear model and how it is applied to navigation and tracking problems. The presentation is from the perspective of a statistician and should be readable to people with a statistical training. The chapter reviews the Kalman filter, the control problem, nonlinear filtering, adaptive estimation (when some of the parameters of the system and observation equations are unknown), smoothing, and prediction.

The dynamic linear model is the most general of those considered in this book and indeed all of the others are special cases.

Other Linear Models

The chapter on other models, Chapter 8, includes many of the multivariate models which are familiar to theoretical and applied statisticians. For example, multivariate regression and design models as well as vector autoregressive and moving average processes are to be considered. For each univariate model considered in this book, there is a corresponding multivariate version.

The linear dynamic model is an example of a multivariate linear model and is to be examined in Chapter 6; however it is so important, so believes the author, that a separate chapter will be devoted to its analysis.

Let us consider the multivariate version of the p-th order autoregressive process, namely

$$y'(t) = \sum_{i=1}^{p} y'(t-i)\theta_i + e'(t) \qquad (2.15)$$

where $t = 1, 2, \ldots, n$, $y(t)$ is a $m \times 1$ observation vector, $y(0)$, $y(-1), \ldots, y(1-p)$ are known $m \times 1$ vectors, the θ_i ($i = 1, 2, \ldots, p$) are unknown $m \times m$ matrices of real numbers where θ_p is nonzero, and the $e(t)$ are independent $N(0, P^{-1})$ random vectors, where P is a $m \times m$ unknown positive definite symmetric precision matrix. This

LINEAR STATISTICAL MODELS 37

model will accommodate the simultaneous analysis of m time series observed at equally spaced points, and as with the univariate AR(p) process, the main objective of our analysis is to develop the joint posterior distribution of the parameters $\theta_1, \theta_2, \ldots, \theta_p$, and P and to predict future observations $y(n+1), y(n+2), \ldots$

It will be shown that the Bayesian analysis of the multivariate AR model is a straightforward generalization of the univariate AR model if one uses the natural conjugate prior density for the parameters. In the univariate version of the model, the conjugate family is the normal-gamma class and the normal-Wishart for the vector version of the model, however, a constraint on the posterior precision matrix of $\theta_1, \theta_2, \ldots, \theta_p$ is imposed in the vector case (m ⩾ 2). See Zellner (1971) for a discussion of this situation.

Since the multivariate autoregression model is also a multivariate linear regression model, the same constraint occurs with that model, but except for this particular nuisance, no difficulties are encountered in the posterior and predictive analysis of the multivariate linear model.

The autoregressive process is, as mentioned earlier, the most often used to model time series data and the vector version is being used more and more along with the vector ARMA model. Jenkins (1979, page 75) gives an interesting example of a bivariate time series

$$y(t) = \begin{pmatrix} y_1(t) \\ y_2(t) \end{pmatrix}, \quad t = 1, 2, \ldots \qquad (2.16)$$

where $y_i(t)$ is the sales of product i, i = 1,2. The products are competitive, and the observations are quarterly and consist of eleven years of data. Since the sales of one product affect those of the other (and conversely), a multivariate ARMA model

$$y'(t) - \sum_{i=1}^{p} y'(t-i)\theta_i = e'(t) - \sum_{j=1}^{q} e'(t-j)\phi_j \qquad (2.17)$$

seems an appropriate tentative guess to explain the data where y(t) is the vector of observations at time t (the t-th quarter) consisting of the sales of the two products at that time, θ_i is a 2 × 2 unknown matrix which explains the dependence of each series on the other at time lag i, the e(t) are bivariate normal vectors with mean zero and unknown precision matrix P, and the ϕ_j's are unknown 2 × 2 moving average matrix coefficients. Also θ_p and ϕ_q are nonzero matrices. Jenkins explains the Box-Jenkins methodology of identification, estimation, diagnostic checking and forecasting in building a model for the sales data.

Box and Tiao (1981) have recently expanded the univariate analysis of Box and Jenkins (1970) to the vector ARMA case where an iterative procedure of identification, estimation, and diagnostic checking is proposed. Unfortunately, the Bayesian analysis of vector ARMA processes (which is also the situation with univariate models) has not progressed as far as the Box-Tiao methodology. With regard to the AR process, the Bayesian analysis is fairly easy to derive, but it is another story with the moving average process because its likelihood function is difficult to work with.

Other models considered in Chapter 8 are the regression model with autocorrelated errors and changing multivariate regression model. The regression model is

$$y_i' = \beta'x_i' + e_i', \quad i = 1, 2, \ldots, n \tag{2.18}$$

where y_i is a m × 1 observation vector, β is a m × 1 unknown parameter vector, x_i a m × m known matrix, and e_i a m × 1 random vector which follows a vector AR(1) model, namely

$$e_i' = e_{t-1}'\theta + \varepsilon_i',$$

where e_0 is known, the ε_i are independent $N(0, P^{-1})$ random m × 1 vectors, θ is a m × m unknown coefficient matrix, and P is an unknown precision matrix.

This model is often used to model time series which have a dependence on a set of m explanatory variables and which have the same correlation structure as an autoregressive time series of order one.

The linear logistic model is in a class by itself because it is quite different from the others since it is used to analyze counting or discrete data. Curiously, the multinormal logit model has not been examined, until recently by Zellner and Rossi (1982) from a Bayesian viewpoint, however, from the classical perspective a lot of work has been accomplished. For example in econometrics, Hausman and McFadden (1981) report that the multinomial logit model is the most used model specification.

The multinomial logit model is the usual multinomial model but with a logit parametrization of the class probabilities. Suppose there are n independent trials and on each exactly one of three events can occur with probabilities p_1, p_2, and p_3 where

$$p_i = e^{\beta Z_i} / \sum_{i=1}^{3} e^{\beta Z_i}, \quad i = 1, 2, 3,$$

$Z_1 = 1$, $Z_2 = 0 = Z_3$, and β is a scalar parameter, then the joint distribution of n_1, n_2, and n_3, where n_i is the frequency of the i-th event in n trials ($n = n_1 + n_2 + n_3$) is multinomial. By reparametrizing the model in terms of $\theta = e^{\beta}/2$, Zellner (1982) and Zellner and Rossi (1982) present a convenient exact small-sample posterior analysis. Their parametrization by θ allows them to identify the inverted Beta as the conjugate class to the model, hence the posterior moments are easily computed. This is an elegant way to solve the problem, and their solution is quite significant since recently very little work on non-normal models has appeared.

The foregoing should give the reader a good idea of what models are going to appear and in what situations they are useful. The linear models appearing in this work are useful in the engineering, physical, social, mathematical, economic, and biological sciences and the following chapters will give the reader a good idea of how to do a Bayesian statistical analysis when one of these linear models is appropriate for the analysis of the data. It is assumed one knows that a particular model is indeed appropriate for a particular experimental situation, and the question of choosing the appropriate model, although very important, is not at issue. What is important is how one implements a Bayesian statistical analysis.

BAYESIAN STATISTICAL INFERENCE

The Bayesian analysis of a statistical problem, using the general linear model, was illustrated in Chapter 1, where the posterior and predictive analysis was done on the basis of either a proper prior density or the proper conjugate prior density. In this section we will discuss just what constitutes our Bayesian analysis when one adopts a linear model for the probability model of the observations, and what Bayesian inference is, as well as what it is not. The results of this section will apply to all the linear models which were introduced in the first part of this chapter.

In what is to follow, we will look at the history of Bayesian inference, the subjective interpretation of probability, the main ingredients of a Bayesian analysis, the various types of prior information, the implementation of prior information, and the advantages and disadvantages of Bayesian inference.

Some History of Bayes Procedures

Our subject's beginning originated with the Rev. Thomas Bayes who published only two papers on the subject. His "An Essay Toward Solving a Problem in the Doctrine of Chances" was published in 1764 and was concerned with the problem of making inferences about the parameter θ of n Bernoulli trials. Bayes assumed θ was uniformly distributed (by construction of a "billiard" table) and presented a way to compute $P(a < \theta < b \mid x = p)$ where x is the number of successes among the n trials and a and b are constants. Of course, this is now recognized as the calculation, a posteriori, that the probability of a success lies in some interval (a,b). Bayes gave two contributions in his paper: (1) the calculation for $P(a < \theta < b \mid x = p)$, called Bayes theorem, and (2) the principle of insufficient reason. The first contribution is widely accepted, however the second is quite controversial.

Some have interpreted the principle of insufficient reason to mean that if nothing is known about the values of θ, then one value of θ is as likely as another, and hence a uniform prior density is appropriate for θ. According to Stigler (1982), this is not what Bayes really meant, but instead knowing nothing about θ really means the marginal distribution of x is uniform.

In any case, such controversy over Bayes' article is not unusual. Beginning with Laplace (1951), many people have commented on the essay of Bayes including Pearson (1920), Fisher (1922), and Jeffreys (1939).

In the essay, Bayes used a uniform prior density for θ but it was constructed to be so, because θ was the horizontal component of the position of a ball tossed at random on a flat table; hence the prior distribution of θ can be given a frequency interpretation and there is very little controversy. The interesting question is that if one cannot use a frequency interpretation for the distribution of θ, a priori, how does one choose a prior for θ and is it appropriate to use a uniform distribution for θ, if one knows "nothing" about the possible values of θ? For example, I have a coin and its probability of landing "heads" on a toss is θ, can I use a uniform prior density for θ if I know "nothing" about θ? There are those that say "no," because if I know nothing about θ, then I know nothing about any function of θ, say $g(\theta) = \theta^2$, $0 \leq \theta \leq 1$; then θ^2 should also have a uniform distribution, which is impossible. For this simple problem, there have been many "solutions." Haldane (1932) proposes

$$\xi(\theta) \propto \theta^{-1}(1-\theta)^{-1}, \quad 0 < \theta < 1,$$

and others, including Jeffreys, use

$$\xi(\theta) \propto \theta^{-1/2}(1-\theta)^{-1/2}, \quad 0 < \theta < 1,$$

to express prior ignorance for θ. Thus we see, from its inception, Bayesian inference is a controversial subject, but not much was heard about it until the twentieth century, where during the last thirty years many advances have been made.

The current activity can be divided into Bayesian decision theory and Bayesian inference, where the former uses three sources of information (in making statistical decisions), namely the sample data s, the prior density $\xi(\theta)$ of the parameters, and the loss function which measures the consequences of making an incorrect decision. On the other hand, Bayesian inference only uses the prior density of the parameters and the sample data because with inference one is not interested in making decisions, only investigating the values of the parameters of the model or making predictions about future observations.

As mentioned earlier, the books by Lindley (1965), Zellner (1971), Box and Tiao (1973), Winkler (1972), and Press (1982) give introductions to Bayesian inference while DeGroot (1970), Ferguson (1967), Raiffa and Schlaifer (1961), Savage (1954) and Berger (1980) deal with decision making.

The early work in Bayesian statistics of this century was done by deFinetti (1930), Jeffreys (1939) and Ramsey (1931/1964). deFinetti and Ramsey develop a formal theory of subjective probability and utility which was continued by Savage (1954), Lindley (1971), and DeGroot, among others, while Jeffreys gave the foundation for Bayesian inference, which was continued by Lindley (1965), Box and Tiao (1973), Winkler (1972), Zellner (1971) and Press (1982). Most of these writers in the latter group employ improper type prior distributions for the parameters, but Raiffa and Schlaifer (1961) introduce conjugate prior distributions, which are proper probability distributions. In recent years, Bayesian statistics has been nurtured and developed by the Seminar on Bayesian Inference in Econometrics, which has been sponsored jointly by the National Bureau of Economic Research and the National Science Foundation. This small group under the leadership of Arnold Zellner has played a significant role in fostering new ideas in Bayesian theory and methodology. One activity they sponsor is the Savage dissertation award for the best doctoral dissertation in Bayesian inference in econometrics and another is the publication of several books devoted to new research. Among these are those edited by Aykac and Brumat (1977), Fienberg and Zellner (1975) and Zellner (1980).

We see Bayesian ideas have come a long way since Bayes, however with regard to statistical practice it will be some time before they are widely accepted for the routine processing of data. One reason Bayesian methodology will have to wait is because non-Bayesian ideas are firmly entrenched, but another is two controversial aspects of Bayes theorem. First is the subjective interpretation of probability, which is so often (but not always) used in Bayesian inference, and the other is the assessment of prior information.

Subjective Probability

The most widely accepted interpretation of probability is the limiting frequency of an event in infinite repetitions of an experiment. Most practicing statisticians use this interpretation and are skeptical of the other interpretations, such as the classical, logical, and subjective interpretations, which is one way to classify the various approaches to probability.

Bernoulli (1713) and DeMorgan (1847) advance the idea of subjective probability in an informal way. For instance DeMorgan says "By degree of probability we really mean, or ought to mean, degree of belief.... Probability then, refers to and implies belief, more or less, and belief is but another name for imperfect knowledge, or it may be, expresses the mind in a state of imperfect knowledge."

Formal theories of subjective probability were formulated by Ramsey (1926) and deFinetti (1937), but, according to Kyburg (1970), it was not until Savage (1954) that this interpretation began to have an impact on statisticians. A more recent formal presentation of subjective probability is given by DeGroot (1970) who axiomatically develops utility and probability.

Why is the Bayesian interested in the subjective interpretation of probability? We saw that in Bayes' original paper, the parameter θ was given a frequency interpretation, because he constructed it that way using a ball-tossing experiment, but in many applications it is difficult to not use a subjective interpretation (non-frequency) for the parameters θ of the probability model. For example, suppose θ is the variance component of a one-way random model, then treating θ as a random variable with values occurring as some experiment is repeated, is for some, to go too far. Barnett (1982), in his book on comparative inference, refers to this idea (giving θ a frequency interpretation) as a super experiment. Thus, in many applications, it appears unreasonable to give θ a frequency interpretation, and one must rely on his subjective interpretation that the probability distribution of θ expresses one's degree-of-belief about the values of the parameter.

If θ is the probability of a coin landing "heads" on one toss, and I believe every possible value of θ is equally likely, then I would use

$$\xi(\theta) = 1, \quad 0 \leq \theta \leq 1$$

as my density for θ, but if I believed the coin is "almost" unbiased, I would use

$$\xi(\theta) \propto \theta^{10}(1-\theta)^{10}, \quad 0 \leq \theta \leq 1$$

to express my belief. Other people would probably use other densities to express their beliefs about the values of θ, but the important thing to observe is that it is possible to express one's degree-of-belief about θ via probability density functions, and in doing so, one is employing a subjective notion of probability. Of course, it is one thing to express one's degree-of-belief about the parameter θ of a binomial experiment and quite another to do it for the variance covariance matrix of a multinormal population.

In the formal theory of subjective probability one must be coherent in the following sense: the individual must not exhibit irrational behavior when assigning his subjective probability to events. He must follow certain rules so that it is not possible to contradict the usual rules of probability. The axioms of subjective probability begin with a preference relation ("at least as likely as," see DeGroot, 1970) between the events and one must be consistent with his preferences. For example, if I prefer event A to event B, and prefer B to event C, then I must prefer A to C. If the axioms of subjective probability are adopted, one will assign probabilities to the events in such a way that the Kolmogorov axioms are satisfied.

Subjective probability is unavoidable, or so it seems, with the Bayesian approach to statistics, and in the remainder of this chapter, we will see how to deal with it when the ideas of prior probability are explained. For an interesting and highly critical account of the frequency interpretation of probability the reader should read chapter VII of Jeffreys (1961).

The Basic Components of Inference

Of the many definitions of statistics, I like Barnett's (1982), who says statistics is "the study of how information should be employed to reflect on, and give guidance for action in, a practical situation involving uncertainty." For this book, if one eliminates "and give guidance for action in" in Barnett's definition, the resulting statement is what this book is about, because this book considers inference (not decision making) and inference is an act of reflection about the parameters of the probability model, among other things.

When one does Bayesian statistical inference, one is using prior information and sample data in order to find the values of the parameters in the model which generated the data.

The components of statistical inference consist of the prior information, the sample data, calculation of the posterior density of the parameters and sometimes calculation of the predictive distribution of future observations. The prior information is expressed by a probability density $\xi(\theta)$, $\theta \in \Omega$ of the parameters θ of the probability model $f(x|\theta)$, $\theta \in \Omega$, $x \in S$, where f is a density of a random variable x and S is the support of f. The information in the data $x_1, x_2, \ldots, x_n$, where $x_1, x_2, \ldots, x_n$ is a random sample from a population with density f is contained in the likelihood function $\mathcal{L}(\theta|x_1, x_2, \ldots, x_n)$, $\theta \in \Omega$, which is the joint density of the sample data, then this is combined with the prior density of θ, by Bayes theorem, and gives the posterior density of θ. Of course, these ideas were shown in Chapter 1, thus one may describe an inference problem in terms of $(S, \Omega, \xi(\theta), f(x|\theta))$ and this problem is solved once the posterior density

$$\xi(\theta|x_1, x_2, \ldots, x_n) \propto \mathcal{L}(\theta|x_1, x_2, \ldots, x_n)\xi(\theta), \quad \theta \in \Omega$$

is calculated.

From the posterior density one may make inferences for θ by examining the posterior density. Some prefer to give estimates of θ, either point or interval estimates which are computed from the posterior distribution. If θ is one-dimensional, a plot of its posterior density tells one the story about θ, but if θ is multidimensional one must be able to isolate those components of θ one is interested in.

The following diagram illustrates the approach to statistical inference which will be adopted in this book:

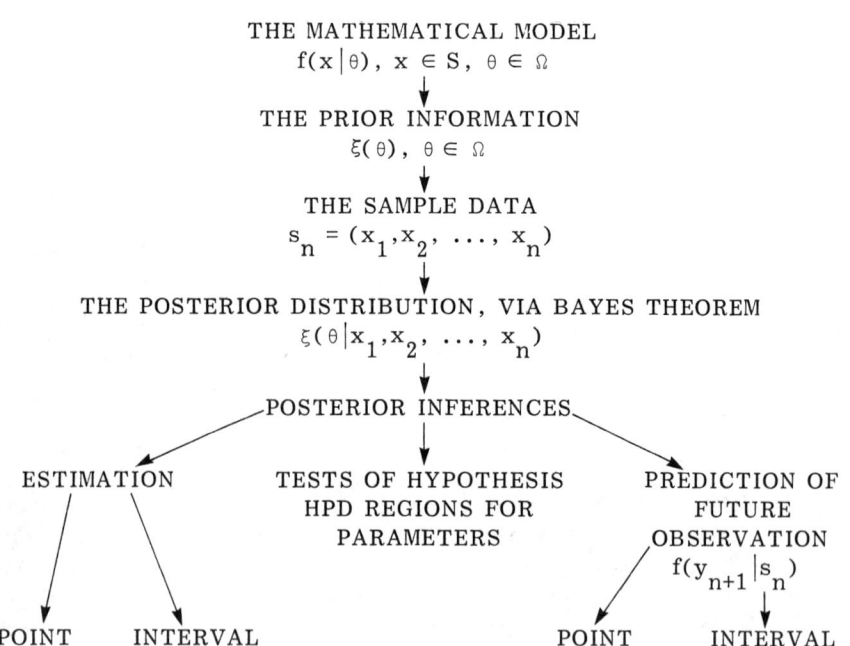

BAYESIAN STATISTICAL INFERENCE

Thus one begins the investigation with a probability model, which tells one how the observations s_n will be distributed (for each value of the parameter), and one's prior (before the data) information is expressed with the density $\xi(\theta)$, and after s_n is observed, the posterior density of θ can be calculated, via Bayes' theorem. The foundation of Bayesian inference is the posterior distribution of θ, because all other aspects of inference depend on this distribution. Perhaps one is more interested in estimation, in which case one must first decide what components of θ are of primary importance and the nuisance parameters eliminated. On the other hand if testing hypotheses is one's main objective, then one must decide how to conduct the test, i.e., to use an HPD region or a posterior odds ratio. Forecasting future observations will be done with the Bayesian predictive density $f(y_{n+1} | s_n)$, where y_{n+1} is a future observation, $y_{n+1} \in S$, and one may provide either point or interval (region) predictions.

It is obvious this is an idealization of the actual way in which one would carry out a statistical analysis. For example, how does one adopt a probability model (say an AR model for a time series) for the observations, and how does one assess the prior information which is available for the parameters? These and other questions are to be examined in the following sections.

One difficulty with advocating such an ideal system of inference is that it ignores the way scientific learning progresses. Box (1980) describes, by means of a Bayesian model, the iterative process of criticism and estimation for building statistical models, and he uses the predictive distribution as a diagnostic check on the probability model. The diagram of inference, presented here, ignores that important aspect of model building, that is, it is assumed the probability model is "correct" and is not to be changed during the analysis. Box's diagnosis should be carried out in practice but will not be adopted here, because it would carry us too far afield from our main goal, which is to demonstrate the technical mechanics of making posterior inferences. In any case, one will benefit from reading the Box paper.

The Probability Model

How does one choose a probability model $f(x|\theta)$ to represent the data generating mechanism? In our context, how does one choose one of the many linear models which are described in the earlier parts of the chapter? This is a separate area of statistics and we will not deal with it here. It will be assumed a particular model is indeed appropriate and that it was chosen either from subject matter considerations or from diagnostic checks. In addition to the classical goodness-of-fit procedures to check model adequacy, there are Bayesian methods of choosing a model from a class of models. It is a direct application of Bayes theorem and given by Zellner (1971, page 292). It will be assumed here that one has correctly chosen a parametric family of densities $\{f(x|\theta); x \in S, \theta \in \Omega\}$ which correctly represents the sample data, and our problem is to find out more about the parameter θ. Or it may be the problem of choosing one particular density (or subset of densities, as in some hypothesis testing problems) from this class.

For example, suppose one is sure an AR(p) process is appropriate for some time series data $\{y(t): t = 1, 2, \ldots, n\}$, then how should one choose p? This identification problem is analyzed in Chapter 5. Here we have assumed the AR(p) class of models is appropriate and one may describe the class by a parametric family of Gaussian densities. Our problem is to learn more about p (the order) as well as the other parameters (the autoregressive coefficients and noise variance) of the model and to forecast future observations.

Prior Information

Of all the various aspects of Bayesian inference, this is the most difficult and controversial. Many object to treating unobservable parameters θ as random variables and giving them subjective probability distributions or densities $\xi(\theta)$; however not many people object to giving observable quantities x a frequency-type probability distribution. This is probably so because one is so used to doing it this way one does not question this part of their model. To assume x is a random variable with a frequency distribution is to assume the existence of an unending sequence of repetitions of the experiment, and of course, since such repetitions in fact do not occur they must be imagined to occur; hence to assume x has a frequency type probability distribution is no more difficult (or is as difficult) than it is to assume θ has some (subjective) probability distribution. In any case to use the Bayesian approach, one must have a probability distribution for θ which expresses one's information about the parameters before the data are observed. Phrased another way, one can say to adopt a frequency probability distribution for x and a subjective probability law for θ are both quite subjective activities. However the probability law of x is more objective in the sense x can be observed and if observed its probability law can be checked against the data.

Choosing a prior density for the parameters has an interesting history. We have seen that Bayes chooses a prior density for the parameter θ of a Bernoulli sequence by constructing a "billiard" table and that $\xi(\theta)$ (uniform over (0,1)) had a frequency interpretation.

In those situations where θ cannot be given a frequency interpretation, how does one choose a prior for θ, if one knows nothing about θ? Stigler (1982) argues that Bayes would choose $\xi(\theta)$ to be uniform because the marginal distribution of x (the total number of successes in n trials) has a discrete uniform distribution over $0, 1, 2, \ldots,$ n, and a continuous uniform prior density for θ implies x has a discrete uniform distribution. This justification for a uniform prior for θ is quite different than the usual interpretation of Bayes' choice of a uniform prior for θ. Many have used the principle of insufficient reason to put a uniform distribution on the parameters. The principle of insufficient reason according to Jeffreys (1961) is "If there is no reason to believe one hypothesis rather than another, the probabilities are equal" and one may choose a uniform prior density for the parameters, however, if one knows nothing about θ, nothing is known about any function of θ, thus any function of θ should have a uniform prior density (according to the principle of insufficient reason), hence one is led to a contradiction in using the principle. Jeffreys has formulated a theory of choosing prior densities based on rules of invariance in situations where "nothing" is known about the parameters. For example, if θ_1 and θ_2 are the mean and standard deviation of a normal population, then Jeffreys' improper prior is

$$\xi(\theta_1, \theta_2) \propto \frac{1}{\theta_2}, \quad \theta_1 \in R, \quad \theta_2 > 0,$$

which implies θ_1 and θ_2 are independent, that θ_1 has a constant marginal density over R and that the marginal density of θ_2 is $\xi_2(\theta_2) \propto 1/\theta_2$, $\theta_2 > 0$. This means every value of θ_1 is equally likely and that smaller values of the standard deviation are more likely than larger values.

This type of prior was used in Chapter 1 and will be used many times in the following chapters. Often Jeffreys' prior leads to confidence intervals which are identical to those constructed from a non-Bayesian theory, but of course the interpretation of the two would be different. To some Bayesians this perhaps is an advantage of the Jeffreys' approach but to others a liability. Whatever the advantages or disadvantages, the Jeffreys improper prior has been used over and over again by Box and Tiao (1971), Zellner (1971), Lindley (1965) and many others, but Lindley now prefers to use proper prior densities instead of Jeffreys' improper prior distributions, because their use can lead to logical inconsistencies, which was demonstrated by Dawid et al. (1973). The issue is still being debated and Jaynes (1980) reportedly refutes the Dawid argument.

Conjugate densities are proper distributions which have the property that the prior and posterior densities both belong to the same parametric family, thus they have the desirable property of mathematical convenience. Raiffa and Schlaifer (1961) developed Bayesian inference and decision theory with conjugate prior densities and their work is a major contribution to statistical theory. DeGroot (1970) also gives an excellent account of conjugate prior distributions and shows how to use them with the usual (popular) population models such as normal, Bernoulli, and Poisson. Press (1982), Savage (1954), and Zellner (1971) also consider conjugate densities.

Considering Bayes' original paper about a Bernoulli population parameter θ, the conjugate class is the two-parameter beta family with density

$$\xi(\theta|\alpha,\beta) \propto \theta^{\alpha-1}(1-\theta)^{\beta-1}, \quad 0 \leq \theta \leq 1, \quad \alpha > 0, \quad \beta > 0$$

and one must choose values for α and β in order to express prior information. This class is quite flexible because the class contains a large number and variety of distributions, but this is not a general property of a conjugate class. The Wishart class is conjugate to the multinormal population with an unknown precision matrix and is not very flexible (as compared to the beta family). See Press (1982) for a discussion of the Wishart and related distributions, but the main problem with employing a conjugate class is that one must choose the parameters (hyperparameters) of the prior distribution. With the beta one must choose two scalar hyperparameters and with the p-dimensional normal distribution, one must find $p + p(p + 1)/2$ hyperparameters.

Still another principle in choosing prior distributions is that of precise measurement, which "is the kind of measurement we have when the data are so incisive as to overwhelm the initial opinion, thus bringing a great variety of realistic initial opinions to practically the same conclusion" (Savage, 1954, page 70). In choosing a prior on the basis of this principle the normalized likelihood function is approximately the same as the posterior distribution, and the prior does not have much of an effect on the posterior distribution.

After the brief history of prior information, we are still left with the problem of how to choose a prior distribution for the parameters of the model, that is, how does one express what one actually knows about θ?

Apparently it is easier to model the observations than it is to model the parameters of the model of the observations. As statisticians we are familiar with choosing various probability models to different experimental or observational conditions, and it becomes somewhat routine in some environments. For example, regression models and models for designed experiments are clearly appropriate in many situations, but when it comes to adopting a model for the parameters of the probability model (such as for the regression coefficients in an autoregressive process) it is another story.

Of course, one should not always use a Jeffreys' improper prior density or a conjugate class, or the principle of precise measurement, but try to accurately represent the prior information.

BAYESIAN STATISTICAL INFERENCE 49

Improper prior densities or a conjugate class of densities are used throughout this book to represent prior information, although this does not mean one should always use these two types of prior information. They are used because they combine nicely with the likelihood function.

Why is it so difficult to choose prior densities for the parameters? Possibly because we are so used to thinking in terms of observable random variables we have not developed a feeling or intuition about what the parameters represent in our probability models.

The approach the author favors is to choose the prior density in terms of the marginal distribution of the observations (predicted or past data). This approach is referred to as the predictive method or the device of imaginary results (Stigler, 1982, page 254) and is exploited by Good (1950, page 35), Geisser (1980), Kadane (1980), Winkler (1977), and Kadane et al. (1980). In a recent paper, Stigler (1982) argues that this "device of imaginary results" was what Bayes did in his justification for the choice of a uniform prior density for θ, where θ is the parameter of a binomial experiment and θ "must" be given a subjective probability distribution.

The device of imaginary results is thoroughly explained by Stigler (1982) and what follows is his description. Consider the usual parametric model $f(x|\theta)$, the conditional density of an observation x when θ is the value of the parameter and Ω is the parameter space, then the marginal density of x is

$$p(x) = \int_\Omega f(x|\theta)\xi(\theta)\, d\theta, \quad x \in S$$

where S is the sample space and ξ the prior density of θ, $\theta \in \Omega$. The device of imaginary results implies that one should choose $\xi(\theta)$ in such a way that it is compatible with one's choice of $p(x)$, $x \in s$. How does one choose $p(x)$? One can either "fit" p to imaginary predictions of future values of the observation or fit p to past data, which were collected under the same circumstances as the future observations will be observed. As Winkler (1980) notes, it is much easier for the experimentor to predict future values of x than it is to think directly in terms of θ.

In the binomial case considered by Bayes (1963) a uniform discrete distribution of x implied $\xi(\theta)$ was uniform on $(0,1)$, that is

$$p(x) = \frac{1}{n+1}, \quad x = 0, 1, 2, \ldots, n$$

implies $\xi(\theta) = 1$, $0 < \theta < 1$, where

$$f(x|\theta) = \binom{n}{x} \theta^x (1-\theta)^{n-x}, \qquad x = 0, 1, 2, \ldots, n.$$

Of course if $\xi(\theta)$ is uniform then $p(x)$ is uniform, but the converse is not necessarily true.

The prior predictive density $p(x)$ allows us to assess prior information about θ from prior knowledge about x instead of contemplating directly the prior density $\xi(\theta)$. If ξ depends on hyperparameters $\alpha \in A$, then the predictive density is

$$p(x|\alpha) = \int_\Omega f(x|\theta) \xi(\theta|\alpha) \, d\theta, \qquad x \in S, \quad \alpha \in A$$

and one may use this integral equation to find values of α which support or fit predicted observations $x_1, x_2, \ldots, x_n$ (sampled from the population with density $p(x|\alpha)$) or to fit past observations which were sampled from a population with density $p(x|\alpha)$.

Suppose in the binomial example

$$\xi(\theta|\alpha) \propto \theta^{\alpha_1 - 1} (1-\theta)^{\alpha_2 - 1}, \qquad 0 \leq \theta \leq 1, \quad \alpha_1 > 0, \quad \alpha_2 > 0$$

then the prior density belongs to the conjugate class, and the prior predictive density of x is

$$p(x|\alpha_1, \alpha_2) \propto \Gamma(\alpha_1 + x) \Gamma(n + \alpha_2 - x), \qquad x = 0, 1, 2, \ldots, n.$$

If one observes values $x_1, x_2, \ldots, x_n$ from this distribution, one may find values of α_1 and α_2 compatible with these predictions by the method of moments or maximum likelihood or some other principle of estimation. One might put a known prior density on α_1 and α_2 and estimate α_1 and α_2 from the conditional distribution of α_1 and α_2 given the predicted observations.

Winkler (1980) shows how this method works with the general linear model of Chapters 1 and 3 and develops the conjugate prior density for the parameters of a multiple linear regression model

$$y = \sum_{i=1}^{k} \theta_i x_i + e,$$

where y is the dependent variable, θ_i is an unknown regression coefficient, x_i is the i-th independent variable, and e is $n(0, \sigma^{-2})$. If σ^2 is known the conjugate class is normal with mean say b and covariance matrix $\sigma^2 V$, where b is k × 1 and V is k × k, then the predictive distribution of y is normal with mean

$$E(y | \sigma^2, x) = b'x,$$

$$Var(y | \sigma^2, x) = \sigma^2(x'Vx + 1),$$

where $x' = (x_1, x_2, \ldots, x_k)$, $b = (b_1, b_2, \ldots, b_k)'$. Winkler shows how one may assess this predictive distribution, thus choosing values for b and V, the hyperparameters of the prior distribution of $\theta = (\theta_1, \theta_2, \ldots, \theta_k)'$. Given x, one predicts the mean and variance of y for $k(k+1)/2$ values of x, where k of the equations (assessments) allows one to find b, and all $k(k+1)/2$ allows one to find V. Winkler also shows how to deal with the case when σ^2 is unknown and Kadane et al. (1980) develop a sophisticated but complicated method of fitting the hyperparameters of prior distributions.

The advantage of the predictive approach is that the experimenter is better able to think about an observable random variable than he is (directly) about an unobservable parameter θ. A disadvantage is that one must usually restrict prior information to a parametric family such as a conjugate class of distributions, and the class may not be as flexible as one desires. Perhaps this can be avoided by a mixture of densities from the conjugate class.

The principle of imaginary results appears to be a very promising way to assess prior information and it should be used in practice to see if experimenters and other users of statistical techniques are willing to adopt it.

One of the latest developments assessing prior information is the idea that the parameters of the model are exchangeable. The idea of exchangeability was introduced by deFinetti (1937) (see Kyburg and Smokler, 1964) and has been applied by Lindley and Smith (1972) and Leonard (1972) for the Bayesian analysis of linear models and binomial probability laws.

Let us consider an example given by Lindley and Smith (1972) where $y_1, y_2, \ldots, y_n$ are independent, normally distributed, have a known variance and $E(y_i) = \theta_i$, $i = 1, 2, \ldots, n$, where the means $\theta_1, \theta_2, \ldots, \theta_n$ are unknown real parameters and very little is known about the means. Since very little is known about the means it perhaps is reasonable to assume the means are exchangeable, that is, one's prior information about θ_1 is the same as that about θ_2, or any other θ_i and that in general one's prior knowledge about $\theta_1, \theta_2, \ldots, \theta_n$ is invariant under any permutation of the indices. An exchangeable sequence can arise as a mixture of independent identically distributed random variables, given some hyperparameter μ, say, that is the density of $\theta = (\theta_1, \theta_2, \ldots, \theta_n)'$ is

$$g(\theta) = \int \prod_{i=1}^{n} p(\theta_i | \mu) \, dQ(\mu),$$

where $p(\theta_i | \mu)$ is the conditional density of θ_i given μ, $Q(\mu)$ the distribution function of μ, and p and Q are arbitrary. Thus $\theta_1, \theta_2, \ldots, \theta_n$ (given μ) constitute a random sample of size n from a population with density p, and μ is a random variable with distribution function $Q(\mu)$. Thus, for example, we might let p be normal with mean μ and known variance τ^{-1} and give μ some known probability distribution, or we could have the prior distribution of μ depend on unknown hyperparameters and put a known distribution on them and continue in a hierarchical fashion.

Lindley and Smith next consider a randomized block layout of b blocks in each of which t treatments are assigned, then the model is

$$y_{ij} = \mu + \alpha_i + \beta_j + e_{ij},$$

where y_{ij} is the observation on the experimental unit of block j to which treatment i was applied, the errors are independent $n(0, \sigma^{-2})$ and one must assign a prior distribution to $\mu, \alpha_1, \alpha_2, \ldots, \alpha_t, \beta_1, \beta_2, \ldots, \beta_b$. If the treatment constants are exchangeable and also the block parameters, and they are independent, then one perhaps would assign $\alpha_i \sim n(0, \sigma_\alpha^{-2})$, $\beta_j \sim n(0, \sigma_\beta^{-2})$ and $\mu \sim n(0, \sigma_\mu^{-2})$, where these distributions are independent. If the variances σ_α^2, σ_β^2, and σ_μ^2 are known, one stops here at this stage and proceeds with the posterior analysis. Also, one might consider these variances as unknown and give them some known distribution, and this is also considered by Lindley and Smith.

The idea of exchangeability has played an important role in the Bayesian analysis of linear models because deFinetti's work provides a mathematical justification for selecting a prior distribution when

BAYESIAN STATISTICAL INFERENCE

a priori symmetry is present about the parameters. It imposes a reasonable restriction on the prior probability distribution when our knowledge of the parameters exhibits this peculiar symmetric property, but it usually doesn't tell us what particular parametric family is the appropriate one to use. For example, in the above two problems exchangeability of the model parameters does not imply a normal prior density for them, although these parameters are indeed exchangeable.

This section has reviewed some of the ways one may express prior information about the parameters. We have looked at vague prior information, conjugate prior densities, the principle of insufficient reason, the principle of imaginary results, and exchangeable prior information. Also, empirical methods of assessing prior information were discussed, thus it is easy to see that much has been developed in the area of prior information but that also much remains to be accomplished.

The Posterior Analysis

Suppose one is now satisfied that the probability model, $f(x|\theta)$, for observations and the prior density, $\xi(\theta)$, for the parameters are appropriate and one must determine the plausible values of θ. One now enters the posterior analysis stage of a Bayesian statistical analysis, where one perhaps might want to estimate θ (or some of its components), test hypotheses about θ, predict future values of the observations, or possibly control the observations at predetermined levels.

Regardless of what particular activity is contemplated, one must first find the posterior density of θ,

$$\xi(\theta|x) \propto f(x|\theta)\, \xi(\theta), \qquad \theta \in \Omega \subseteq R^p$$

which is found by Bayes theorem. Often all the components of θ are of interest, but sometimes some of these components θ_1 are regarded as nuisance parameters and the remaining $\theta_2 (q \times 1)$ are of primary interest. How should one make inferences about θ_2? The Bayesian will use

$$\xi_2(\theta_2|x) = \int_{\Omega_1} \xi(\theta_1, \theta_2|x)\, d\theta_1, \qquad \theta_2 \in \Omega_2, \qquad \Omega = \Omega_1 \times \Omega_2,$$

where θ_1 is $(p-q) \times 1$. ξ_2 is called the marginal posterior density of θ_2, and as with any posterior density function it may be used to estimate and test hypotheses concerning θ_2. To estimate θ_2, either the marginal mean, median, or mode, or any other typical or representative value of θ_2 can be computed from the marginal posterior density of θ_2. Of course these estimates are summary characteristics of $\xi(\theta_2|x)$ and only tell us part of the story about the parameter and they are not a substitute for the entire distribution.

Consider an experiment which has been conducted according to a randomized block design (see Chapter 3), then the primary parameters are the treatment effects but the block effects and error variance are regarded as nuisance parameters. Then the Bayesian would test for significance of treatment effects by referring to the marginal posterior distribution of those parameters.

To test hypotheses about the parameters θ_2, one perhaps would find a $1 - \gamma$ HPD, $0 \leq \gamma \leq 1$ (highest posterior density) region $R_\gamma(\theta_2)$ from $\xi(\theta_2|x)$. Such a region has the property that if $\theta_2' \in R_\gamma(\theta_2)$ and $\theta_2'' \notin R_\gamma(\theta_2)$, then $\xi(\theta_2'|x) > \xi(\theta_2''|x)$, that is parameter values inside the region have larger posterior probability density than those excluded from the region which must satisfy

$$1 - \gamma = \int_{R_\gamma(\theta_2) \subseteq \Omega_2} \xi(\theta_2|x) \, d\theta_2,$$

that is the HPD region has posterior probability content $1 - \gamma$. Box and Tiao (1971) and Chapter 1 give other properties of HPD regions and they are used in Chapter 3. To test the hypothesis $H_0: \theta_2 = \theta_{20}$ (known) versus $H_a: \theta_2 \neq \theta_{20}$ one rejects H_0 if θ_{20} (the hypothesized value) is excluded from the region. In the case of a randomized block design, one tests for treatment effects by finding an HPD region for the treatment contrasts.

The commonly used other way to test hypotheses is based on the posterior odds ratio but is not implemented in this book. DeGroot (1970), Zellner (1971), and others use the odds ratio, however Box and Tiao (1971) do not. Suppose θ_2 is scalar, that Ω_2 is some nondegenerate interval of the real line and that $H_0: \theta_2 \in (a,b) \subset \Omega_2$ while $H_a: \theta_2 \notin (a,b)$, $\theta_2 \in \Omega_2$, then the posterior odds ratio is

$$O(H_0|H_a, x) = \frac{P_\gamma\{\theta_2 \in (a,b)|x\}}{P_\gamma\{\theta_2 \in \Omega_2 - (a,b)|x\}}$$

where

$$P_\gamma\{\theta_2 \in (a,b) | x\} = \int_a^b \xi(\theta_2|x) \, d\theta,$$

and the larger the ratio the more the indication the null hypothesis is true.

If the null hypothesis is $H_0: \theta_2 = \theta_{20}$ versus $H_a: \theta_2 \neq \theta_{20}$, Jeffreys (1939/1961) expresses prior information about θ in terms of a mixed (part discrete, part continuous) distribution made up of a discrete prior probability of $\xi(\theta_{20})$ at $\theta_2 = \theta_{20}$ and a probability density $\xi(\theta)$, $\theta \neq \theta_{20}$ for the other values of θ, then the posterior odds ratio for the null versus the alternative is

$$\xi(\theta_{20}|x) / \int_{\Omega_2 - \{\theta_{20}\}} \xi(\theta|x) \, dx = O(H_0|H_a, x).$$

For an up-to-date account of Bayesian hypothesis testing see Bernardo (1980) and Barnett (1982). Since the posterior odds ratio is still being developed, it will not be used to test hypotheses but instead the HPD region method which was formulated by Lindley (1965) will be employed. Lindley advocates the HPD region only when a vague prior density is appropriate. However, it is not an unreasonable procedure with any prior probability density function, such as a member of a conjugate class and it is done this way, sometimes (see Chapter 3) in this book.

For almost all the linear models of this book, estimation is accomplished by formulas for the posterior mean and variance of the parameters of the model. For example, Chapter 4 examines the random and mixed linear models where the marginal distribution of each variance component is found (approximately) and formulas for the mean and variance are derived. In this way each component can be estimated and an idea of the uncertainty in the estimate is given by the marginal posterior variance. Unfortunately, it was not possible to find estimates for the fixed effects of the model because their marginal posterior distribution cannot be determined analytically and must be determined numerically as was done by Box and Tiao (1973).

Another example of estimation is the Kalman filter, which is the mean of the marginal normal posterior distribution of the successive states of a linear dynamic system. This particular method of estimation is explained in Chapter 6.

Sometimes all of the parameters of the model are nuisance parameters, in the sense one is mainly interested in predicting future observations or controlling the future values of the dependent variable of a regression analysis or time series.

Bayes Forecasting

Suppose one observes a series of observations y(1),y(2), ..., y(n) and one wants to forecast the next observation y(n + 1) and that the generating mechanism of the series is known and consequently one "knows" the joint density f[y(1),y(2), ..., y(n)|θ], θ ∈ Ω, of the observations and the prior density for θ, say ξ(θ). Then how does the Bayesian predict the next observation y(n + 1)? The standard way is to find the conditional distribution of y(n + 1) given y(1),y(2), ..., y(n) and this is called the Bayesian predictive distribution of y(n + 1) and is

$$f[y(n+1)|y(1), \ldots, y(n)] = \int_\Omega f[y(n+1)|y(1),y(2), \ldots, y(n),\theta]$$

$$\times \xi(\theta|y(1), \ldots, y(n)] \, d\theta,$$

where the first factor of the integrand is the conditional density of y(n + 1) given θ and the previous observations, and the second factor is the posterior density of θ. The predictive density is the average of the conditional predictive density of y(n + 1), given θ, with respect to the posterior distribution of θ, that is if one knew θ, one would predict y(n + 1) with f[y(n + 1)|y(1),y(2), ..., y(n),θ] but since θ is unknown this density is averaged over the posterior distribution of θ and a Bayesian forecast consists of an average of conditional predictions.

We have seen how this approach is used with the general linear model of Chapter 1. In what is to follow, the Bayesian predictive density is found for the time series models of Chapter 5 and the linear dynamic model of Chapter 6. Often with linear dynamic systems it is important to control the states (parameters) of the system at predetermined levels and to do this one must be able to forecast the future observations at each stage in the evolution of the system. The Bayesian predictive density plays an important role in the control and monitoring of physical and economic systems and its use will be demonstrated in Chapters 5 and 6.

Press (1982) and Zellner (1971) consider a control problem which is not discussed in this book. Suppose

$$Y = X\theta + Z\phi + e$$

is a linear model, where Y is the vector of observations, X is a matrix consisting of the observations on independent variables which cannot be controlled, θ is an unknown parameter vector, Z a matrix consisting of the values of control variables, φ another unknown parameter vector, and e an error vector. Under the assumption of normally distributed errors and a vague prior for the parameters, Zellner shows how to choose the control variable Z so that Y is "close" to some predetermined value. He adopts three control strategies to do this but in all three the Bayesian predictive density plays a crucial role. These problems, although important, will not be examined because the author is unable to add anything new and Zellner's account is excellent.

The type of control problem which occurs in Chapter 6 is similar to the following problem. Suppose $\{y(t): t = 1, 2, \ldots, n, \ldots\}$ is a time series which follows the law

$$y(t) = \theta(t)y(t - 1) + \varepsilon(t)$$

and

$$\theta(t) = G\theta(t - 1) + H(t)x(t - 1) + e(t)$$

where $t = 1, 2, \ldots$. This is a dynamic AR(1) process where the time dependent autoregressive coefficients follow an AR(1) law. The observation errors $\varepsilon(t)$ and the system errors $e(t)$ are independent normally distributed random variables, $y(0)$ is known, G is known, and $H(t)$ is known, and at the initial stage, assume $\theta(0)$ is known. Among these and other assumptions, the problem is to control $\theta(1)$ at some level, say $T(1)$, but to do this one must have an idea of the value of $\theta(1)$ (as a function of $H(0)$), thus one must predict $\theta(1)$ (one-step-ahead). Since the posterior mean of $\theta(1)$ depends on $y(1)$ to predict $\theta(1)$, one averages this posterior mean over the Bayesian predictive distribution of $y(1)$, because $y(1)$ has yet to be observed. This procedure is applied at each stage in the evolution of the system.

Linear dynamic systems are very interesting models to study from a Bayesian perspective because in order to analyze them one must use the complete arsenal of Bayesian methodology.

Once the Bayesian predictive density is found, one predicts the future value $y(n + 1)$ by examining this distribution as one would do in making inferences for the parameters from their posterior distribution. If one wants one value to forecast $y(n + 1)$, perhaps the mean, median, or mode will do, and if one wants an interval forecast, one could construct an HPD (highest predictive density?) for $y(n + 1)$, analogous to the way one would find a region of highest posterior density for the parameters of the model.

The Bayesian predictive distribution for $y(n + 1)$ turns out to be a univariate, multivariate, or matrix t (depending on the dimension of $y(n + 1)$) for most of the linear models, with the exception of the linear models exhibiting structural change, where the predictive distribution is a mixture of t distributions. Other exceptions are the moving average process, the ARMA model and linear dynamic models.

LINEAR MODELS AND BAYESIAN INFERENCE

SUMMARY

This chapter gives the reader a chance to see what this book is about. First, a description of each class of models is given with some applications provided illustrating their use. For example, we have seen that linear dynamic models are sophisticated ways for modeling the monitoring and control of many physical and economic variables and that models for structural change give one a degree of flexibility not possessed by the others.

Secondly, we see the way a Bayesian would analyze data which were generated by one of the linear models. The components of an analysis, namely, the probability model, prior information, the posterior and predictive analysis were carefully explained. Problems with choosing prior information were discussed and it was decided that the "principle of imaginary results" was a promising approach to assessing prior information, but that there was no easy and "quick" solution to this vexing problem. Further research along the lines suggested by Winkler (1980) and Kadane et al. (1980) should enhance the way one expresses their prior information about the parameters, and if these ideas can be made operational so that they become standard in the routine problems of regression, the analysis of variance, and time series, the Bayesian methodology of this book will easily be implemented.

EXERCISES

1. If $x_1, x_2, \ldots, x_n$ is a random sample from a population with a mass function

    $$f(x|\theta) = \theta^x (1-\theta)^{1-x}, \quad x = 0, 1, \quad 0 \leq \theta \leq 1$$

    and

    $$\xi(\theta) \propto \theta^{\alpha-1}(1-\theta)^{\beta-1}, \quad 0 \leq \theta \leq 1, \quad \alpha < 0, \beta > 0$$

    is the prior density of θ, find:
    (a) The posterior density of θ,
    (b) The predictive distribution of x_{n+1}, and
    (c) The joint predictive distribution of x_{n+1}, x_{n+2}.
    (d) If $n = 4$, $x_1 = 1$, $x_2 = 1$, $x_3 = 0$, $x_4 = 0$, $\alpha = 1 = \beta$, plot the posterior density of θ and predictive mass function of x_{n+1}.

EXERCISES

2. Consider an exponential population with density

$$f(x|\theta) = \theta e^{-\theta x}, \quad x > 0, \quad \theta > 0$$

and let $x_1, x_2, \ldots, x_n$ be a random sample.
   (a) Examine the likelihood function for θ (based on $x_1, x_2, \ldots, x_n$) as a function of θ, where $\theta > 0$. What type of function is it?
   (b) Upon examining the likelihood function, what is a conjugate prior density for θ?
   (c) Using the conjugate prior for θ, derive the posterior density of θ and the predictive density of x_{n+1}.
   (d) Suppose, a priori, that previous values of x have been observed, namely $x_1^* = 1$, $x_2^* = 6$, and $x_3^* = 4$. What values would you use for the hyperparameters of the conjugate prior density?

3. Let $x_1, x_2, \ldots, x_n$ be a random sample from a Poisson population

$$f(x|\theta) = e^{-\theta} \theta^x / x!, \quad x = 0, 1, 2, \ldots$$

and suppose the mean θ has prior density

$$\xi(\theta) \propto \theta^{\alpha-1} e^{-\theta\beta}, \quad \alpha > 0, \quad \beta > 0, \quad \theta > 0.$$

   (a) Is the above a conjugate prior density?
   (b) Using the above prior density, find a HPD region for θ.
   (c) Find the predictive mass function of x_{n+1}.
   (d) Give an interval forecast of x_{n+1}.

4. Let $x_1, x_2, \ldots, x_n$ be a random sample of size n from a population with density

$$f(x|\theta_1, \theta_2) = (\theta_2 - \theta_1)^{-1}, \quad \theta_1 < x < \theta_2$$

where θ_1 and θ_2 are unknown real parameters. Give a conjugate prior density for θ_1 and θ_2 and use it to find the predictive density of x_{n+1}. Refer to Chapter 9 of DeGroot (1970).

5. If x and y have a normal-gamma distribution, that is the conditional distribution of x given y is normal and the marginal distribution

of y is gamma, find the conditional distribution of y given x. Also, find the upper .01 point of the marginal distribution of x. Refer to the Appendix.

6. If $x_1, x_2, \ldots, x_n$ is a random sample of size n from a bivariate normal population with mean vector μ and prediction matrix P (2 × 2), and if the prior distribution of μ and P is a normal-Wishart, find the posterior distribution of the correlation coefficient.

7. The gamma population has density

$$f(x|\alpha, \beta) = \frac{\Gamma^{-1}(\alpha)}{\beta^{-\alpha}} x^{\alpha-1} e^{-x\beta}, \quad x > 0, \quad \beta > 0$$

where Γ is the gamma function.
   (a) Find a conjugate prior density for α and β.
   (b) If α is known, what is the conjugate class of prior densities for β?
   (c) If α is known, and $x_1^*, x_2^*, \ldots, x_m^*$ are previous values of x (the population random variable), how would you choose values for the parameters of the conjugate prior density of β?

8. Is the sampling distribution of an estimator of θ irrelevant in making inferences about θ? Is the posterior distribution of θ relevant in making inferences about θ? Which approach is more relevant? Explain your answers carefully.

9. What are the controversial aspects of the Bayesian approach to statistical analysis?

10. Why isn't the Bayesian approach to statistics more popular? Is there an inherent defect in the Bayesian analysis of a statistical problem?

REFERENCES

Aykac, Ahmet and Carlo Brumat (1977). *New Developments in the Applications of Bayesian Methods.* North-Holland Publishing Company, New York.

Barnett, Vic (1982). *Comparative Statistical Inference,* second edition. John Wiley and Sons, New York.

Bayes, T. (1963). "An essay toward solving a problem in the doctrine of chances," Philosophical Transactions Royal Society, Vol. 53, pp. 320-418.

Berger, James O. (1980). *Statistical Decision Theory, Foundations, Concepts, and Methods.* Springer-Verlag, New York.

REFERENCES

Bernardo, J. M. (1980). "A Bayesian analysis of classical hypothesis testing," in *Bayesian Statistics: Proceedings of the First International Meeting, Valencia, 1979,* edited by Bernardo, DeGroot, Lindley and Smith.

Bernoulli, J. (1713). *Ars Conjectandi.* Basel.

Box, G. E. P. (1980). "Sampling and Bayes inference in scientific modeling and robustness" (with discussion), Journal of the Royal Statistical Society, Series A, Vol. 143, part 4, pp. 383-430.

Box, G. E. P. and Gwilym M. Jenkins (1970). *Time Series Analysis, Forecasting, and Control,* Holden-Day, San Francisco.

Box, G. E. P. and G. C. Tiao (1973). *Bayesian Inference in Statistical Analysis,* Addison-Wesley, Reading, Ma.

Box, G. E. P. and G. C. Tiao (1981). "Modeling multiple time series," Journal of the American Statistical Association, Vol. 76, No. 376, pp. 802-816.

Broemeling, Lyle (1982). "Introduction," Journal of Econometrics, Vol. 19, pp. 1-5.

Davies, O. L. (1967). *Statistical Methods in Research and Production,* third edition, Oliver and Boyd, London.

Dawid, A. P., Stone, M. and J. W. Zedek (1973). "Marginalization paradoxes in Bayesian structural inference" (with discussion), Journal of the Royal Statistical Society, B, Vol. 35, pp. 189-233.

deFinetti, B. (1930). "Sulla proprieti conglomerativa delle probabilita subordinate," Rend R. Inst. Lombardo (Milano), Vol. 63, pp. 414-418.

deFinetti, B. (1937). "Foresight: its logical laws, its subjective sources," Annals de l'Institute Henri Poincare. Reprinted (in translation) in Kyburg and Smokler (1964).

DeGroot, M. H. (1970). *Optimal Statistical Decisions,* McGraw-Hill, New York.

DeMorgan, August (1847). *Formal Logic,* Taylor and Walton, London.

Fienberg, S. E. and A. Zellner (1975). *Studies in Bayesian Econometrics and Statistics.* North-Holland Publishing Co., New York.

Fisher, R. A. (1922). "On the mathematical foundations of theoretical statistics," Phil. Transactions Roy. Soc. London, A, Vol. 222, pp. 309-362.

Ferguson, Thomas, S. (1967). *Mathematical Statistics, A Decision Theoretic Approach,* Academic Press, New York.

Geisser, S. (1980). "A predictivistic primer," in *Bayesian Analysis in Econometrics and Statistics,* edited by A. Zellner, North-Holland, Amsterdam.

Good, I. J. (1950). *Probability and the Weighing of Evidence,* Griffin, London.

Holbert, Don (1982). "A Bayesian analysis of a switching linear model," Journal of Econometrics, Vol. 19, pp. 77-87.

Jaynes, Edwin T. (1980). "Marginalization and prior probabilities," in *Bayesian Analysis in Econometrics and Statistics*, edited by A. Zellner, North Holland Publishing Co., Amsterdam.

Jazwinski, Andrew H. (1970). *Stochastic Processes and Filtering Theory*, Academic Press, New York.

Jeffreys, Harold (1939/1961). *Theory of Probability*, Oxford at the Clarendon Press, first and third editions.

Jenkins, Gwilym M. (1979). *Practical Experiences with Modeling and Forecasting Time Series*, Gwilym Jenkins and Partners (overseas) Ltd.

Kadane, J. B. (1980). "Predictive and structural methods for eliciting prior distributions," in *Bayesian Analysis in Econometrics and Statistics*, edited by A. Zellner, North-Holland, Amsterdam.

Kadane, J. B., J. M. Dickey, R. I. Winkler, W. S. Smith, and S. C. Peters (1980). "Interactive elicitation of opinion for a normal linear model," Journal of American Statistical Association, Vol. 75, pp. 845-854.

Kalman, R. E. (1960). "A new approach to linear filtering and prediction problems," Transactions of ASME, Series D, Journal of Basic Engineering, Vol. 83, pp. 95-108.

Kyburg, Henry E., Jr. (1970). *Probability and Inductive Logic*. The Macmillan Company, Collier-Macmillan Limited, London.

Kyburg, H. E., Jr., and Smokler, H. E. (eds.) (1964). *Studies in Subjective Probability*, John Wiley and Sons, New York.

Laplace, P. S. de (1951). *A Philosophical Essay on Probabilities*. (An English translation by Truscott and Emory of 1820 edition.) Dover, New York.

Leonard, T. (1972). "Bayesian methods for binomial data," Biometrika, Vol. 59, pp. 581-589.

Lindley, D. V. (1965). *Introduction to Probability and Statistics from a Bayesian Viewpoint, Part 2, Inference*. Cambridge University Press.

Lindley, D. V. (1971). *Making Decisions*, Wiley-Interscience, John Wiley and Sons Ltd., London.

Lindley, D. V. and A. F. M. Smith (1972). "Bayes estimates for the linear model," Journal of the Royal Statistical Society, B, Vol. 34, pp. 1-42.

McGee, V. E. and W. T. Carleton (1970). "Piecewise regression," Journal of the American Statistical Association, Vol. 65, pp. 1109-1124.

Monahan, J. F. (1983). "Fully Bayesian analysis of ARMA time series models," Journal of Econometrics, Vol. 21, pp. 307-331.

Pearson, K. (1920). "Note on the fundamental problem of practical statistics," Biometrika, Vol. 13, pp. 300-301.

Poirier, D. J. (1976). *The Econometrics of Structural Change*, North-Holland Publishing Company, Amsterdam.

REFERENCES

Press, S. James (1982). *Applied Multivariate Analysis: Using Bayesian and Frequentist Methods of Inference,* second edition, Robert E. Krieger Publishing Company, Malabar, Florida.

Raiffa, H. and R. Schlaifer (1961). *Applied Statistical Decision Theory,* Division of Research, Graduate School of Business Administration, Harvard University, Cambridge, Massachusetts.

Rajagopalan, Muthiya (1980). Bayesian Inference for the Variance Components in Mixed Linear Models. Ph.D. dissertation, December, 1980, Oklahoma State University, Stillwater, Oklahoma.

Ramsey, F. P. (1931/1964). "Truth and probability," in Kyburg and Smokler (1964), Reprinted from the *Foundations of Mathematics and other Essays,* Kagan, Paul, Trench, London, 1931.

Savage, L. J. (1954). *The Foundation of Statistics.* John Wiley and Sons, Inc., New York.

Stigler, Stephen M. (1982). "Thomas Bayes and Bayesian inference," Journal of the Royal Statistical Society, A, Vol. 145, part 2, pp. 250-258.

Winkler, R. L. (1972). *Introduction to Bayesian Inference and Decision,* Holt, Rinehart and Winston, New York.

Winkler, R. L. (1980). "Prior distributions and model building in regression analysis," in *New Developments in the Applications of Bayesian Methods,* edited by Aykac and Brumat, North-Holland Publishing Company, Amsterdam.

Zellner, A. (1971). *An Introduction to Bayesian Inference in Econometrics,* John Wiley and Sons, Inc., New York.

Zellner, Arnold (1980). *Bayesian Analysis in Econometrics and Statistics, Essays in Honor of Harold Jeffreys,* North-Holland Publishing Company, New York.

3
THE TRADITIONAL LINEAR MODELS

INTRODUCTION

A first one-semester course in linear models introduces the student to simple and multiple regression models and fixed models of designed experiments, and if time permits to random and mixed models of designed experiments. These are the so-called standard or traditional linear models that the statistician is expected to know and be able to use.

The Bayesian analysis of the general linear model was introduced in Chapter 1 and will be used to analyze the regression models and fixed models for designed experiments, and these are to be studied in this chapter. The random and fixed models will be examined in the following chapter.

First, elementary models for one and two normal populations are given and the Bayesian approach is illustrated with the prior, posterior, and predictive analysis.

Secondly, simple and multiple linear regression models are analyzed, then nonlinear regression models, and finally models for designed experiments including models containing concomitant variables are introduced and Bayesian inferences are provided and illustrated with numerical examples.

Prior information about the parameters of these models is specified by either a vague improper Jeffreys' density or a member of a class which is conjugate to the model. We have seen the conjugate class is the normal-gamma family of distributions and that the four parameters of this distribution must be specified from one's prior knowledge of the future experiment. This chapter will use Winkler's (1977) method of the prior density of a future observation to set the values of the parameters of the normal-gamma prior density.

Once having determined the prior distribution, the joint posterior distribution of all the parameters and the marginal distribution of certain subsets of the parameters will be found. Of course, from Chapter 1 the relevant distribution theory has been derived and all that is needed is to find the parameters of the various posterior distributions. For example, in a simple linear regression model, the intercept and slope parameters will have a bivariate t distribution, while the marginal distribution of these parameters will be univariate t distributions. Having the sample information and the parameters of the prior distribution will allow one to know the precision matrix, degrees of freedom, and mean vector of the joint posterior distribution of the slope and intercept coefficients of the model. Thus, this will allow one to test hypotheses about the parameters, using HPD regions, and construct confidence intervals for these parameters as well as others such as the average response of the dependent variable at selected values of the regressor variables.

To predict a future value of the dependent variable, the Bayesian predictive density will give point and interval forecasts.

This type of analysis will be repeated for multiple linear regression models and those for some designed experiments and accompanied by numerical examples which will provide tables and graphs of posterior and predictive densities.

The reader should consult Box and Tiao (1973), Lindley (1965) and Zellner (1971) for a similar presentation along the lines developed in this chapter.

PRIOR INFORMATION

Two types of prior information will be used in this chapter. First, Jeffreys' improper density

$$\xi(\theta,\tau) \propto 1/\tau, \quad \theta \in R^p, \quad \tau > 0 \qquad (1.13)$$

is to be used when very little prior information is available for the parameters θ and τ of the linear model, and when one is more informed, a priori, recall

$$\xi(\theta,\tau) = \xi_1(\theta|\tau)\xi(\tau), \quad \tau > 0, \quad \theta \in R^p \qquad (1.5)$$

where

PRIOR INFORMATION 67

$$\xi_1(\theta|\tau) \propto \tau^{p/2} \exp -\frac{\tau}{2}(\theta - \mu)'P(\theta - \mu), \quad \theta \in R^p \quad (1.6)$$

and

$$\xi_2(\tau) \propto \tau^{\alpha-1} e^{-\tau\beta}, \quad \tau > 0 \quad (1.7)$$

is the normal-gamma density with known parameters $\mu \in R^p$, P a symmetric positive definite matrix, $\alpha > 0$, and $\beta > 0$. Of course the difficulty with using this conjugate class is that the parameters need to be specified, thus how does one know what values to assign to the hyperparameters?

Consider one way advocated by Winkler (1977) but what is to follow is my interpretation. Thus consider the model

$$Y = X\theta + e \quad (1.1)$$

which describes the experiment which will be run in the future, where Y is $n \times 1$, X is $n \times p$, θ is $p \times 1$, and e is $n(0, \tau I_n)$. Also consider the model

$$Z = X^*\theta + \varepsilon, \quad (3.1)$$

where Z is $m \times 1$, X^* is $m \times p$, θ is $p \times 1$, and $\varepsilon \sim N(0, \tau^{-1} I_m)$, and where θ and τ are the parameters of both models. Note θ and τ have the same meaning in both models. The latter model (3.1) thus describes a related experiment (related by the parameters θ and τ) which perhaps is a past experiment with past data Z when the design matrix was X^*. On the other hand (3.1) is perhaps a hypothetical experiment with hypothetical observations Z when the design matrix is X^*. The hypothetical data Z are supplied by the experimenter and in this situation, if Z is the experimenter's guess or prediction of Y, then the design matrix is $X(= X^*)$.

What is the predictive prior density of Z? It is the marginal density of Z, which will depend on the hyperparameters μ, P, α, and β. The conditional density of Z given θ and τ is from (3.1).

$$g(z|\theta, \tau) \propto \tau^{m/2} \exp -\frac{\tau}{2}(z - X^*\theta)'(z - X^*\theta), \quad z \in R^m \quad (3.2)$$

thus the marginal density of Z is

$$h(z|\mu, P, \alpha, \beta) \propto \frac{1}{[(z - A^{-1}B)'A(z - A^{-1}B) + C - B'A^{-1}B]^{(m+2\alpha)/2}},$$

$$z \in R^m, \quad (3.3)$$

where

$$A = I - X^*(X^{*'}X^* + P)^{-1}X^{*'}, \qquad (3.4)$$

$$B = X^*(X^{*'}X^* + P)^{-1}P\mu, \qquad (3.5)$$

and

$$C = \mu'P\mu + 2\beta - \mu'P(X^{*'}X^* + P)^{-1}P\mu. \qquad (3.6)$$

Since the prior predictive density (3.3) of Z is a m-variate multivariate t with 2α degrees of freedom, location

$$E(Z) = A^{-1}B$$

and precision matrix

$$P(Z) = \frac{(2\alpha)A}{C - B'A^{-1}B}$$

and since these moments depend on the hyperparameters, one may choose them in such a way that the equations

$$A^{-1}B = X^*(X^{*'}X^*)^{-1}X^{*'}Z \qquad (3.7)$$

and

$$\frac{(2\alpha)A}{C - B'A^{-1}B} = \frac{Z'[I - X^*(X^{*'}X^*)^{-1}X^{*'}]ZI_m}{(m - p)} \qquad (3.8)$$

are satisfied; however, the solution is not unique. The right-hand side of these two equations were obtained from (3.1) as follows. From (3.1) $E(Z|\theta) = X^*\theta$ and this is estimated by the right-hand side of (3.7), since the usual estimate of θ is $(X^{*'}X^*)^{-1}X^{*'}Z$. Also from (3.1), the dispersion matrix of Z is τI_m, thus its precision matrix is $\tau^{-1}I_m$ and τ^{-1}, the variance is usually estimated by the residual sum of squares of (3.8).

Still a much easier way to set the values of the hyperparameters is to notice that from (3.1), the usual estimators of θ and τ^{-1} are

$$\theta^* = (X^{*'}X^*)^{-1}X^{*'}Z \qquad (3.9)$$

and

$$(\tau^{-1})^* = \frac{Z'Z - Z'X^*(X^{*'}X^*)^{-1}X^{*'}Z}{(m-p)}, \quad (3.10)$$

and since the prior mean of θ is μ, choose $\mu = \theta^*$. In the same way, since the prior mean of τ^{-1} is $\beta/(\alpha - 1)$, $(\alpha > 1)$, choose α and β such that $\beta/(\alpha - 1) = (\tau^{-1})^*$. Of course the choice of α and β is not unique. The prior dispersion of θ is $\beta P(\alpha - 1)^{-1}$, thus choose P so that $(\tau^*)^{-1}P = (\tau^*)^{-1}(X^{*'}X^*)^{-1}$ or let $P = (X^{*'}X^*)^{-1}$. This determines values of the hyperparameters from the past or hypothetical experiment (3.1), although the choice of α and β is not unique. Is there a way to choose α and β uniquely?

Of course it is important that Z be known, and this is not a problem if Z is from a past experiment, however if Z must be chosen hypothetically as the future value of Y, more subjectivity is involved because if Z is the future value of Y, Z must be supplied by the experimenter in the hypothetical future experiment

$$Z = X\theta + e,$$

which is the experiment to be conducted where Y (in place of Z) will be observed.

As Winkler emphasizes, the advantage of this method is that the values of the hyperparameters are determined indirectly in terms of future (or past) values of Y and not directly in terms of the hyperparameters. It is much easier to "guess" future values of Y or to observe past values of Y than it is to "guess" the values of the hyperparameters.

For additional information on choosing the hyperparameters of the conjugate prior density (1.5), Zellner (1980) uses a g-prior distribution, while Kadane et al. (1980) choose the hyperparameters from the quantiles of the prior predictive distribution.

To recapitulate one may either use an improper prior density or a conjugate prior density with which to express one's prior information about the parameters, and in the latter case, one may use the prior predictive density to set the values of the hyperparameters.

Consider a $n(\theta, \tau)$ population and suppose a normal-gamma conjugate prior $\xi(\theta, \tau)$ is appropriate, then how does one choose the hyperparameters when m past observations $Z_1, Z_2, \ldots, Z_m$ are available? From the past data, θ and τ^{-1} would be estimated by $\theta^* = \bar{z}$ and $(\tau^{-1})^* = \Sigma_1^m (z_i - \bar{z})^2/(m-1)$ then the hyperparameters could be chosen as follows: μ by $\bar{z}$, α and β to satisfy $\beta(\alpha-1)^{-1} = (\tau^{-1})*$ and P (scalar) by m^{-1}, where $\alpha > 1$.

NORMAL POPULATIONS

Bayesian inferences for one normal population were discussed in Chapter 1 so now several normal populations will be examined.

Suppose one has two normal populations, $n(\theta_1, \tau_1^{-1})$ and $n(\theta_2, \tau_2^{-1})$; then four cases arise, namely:

(1) $\theta_1 = \theta_2$, $\tau_1 = \tau_2$

(2) $\theta_1 \neq \theta_2$, $\tau_1 = \tau_2$

(3) $\theta_1 = \theta_2$, $\tau_1 \neq \tau_2$

(4) $\theta_1 \neq \theta_2$, $\tau_1 \neq \tau_2$.

In the first case the populations are identical and there is actually only one population, therefore only the three remaining cases are of interest.

Two Normal Populations, Distinct Means

Let $S_1 = (x_{11}, x_{12}, \ldots, x_{1n_1})$ be a random sample from a $n(\theta_1, \tau^{-1})$ population and suppose $S_2 = (x_{21}, x_{22}, \ldots, x_{2n_2})$ is a random sample from the second $n(\theta_2, \tau^{-1})$ population, then the likelihood function of θ_1, θ_2 and τ is

$$L(\theta, \tau | s) \propto \tau^{(n_1+n_2)/2} \exp\left\{-\frac{\tau}{2}\left[\sum_{i=1}^{n_1}(x_{1i}-\theta_1)^2 + \sum_{i=1}^{n_2}(x_{2i}-\theta_2)^2\right]\right\},$$

$$\theta \in R^2, \quad \tau > 0 \quad (3.11)$$

where $s = (S_1, S_2)$ and $\theta = (\theta_1, \theta_2)'$ and if one uses the normal (bivariate)-gamma prior density of (θ, τ) suggested by (3.11), with parameters $\mu(2 \times 1)$, $P(2 \times 2)$, α and β, the joint posterior density of (θ, τ) is

$$\xi(\theta,\tau|s) \propto \tau^{(n+2\alpha+2)/2-1} \exp -\frac{\tau}{2}[(\theta - A^{-1}B)'A(\theta - A^{-1}B)$$
$$+ C - B'A^{-1}B], \qquad (3.12)$$

where $n = n_1 + n_2$,

$$A = \begin{pmatrix} n_1 & 0 \\ 0 & n_2 \end{pmatrix} + P,$$

$$B = \begin{pmatrix} \Sigma x_{1i} \\ \Sigma x_{2i} \end{pmatrix} + P\mu,$$

and

$$C = 2\beta + \Sigma x_{1i}^2 + \Sigma x_{2i}^2 + \mu'P\mu.$$

Furthermore, the marginal posterior density of θ is a bivariate t density

$$\xi(\theta|s) \propto [(\theta - A^{-1}B)'A(\theta - A^{-1}B) + C - B'A^{-1}B]^{-(n+2\alpha+2)/2},$$
$$\theta \in R^2, \qquad (3.13)$$

thus θ has $n + 2\alpha$ degrees of freedom, location vector $A^{-1}B$ and precision matrix

$$P(\theta|s) = \frac{(n + 2\alpha)A}{C - B'A^{-1}B}. \qquad (3.14)$$

In addition the marginal posterior distribution of τ is a gamma with parameters

$$\alpha' = (n + 2\alpha)/2$$

and

$$\beta' = (C - B'A^{-1}B)/2.$$

How does one make posterior inferences about θ and τ? First consider θ, then the mean $A^{-1}B$ jointly estimates θ_1 and θ_2, but suppose one wants to estimate θ_1, ignoring θ_2. If $\theta|s$ has a bivariate t distribution then $\theta_1|s$ has a univariate t distribution (see DeGroot, Chapter 5, and the Appendix of this book) with $n + 2\alpha$ degrees of freedom, location parameter which is the first component of $A^{-1}B$, and precision $P_1(\theta_1|s) = P_{11}(\theta|s) - P_{12}(\theta|s)P_{22}^{-1}(\theta|s)P_{21}(\theta|s)$, where p_{ij} is the ij-th component of $p(\theta|s)$ given by (3.14).

Often $\gamma = \theta_1 - \theta_2$ is the parameter of interest, because γ measures the difference in the means of the two normal populations and if $\gamma = 0$, the two populations are identical. What is the posterior distribution of γ? Since $\theta|s$ is a bivariate t, $\gamma|s$ has a univariate t distribution with $n + 2\alpha$ degrees of freedom, location $(1, -1)A^{-1}B$, and precision $[(1, -1)P^{-1}(\theta|s)(1, -1)']^{-1} = P(\gamma|s)$. These results can be verified by referring to DeGroot, Chapter 5, or to the Appendix of this book.

Remember the hyperparameters μ, P, α and β must be known in order to do the posterior analysis and they can be determined by using the methods of the previous section of this chapter. On the other hand, if very little is known about the parameters, before the experiment, one perhaps would use the Jeffreys' prior (1.13) which results in a simplified posterior analysis. For example the posterior distribution of θ is a bivariate t with $n - 2$ degrees of freedom, location vector $E(\theta|s) = \begin{pmatrix} \bar{x}_1 \\ \bar{x}_2 \end{pmatrix}$, and precision matrix

$$P(\theta|s) = \frac{(n_1 + n_2 - 2)\begin{pmatrix} n_1 & 0 \\ 0 & n_2 \end{pmatrix}}{(n_1 - 1)s_1^2 + (n_2 - 1)s_2^2},$$

where

$$s_i^2 = \sum_{j=1}^{n_i} (x_{ij} - \bar{x}_i)^2 (n_i - 1)^{-1}, \quad i = 1, 2$$

NORMAL POPULATIONS 73

and $\bar{x}_1$ and $\bar{x}_2$ are the sample means and s_1^2 and s_2^2 the sample variances.

The posterior analysis for θ using Jeffreys' prior density is based on Theorem 1.1 of Chapter 1. One may obtain the same results by letting $\alpha \to -1$, $\beta \to 0$, $P \to 0(2 \times 2)$ in the joint posterior distribution of θ with density (3.12).

Regions of highest posterior density, HPD, may be found for θ, θ_i, $i = 1, 2$, and γ. First for θ, from (3.12) one may show (see Chapter 1, 1.28)

$$F(\theta) = \frac{2^{-1}(\theta - A^{-1}B)'A(\theta - A^{-1}B)(n + 2\alpha)}{C - B'A^{-1}B} \qquad (3.15)$$

has an F-distribution with 2 and $n + 2\alpha$ degrees of freedom and one may plot 99%, 95% regions for θ. If θ_1 or θ_2 or γ is the parameter of interest, one may construct an HPD interval using the Student t tables. How would one do this?

Since θ_1 and θ_2 are the primary parameters of interest, Bayesian inferences for τ will not be discussed.

Two Normal Populations with a Common Mean

Consider two normal populations $n(\theta, \tau_1^{-1})$ and $n(\theta, \tau_2^{-1})$ with a common mean θ but distinct precisions τ_1 and τ_2, then on the basis of random samples $S_i = (x_{i1}, x_{i2}, \ldots, x_{in_i})$, $i = 1, 2$, how does one make inferences about the common mean? Suppose given θ, τ_1 and τ_2 that S_1 and S_2 are independent, then the likelihood function for θ and $\tau = (\tau_1, \tau_2)$ is

$$L(\theta, \tau | s) \propto \prod_{i=1}^{2} \tau_i^{n_i/2} \exp -\frac{\tau_i}{2} \sum_{j=1}^{n_i} (x_{ij} - \theta)^2, \quad \theta \in R, \quad \tau_i > 0,$$

(3.16)

where $s = (S_1, S_2)$. The conjugate prior density is

$$\xi(\theta, \tau) \propto \prod_{i=1}^{2} \tau_i^{\alpha_i - 1} e^{-\tau_i \beta_i} \tau_i^{1/2} e^{-\tau_i P_i (\theta - \mu_i)^2/2},$$

74   THE TRADITIONAL LINEAR MODELS

$$\theta \in R, \quad \tau_i > 0, \tag{3.17}$$

where $\alpha_i > 0$, $\beta_i > 0$, $P_i > 0$, and $\mu_i \in R$, and is the product of two normal-gamma densities, one for each population. This implies the marginal prior density of θ is

$$\xi(\theta) \propto \prod_{i=1}^{2} [2\beta_i + P_i(\theta - \mu_i)^2]^{-(2\alpha_i+1)/2}, \quad \theta \in R,$$

which is the product of two t densities on θ and is called a 2/0 univariate t density with parameters α_i, β_i, P_i, and μ_i, $i = 1,2$. This is a very complicated density because its normalizing constant and moments are unknown, thus it is very difficult to assign values to the hyperparameters. How would one assign values to the hyperparameters? One solution is to consider each population separately, so for the $n(\theta, \tau_1^{-1})$ population, one may set the values of μ_1, P_1, α_1, and β_1 and then repeat the process for the second population. Is this a legitimate procedure?

The marginal prior density of τ_1 and τ_2 is

$$\xi(\tau) \propto \prod_{i=1}^{2} \tau_i^{\alpha_i-1} e^{-\tau_i \beta_i} \frac{\tau_1^{1/2} \tau_2^{1/2}}{(\tau_1 P_1 + \tau_2 P_2)^{1/2}}$$

$$\times e^{-\tau_1 \tau_2 P_1 P_2 (\mu_1 - \mu_2)^2 / (\tau_1 P_1 + \tau_2 P_2)}, \quad \tau_1 > 0, \quad \tau_2 > 0,$$

$$\tag{3.18}$$

thus τ_1 and τ_2 are not independent, a priori, even if $\mu_1 = \mu_2$. Since the populations have a common mean, it is probably wise to let μ_1 and μ_2 have a common value, in which case (3.18) simplifies but is not the density of independent precision components.

Proceeding with the posterior analysis, we have

$$\xi(\theta, \tau | s) \propto \prod_{i=1}^{2} \tau_i^{(2\alpha_i + n_i + 1)/2 - 1} \exp -\frac{\tau_i}{2} \left\{ 2\beta_i + \sum_{j=1}^{n_i} (x_{ij} - \theta)^2 \right.$$

$$\left. + P_i(\theta - \mu_i)^2 \right\}, \tag{3.19}$$

NORMAL POPULATIONS

where $\theta \in R$, $\tau_i > 0$, for the joint posterior density of the parameters.

Completing the square on θ and integrating with respect to τ_1 and τ_2 gives

$$\xi(\theta|s) \propto \prod_{i=1}^{2} \left\{ 2\beta_i + \sum_{j=1}^{n_i} (x_{ij} - \bar{x}_i)^2 + n_i P_i (\bar{x}_i - \mu_i)^2 / (n_i + P_i) \right.$$

$$\left. + [\theta - (n_i + P_i)^{-1}(n_i \bar{x}_i + P_i \mu_i)]^2 (n_i + P_i) \right\}^{-(n_i + 2\alpha_i + 1)/2},$$

$$\theta \in R \quad (3.20)$$

for the marginal posterior density of θ and is a 2/0 scalar poly-t density, which, as we have seen, is the same form as the prior density of θ. For the properties of this distribution, see Box and Tiao (1973), Zellner (1971), Dreze (1977), Sedory (1980), and Yusoff (1982). Since θ is scalar, the density (3.20) is easily plotted and its moments calculated by numerical integration; however if θ is a vector, it may be necessary to use approximations to the density and this has been done by Box and Tiao (1973), Zellner (1971), Dreze (1977), and Yusoff (1982).

What is the marginal posterior density of τ_1 and τ_2? If θ is eliminated from the joint density,

$$\xi(\tau|s) \propto \prod_{i=1}^{2} \tau_i^{(n_i + 2\alpha_i)/2 - 1} \exp{-\frac{\tau_i}{2} \sum_{j=1}^{n_i} (x_{ij} - \bar{x}_i)^2} \frac{\exp - A/2}{[\Sigma \tau_i (n_i + P_i)]^{1/2}},$$

$$\tau_i > 0 \quad (3.21)$$

is the density of the precision components, where

$$\sum_{1}^{2} \tau_i (n_i + P_i) A = \sum_{1}^{2} \tau_i^2 n_i P_i (\bar{x}_i - \mu_i)^2 + \tau_1 \tau_2 \{n_1 n_2 (\bar{x}_1 - \bar{x}_2)^2$$

$$+ n_1 P_2 (\bar{x}_1 - \mu_2)^2 + n_2 P_1 (\bar{x}_2 - \mu_1)^2$$

$$+ P_1 P_2 (\mu_1 - \mu_2)^2\},$$

THE TRADITIONAL LINEAR MODELS

which is some type of bivariate-gamma distribution and it is necessary to employ numerical integration to calculate the joint and marginal properties of the distribution.

Thus with a common mean for the two populations, it is very difficult to do the posterior analysis because the marginal posterior distributions for θ, then for τ_1 and τ_2 are not standard densities.

For example, if the precisions are equal and the means are distinct, it has been shown the posterior analysis results in standard well-known distributions for the parameters.

How should inferences be made for this problem? On the one hand, for θ, one must use the poly-t density, and on the other, to make inferences for τ_1 and τ_2, one must use the bivariate type gamma density (3.21), and both densities must be numerically determined. Is there some other way by which posterior inferences for θ and τ are possible?

Consider the conditional posterior densities for the parameters, for example, for θ given τ_1 and τ_2, and then for τ given θ. The former distribution is normal with mean

$$E(\theta|\tau,s) = \left[\sum_1^2 \tau_i(n_i + P_i)\right]^{-1}\left[\sum_1^2 \tau_i(n_i\bar{x}_i + P_i\mu_i)\right]$$

and precision

$$P(\theta|\tau,s) = \sum_1^2 \tau_i(n_i + P_i)$$

and the conditional distribution of τ_1 and τ_2, given θ, is that of independent gamma random variables, such that $\tau_i|\theta$ is gamma with parameters $(2\alpha_i + n_i + 1)/2$ and

$$\frac{2\beta_i + \sum_{j=1}^{n_i}(x_{ij} - \bar{x}_i) + n_i(\theta - \bar{x}_i)^2 + P_i(\theta - \mu_i)^2}{2}.$$

Although the marginal posterior distributions of the parameters are intractable, the corresponding conditional distributions are well-known. How can these be used for inferences about the parameters?

Perhaps the mode of all the parameters can be found from the modes of the conditional densities, which are

$$M(\theta|\tau,s) = \frac{\sum_{1}^{2} \tau_i(n_i \bar{x}_i + P_i\mu_i)}{\sum_{1}^{2} \tau_i(n_i + P_i)} \qquad (3.22)$$

and

$$M(\tau_i|\tau_j, i \ne j, \theta) = \frac{n_i + 2\alpha_i - 1}{2\beta_i + \sum_{j=1}^{n_i} (x_{ij} - \bar{x}_i)^2 + n_i(\theta - \bar{x}_i)^2 + P_i(\theta - \mu_i)^2} \qquad (3.23)$$

for $i = 1, 2$.

Using an approach of Lindley and Smith (1972), the mode of the joint density of θ, τ_1, and τ_2, perhaps, can be found by solving simultaneously the above three modal equations for the parameters. An explicit solution cannot be found and an iterative technique can be tried by starting the process with some trial values for τ_1 and τ_2 (say the sample precisions) and finding the conditional mode of θ from (3.22) which is then substituted into the equations, (3.23), for the conditional modes of τ_1 and τ_2. The process is repeated until the estimates stabilize and then the final estimates are a solution to the modal equations, and the solution, perhaps, is the joint mode of the joint posterior density.

Since the joint density may have multiple modes, the iterative technique may not converge to the absolute maximum but instead to a local maximum, depending on the starting value of the iterative procedure.

Of course, the joint mode of the parameters, if it can be found, is a useful characteristic of the posterior distribution since it gives one a point estimate of the parameters, but it does not tell us anything about the precision of the posterior distribution.

Suppose θ is the primary parameter of interest and τ_1 and τ_2 are nuisance parameters, then how should one estimate θ? There are several possibilities. Suppose we obtain the marginal posterior mean and mode of θ from the 2/0 poly-t density (3.20) via numerical integration and the θ component of the joint posterior mode, which is found by an iterative solution to the modal equations (3.22) and (3.23) where the sample precisions are used as starting values. This will give three point estimates of θ.

Suppose the sample values are $n_1 = n_2 = 10$ and $\Sigma_1^{10}(x_{1j} - \bar{x}_1)^2 = 31.7882$, $\Sigma_1^{10}(x_{2j} - \bar{x}_2)^2 = 9.9999$, $\bar{x}_1 = -.1499$ and $\bar{x}_2 = .3611$, thus there are two normal populations with a common mean and distinct variances and the two sample means are $-.1499$ and $.3611$. The sample precisions are $.28312$ and $.9$ respectively and are used as the starting values of the iterative solution to the modal equations.

One must choose the hyperparameters α_1, β_1, α_2, β_2, P_1, P_2, μ_1, and μ_2 of the prior distribution, and for fifty-two combinations of these parameters, Table 3.1 lists the mode of the marginal density of θ, the θ component of the solution to the modal equations, and the mean of the marginal density of θ.

We see, first of all, that for each combination of the eight hyperparameters, the three estimates of θ are very close. The marginal mean and mode of θ are not the same because the marginal density of θ, (3.20), is usually not symmetric and the θ component of the joint modal estimate is not the same as that calculated from the marginal density of θ, nevertheless, the three estimates are close to one another.

Table 3.1 reveals the sensitivity of the three posterior estimates to the values of the parameters of the prior distribution. For example, the range of the largest to the smallest values of each estimate is 5.4087, 5.3870, and 5.3865 for the marginal mode, θ component of iterative solution, and posterior marginal mean, respectively, and the last has the smallest range.

The three estimates appear to be sensitive to μ_1 and μ_2, the two means of the prior distribution of θ. As μ_1 and μ_2 increase, so do the posterior mean of θ and the other two estimates and this was not unexpected.

In order to arrive at definite conclusions about this way of estimation, more numerical studies need to be conducted.

Another way to do the posterior analysis is with an improper prior density

$$\xi(\theta, \tau_1, \tau_2) \propto \tau_1^{-1} \tau_2^{-1}, \quad \theta \in R, \quad \tau_1 > 0, \tau_2 > 0 \qquad (3.24)$$

for the parameter. If this is done, one will see that the marginal posterior density of θ is a 2/0 poly-t, of the form (3.20), and the marginal posterior density of τ is a bivariate type gamma density, (3.21).

Also, the analysis of this section is easily extended to $k(\geq 2)$ normal populations with the same mean but with distinct variances.

Table 3.1
Comparison of the Mode (from Marginal Distribution),
Mode (by Iteration), and the Mean

α_1	α_2	β_1	β_2	ξ_1	ξ_2	μ_1	μ_2	Mode (marginal)	Mode (iterative)	Mean
0	0	0	0	0	0	0	0	.2458	.2418	.2261
0	0	0	0	0	0	2	2	.2458	.2418	.2261
0	0	0	0	0	0	4	4	.2458	.2418	.2261
0	0	0	0	0	0	6	6	.2458	.2418	.2261
0	0	0	0	0	0	8	8	.2458	.2418	.2261
0	0	0	0	0	0	10	10	.2458	.2418	.2261
0	0	0	0	2	2	0	0	.2060	.1995	.1867
0	0	0	0	4	4	0	0	.1661	.1698	.1589
0	0	0	0	6	6	0	0	.1528	.1478	.1384
0	0	0	0	8	8	0	0	.1262	.1308	.1225
0	0	0	0	10	10	0	0	.1130	.1173	.1099
0	0	0	0	12	12	0	0	.1130	.1064	.0997
0	0	0	0	14	14	0	0	.0997	.0973	.0911
0	0	0	0	16	16	0	0	.0864	.0896	.0840
0	0	0	0	18	18	0	0	.0864	.0831	.0779
0	0	1	3	0	0	0	0	.1894	.1989	.1867
0	0	2	6	0	0	0	0	.1528	.1679	.1592
0	0	3	9	0	0	0	0	.1528	.1443	.1389
0	0	4	12	0	0	0	0	.1162	.1260	.1231
0	0	5	15	0	0	0	0	.1162	.1113	.1105
0	0	6	18	0	0	0	0	.1162	.0992	.1001
0	0	7	21	0	0	0	0	.0797	.0891	.0915
0	0	8	24	0	0	0	0	.0797	.0806	.0841
0	0	9	27	0	0	0	0	.0797	.0733	.0778
2	2	0	0	0	0	0	0	.2458	.2418	.2305
4	4	0	0	0	0	0	0	.2458	.2418	.2330

Table 3.1 (Continued)

α_1	α_2	β_1	β_2	ξ_1	ξ_2	μ_1	μ_2	Mode (marginal)	Mode (iterative)	Mean
6	6	0	0	0	0	0	0	.2458	.2418	.2346
8	8	0	0	0	0	0	0	.2458	.2418	.2357
10	10	0	0	0	0	0	0	.2458	.2418	.2365
12	12	0	0	0	0	0	0	.2458	.2418	.2371
14	14	0	0	0	0	0	0	.2458	.2418	.2376
16	16	0	0	0	0	0	0	.2458	.2418	.2380
2	2	1	1	3	3	−8	−8	−1.7741	−1.7634	−1.7636
2	2	1	1	3	3	−6	−6	−1.2957	−1.2967	−1.2974
2	2	1	1	3	3	−4	−4	−.8173	−.8225	−.8245
2	2	1	1	3	3	−2	−2	−.3389	−.3299	−.3349
2	2	1	1	3	3	0	0	.1794	.1745	.1664
2	2	1	1	3	3	2	2	.6179	.6262	.6187
2	2	1	1	3	3	4	4	1.0565	1.0572	1.0520
2	2	1	1	3	3	6	6	1.4950	1.4999	1.4966
2	2	1	1	3	3	8	8	1.9336	1.9512	1.9489
2	2	1	1	0	0	1	1	.2359	.2299	.2191
2	2	1	1	3	3	1	1	.4186	.4078	.3996
2	2	1	1	6	6	1	1	.5382	.5189	.5124
2	2	1	1	9	9	1	1	.5781	.5949	.5895
2	2	1	1	12	12	1	1	.6578	.6502	.6455
2	2	1	1	15	15	1	1	.6977	.6922	.6881
2	2	1	1	18	18	1	1	.7375	.7252	.7215
2	2	1	3	2	2	2	2	.4983	.4947	.4877
4	4	2	6	4	4	4	4	1.2425	1.2501	1.2477
6	6	3	9	6	6	6	6	2.3322	2.3318	2.3301
8	8	4	12	8	8	8	8	3.6346	3.6236	3.6229

Two Normal Populations with Distinct Means and Precisions

The Behrens-Fisher problem is the problem of comparing the means θ_1 and θ_2 of two normal populations which have distinct variances or precisions τ_1 and τ_2. If the precisions are equal, the two-sample t-test is the way to test the hypothesis the two means are equal, and the Bayesian analysis of this problem was discussed in the section on two normal populations with distinct means (see pp. 70-73). When the precisions are unequal, various tests such as that of Scheffé (1943) have been proposed.

Suppose $S_i = (x_{i1}, x_{i2}, \ldots, x_{in_i})$ is a random sample of size n_i from a $n(\theta_i, \tau_i^{-1})$ population, where $i = 1, 2$, then how does one provide posterior inferences for the parameters? The likelihood function for $\theta = (\theta_1, \theta_2)$ and $\tau = (\tau_1, \tau_2)$ is

$$L(\theta, \tau | s) \propto \prod_1^2 \tau_i^{n_i/2} e^{-\tau_i/2 \sum_{j=1}^{n_i} (x_{ij} - \theta_i)^2},$$

$$\theta \in R^2, \quad \tau_1 > 0, \quad \tau_2 > 0 \tag{3.25}$$

where $s = (S_1, S_2)$, which implies

$$\xi(\theta, \tau) \propto \prod_1^2 \tau_i^{\alpha_i - 1} e^{-\tau_i \beta_i} \tau_i^{1/2} e^{-(\tau_i/2) P_i (\theta_i - \mu_i)^2},$$

$$\theta \in R^2, \quad \tau_1 > 0, \quad \tau_2 > 0 \tag{3.26}$$

is the conjugate density, a product of two normal-gamma densities, thus, a priori, (θ_1, τ_1) and (θ_2, τ_2) are independent. It follows, a posteriori, that (θ_1, τ_1) and (θ_2, τ_2) are independent with density,

$$\xi(\theta, \tau | s) \propto \prod_{i=1}^2 \tau_i^{(n_i + 2\alpha_i + 1/2) - 1} \exp - \frac{\tau_i}{2} \left\{ 2\beta_i + \sum_{j=1}^{n_i} (x_{ij} - \bar{x}_i)^2 \right.$$

$$\left. + P_i(\theta_i - \mu_i)^2 + n_i(\theta_i - \bar{x}_i)^2 \right\}, \tag{3.27}$$

where $\tau_i > 0$, $\theta \in R^2$, hence (θ_i, τ_i) has a normal-gamma distribution with parameters α_i, β_i, μ_i, and P_i, $i = 1,2$. The joint marginal p.d.f. of θ_1 and θ_2 is

$$\xi(\theta|s) \propto \prod_{i=1}^{2} \left\{ 2\beta_i + \sum_{j=1}^{n_i} (x_{ij} - \bar{x}_i)^2 + \frac{n_i P_i (\mu_i - \bar{x}_i)^2}{(n_i + P_i)} \right. \quad (3.28)$$

$$\left. + [\theta_i - (n_i + P_i)^{-1}(P_i \mu_i + n_i \bar{x}_i)]^2 (n_i + P_i) \right\}^{-(n_i + 2\alpha_i + 1)/2}$$

and θ_i has a univariate t distribution with $n_i + 2\alpha_i$ degrees of freedom, location

$$E(\theta_i|s) = (n_i + P_i)^{-1}(P_i \mu_i + n_i \bar{x}_i)$$

and precision

$$P(\theta_i|s) = \frac{(n_i + 2\alpha_i)(n_i + P_i)}{2\beta_i + \sum_{1}^{n_i}(x_{ij} - \bar{x}_i)^2 + n_i P_i (\mu_i - \bar{x}_i)^2 (n_i + P_i)^{-1}},$$

$$i = 1, 2.$$

Furthermore, τ_1 and τ_2 are independent and τ_i is gamma with parameters $\alpha' = (n_i + 2\alpha_i)/2$ and β', where

$$2\beta' = 2\beta_i + \sum_{j=1}^{n_i}(x_{ij} - \bar{x}_i)^2 + n_i P_i (\mu_i - \bar{x}_i)^2 (n_i + P_i)^{-1}, \quad i = 1, 2.$$

In this type of problem, often $\gamma = \theta_1 - \theta_2$ is the parameter of interest and τ_1 and τ_2 are nuisance parameters. How would one find the posterior density of γ? See Box and Tiao (1973, pages 104–109) where they explain an approximation of Patil (1964) to the posterior density of γ and illustrate the approximation with a "textile" example. When reading Box and Tiao the reader should remember, they are using an improper prior density

NORMAL POPULATIONS

$$\xi(\theta,\tau) \propto \tau_1^{-1}\tau_2^{-1}, \quad \theta \in R^2, \quad \tau_1 > 0, \tau_2 > 0$$

for the parameters and not the conjugate prior density (3.26); however, their analysis can be duplicated by letting $\alpha_i \to -1/2$, $\beta_i \to 0$, and $P_i \to 0$, $i = 1,2$, in the posterior density (3.27). When letting the hyperparameters approach these limits in the posterior density, the normalizing constant of the density should be ignored.

Now returning to the distribution of γ, let $\xi_i(\theta_i|s)$ be the posterior density of θ_i, then we have seen ξ_i is a t density with $n_i + 2\alpha_i$ degrees of freedom, location $E(\theta_i|s)$ and precision $P(\theta_i|s)$, $i = 1,2$, hence the posterior density of γ is

$$g(\gamma|s) = \int_R \xi_1(\gamma + \Delta|s)\xi_2(\Delta|s)d\Delta, \quad \gamma \in R \qquad (3.29)$$

and the integral cannot be expressed in terms of simple functions and must be computed by numerical integration. Patil's approximation to $g(\gamma|s)$ avoids the numerical integration of (3.29) and expresses $g(\gamma|s)$ as a t distribution. Box and Tiao compare the Patil approximation to the exact distribution of $\gamma|s$ and show the approximation works well.

How would one develop a normal approximation to the distribution of $\gamma|s$?

Concluding Remarks

This chapter begins with the Bayesian analysis of two normal populations, where either the means are distinct or the precisions are distinct, or both.

In the case the means are distinct, but the precisions are equal, the marginal posterior distribution of the means is a bivariate t distribution with density (3.13) and the marginal posterior distribution of the common precision parameter is a gamma. Also, the marginal posterior distributions of θ_1, θ_2, and $\gamma = \theta_1 - \theta_2$ are univariate t densities and Bayesian inferences for these parameters are easily obtained.

If the precision parameters are distinct and the means are equal, the marginal posterior density of the common mean is a univariate 2/0 poly-t density (3.20), and the joint posterior density of the precisions is given by (3.21), which is a bivariate-type gamma density. A numerical example gave three estimates of the common mean for various combinations of the parameters of the prior distribution and the estimates are listed in Table 3.1. One of these estimates was calculated from the θ component of the mode of the posterior distribution of all the parameters. An iterative procedure of Lindley and Smith (1972) produced this modal estimate of θ while the other estimates, the marginal mode and mean, were calculated from the marginal posterior density of θ.

This section was concluded with a Bayesian analysis of the Behrens-Fisher problem. Using a conjugate prior density for the parameters, it was shown that the joint posterior density (3.27) is such that (θ_1, τ_1) and (θ_2, τ_2) are independent and each has a normal-gamma distribution, however the distribution of $\gamma = \theta_1 - \theta_2$ must be evaluated numerically. The marginal posterior distributions for τ_1 and τ_2 are each gamma, which follows from (3.27) but inferences for these parameters were not made. How would one compare τ_1 and τ_2? Perhaps the ratio of these two parameters could be used.

LINEAR REGRESSION MODELS

One of the standard models the statistician must know how to use is the linear regression model. With such models one attempts to build a connection between a dependent variable Y and a set of m (m $\geq$ 1) independent variables. Often, but not always, the average value of Y is a linear function of a set of p unknown parameters θ, thus if there are n observations $(Y_i, x_{i1}, x_{i2}, \ldots, x_{im})$ on these m + 1 variables, the model relating Y to m independent variables is

$$Y = x\theta + e, \qquad (3.30)$$

where Y is the n × 1 observation vector, x is the n × p, (p = m + 1), matrix of values of the independent variables, θ a p × 1 parameter vector, and e is a n × 1 vector of unobservable random variables with zero means. If, in addition, the n observations on y are independent and normally distributed with a common precision τ, model (3.30) is the general linear model of Chapter 1. The prior, posterior and predictive analysis for such a model was developed in Chapter 1.

In this section, a detailed Bayesian analysis of the linear regression model will be given. That is, using the normal-gamma prior density for θ and τ:

LINEAR REGRESSION MODELS

(i) Point estimates for θ and τ will be found.
(ii) HPD interval estimates for the parameters will be constructed.
(iii) Point and interval estimates of $E(Y|x)$ for a given set X of values of the independent variables will be constructed.
(iv) Prediction intervals for future Y values for a given set of values of the independent variables are to be derived.
(v) Numerical examples will illustrate the above four techniques.

Of course, not all regression studies involve the general linear model (3.30), since sometimes the $E(Y|x)$ is a nonlinear function of x and often the observations on Y are correlated, that is the precision matrix $P(Y|x)$ is not τI_n, but is nondiagonal. Nonlinear regression models and models with correlated observations will be presented later.

Simple Linear Regression

Let

$$y_i = \theta_1 + \theta_2 x_i + e_i, \quad i = 1, 2, \ldots, n$$

where y_i is the i-th observation on a dependent variable Y, x_i is the i-th observation on an independent variable x, θ_1 and θ_2 are unknown intercept and slope coefficients respectively, and e_i is the i-th error term. If the e_i are n.i.d. $(0, \tau^{-1})$ variables, then the general linear model and the theory of Chapter 1 apply. In terms of this model Y is the n × 1 vector of observations of the dependent variable, x is a n × 2 matrix

$$x = \begin{pmatrix} 1 & x_1 \\ 1 & x_2 \\ \cdot & \cdot \\ \cdot & \cdot \\ \cdot & \cdot \\ 1 & x_n \end{pmatrix}$$

$\theta = (\theta_1, \theta_2) \in R^2$, and e is the n × 1 vector of errors e_i, where $e \sim n(0, \tau^{-1} I_n)$.

If one uses a normal-gamma prior density for θ and τ, the Bayesian analysis was developed in Chapter 1.

First, how does one develop point and interval estimates for θ_1, θ_2, and τ? From Chapter 1 the marginal posterior density of θ is a t with $n + 2\alpha$ degrees of freedom, location vector

$$\mu^* = (x'x + P)^{-1}(x'y + P\mu) \qquad (3.31)$$

and precision matrix

$$P(\theta|s) = \frac{(x'x + P)(n + 2\alpha)}{2\beta + y'y - (x'y + P\mu)'(x'x + P)^{-1}(x'y + P\mu)} \qquad (3.32)$$

where

$$x'x = \begin{pmatrix} n & \sum_{1}^{n} x_i \\ \sum_{1}^{n} x_i & \sum_{1}^{n} x_i^2 \end{pmatrix},$$

and

$$x'y = \sum_{1}^{n} x_i y_i.$$

If one uses an improper prior density

$$\xi(\theta, \tau) \propto 1/\tau, \qquad \tau > 0, \qquad \theta \in R^2, \qquad (3.33)$$

the marginal posterior density of θ is a t with $n - 1$ degrees of freedom, location vector

$$\hat{\theta} = (x'x)^{-1} x'y \qquad (3.34)$$

$$= \begin{pmatrix} \hat{\theta}_1 \\ \hat{\theta}_2 \end{pmatrix},$$

LINEAR REGRESSION MODELS

where

$$\hat{\theta}_1 = \bar{y} - \hat{\theta}_2 \bar{x}$$

and

$$\hat{\theta}_2 = \frac{\sum_{1}^{n}(x_i - \bar{x})(y_i - \bar{y})}{\sum_{1}^{n}(x_i - \bar{x})^2},$$

which is the usual least squares estimator of θ, and the precision matrix of θ is

$$P^*(\theta|s) = \frac{x'x}{s^2}, \qquad (3.35)$$

where

$$s^2 = \frac{y'y - y'x(x'x)^{-1}x'y}{(n-2)}$$

$$= \sum_{i=1}^{n}(y_i - \hat{\theta}_1 - \hat{\theta}_2 x_i)^2/(n-2),$$

and s^2 is the usual unbiased estimator of $\sigma^2 = 1/\tau$. The dispersion matrix of the marginal posterior distribution of θ is $s^2(x'x)^{-1}(n-1) \cdot (n-3)^{-1}$, $n > 3$.

We see if one uses an improper prior density for the parameters, the Bayesian analysis closely resembles the least squares or maximum likelihood approach to estimating the parameters.

To estimate θ from the joint posterior distribution, the vector $\hat{\theta}$ of least squares estimators seems appropriate, if one uses an improper prior density (3.33), but if one uses a conjugate prior density, then μ^*, the mean of the marginal posterior density of θ seems appropriate. By letting $\alpha \to -1$, $\beta \to 0$, $P \to 0$ (2×2) in the joint posterior density of θ and τ, which corresponds to the conjugate prior density, one will obtain the joint posterior density of θ and τ corresponding to the improper prior density (3.33). In particular, $\mu^* \to \hat{\theta}$ (the least squares estimator of θ) as $P \to 0$ and as $\alpha \to -1$ also if $P \to 0$, and $\beta \to 0$, $P(\theta|s) \to (x'x)/s^2$, the precision matrix of the marginal posterior distribution of θ corresponding to the improper prior density (3.33).

If one is interested in τ, how should this parameter be estimated? We know the marginal posterior density of τ is

$$\xi(\tau|s) \propto \tau^{(n+2\alpha)/2-1} e^{-(\tau/2)\{2\beta+y'y-(x'y+P\mu)'(x'x+P)^{-1}(x'y+P\mu)\}},$$

$$\tau > 0 \quad (3.36)$$

if one uses a conjugate prior density with hyperparameters μ, P, α, and β. Since $\tau|s$ has a gamma density, the mean and mode of $\tau|s$ are distinct, thus if one uses the mean to estimate τ, the estimator is

$$E(\tau|s) = \frac{n + 2\alpha}{2\beta + y'y - (x'y + P\mu)'(x'x + P)^{-1}(x'y + P\mu)} \quad (3.37)$$

whereas the marginal posterior mode of τ is

$$M(\tau|s) = \frac{n + 2\alpha - 2}{2\beta + y'y - (x'y + P\mu)'(x'x + P)^{-1}(x'y + P\mu)} \quad (3.38)$$

What is the marginal posterior mean of τ^{-1} (which is the variance σ^2 along the regression line) and what is $E(\tau|s)$ if an improper prior density is used to express one's prior information about the parameters?

The marginal posterior densities for θ and τ are also useful in obtaining HPD regions. For example, what is the HPD interval for θ_2? Since $\theta_2|s$ has a univariate t density, one may use Student's t-tables to find these intervals for θ_2, as well as for θ_1. With regard to τ, HPD intervals are more difficult to construct because the gamma marginal posterior distribution is not symmetric and numerical integration is necessary in order to find an HPD interval for τ. An approximate HPD interval is easily constructed from the chi-square tables since, a posteriori, $[2\beta + y'y - (x'y + P\mu)'(x'x + P)^{-1}(x'y + P\mu)]\tau/s$ has a chi-square distribution with $n + 2\alpha$ degrees of freedom.

Often, one is interested in estimating the average value of the dependent variable Y when the independent variable $x = x^*$, where x^* is known. Now

$$E(Y|s^*) = (1, x^*) \begin{pmatrix} \theta_1 \\ \theta_1 \end{pmatrix} = \gamma$$

LINEAR REGRESSION MODELS

is the average value of Y when $x = x^*$, and since θ has a bivariate t distribution with $n + 2\alpha$ degrees of freedom, location vector μ^* and precision matrix $P(\theta|s)$, γ has a t distribution with $n + 2\alpha$ degrees of freedom, location $(1,x^*)\mu^*$ and precision

$$P(\gamma|s) = [(1,x^*)P^{-1}(\theta|s)(1,x^*)']^{-1} \qquad (3.39)$$

and one would usually estimate γ by $(1,x^*)\mu^*$ and an HPD interval for γ is easily found from the Student t tables. For example, a $1 - \Delta$ ($0 \leq \Delta \leq 1$) HPD interval for γ is

$$(1,x^*)\mu^* \pm t_{\Delta/2,n+2\alpha} P(\gamma/s)^{-1/2}, \qquad (3.40)$$

where $P(\gamma|s)$ is given by (3.39).

The analysis of a simple linear regression model is completed with the development of a forecasting procedure to predict a future value w of the dependent variable Y when the regressor $x = y_{n+1}$, and x_{n+1} is known. Recall from the section on predictive analysis in Chapter 1 (see pp. 14-18) the derivation of the Bayesian predictive density, with which a future value of Y is to be predicted. One obtains the future predictive density of Y if one substitutes the appropriate quantities into formula (1.39), which is based on the conjugate prior density with hyperparameters α, β, μ, and P.

Use the following substitutions: $k = 1$, $z = (1,x_{n+1})$, x is the $n \times 2$ matrix following formula (3.30), and Y is the $n \times 1$ vector of observations. The predictive density of $Y_{n+1} = w$ is a t-distribution with $n + 2\alpha$ degrees of freedom, location $A^{-1}B$ and precision $(n + 2\alpha)A \cdot (C - B'A^{-1}B)^{-1}$, where A, B, and C are explained by formula (1.39).

Notice, letting $\beta \to 0$, $P \to 0(2 \times 2)$ and $\alpha \to -1$,

$$A \to 1 - (1,x_{n+1})(x'x + z'z)^{-1} \begin{pmatrix} 1 \\ x_{n+1} \end{pmatrix} \qquad (1.41)$$

$$B \to (1,x_{n+1})(z'z + x'x)^{-1}x'y \qquad (1.42)$$

$$C \to \sum_{1}^{n} y_i^2 - y'x(z'z + x'x)^{-1}x'y, \qquad (1.43)$$

where

$$z'z = \begin{pmatrix} 1 & x_{n+1} \\ x_{n+1} & x_{n+1}^2 \end{pmatrix}$$

and

$$x'y = \begin{pmatrix} \sum_{1}^{n} y_i \\ \sum_{1}^{n} x_i y_i \end{pmatrix}.$$

The predictive density of w which corresponds to the improper prior density is a univariate t density with $n - 2$ ($n > 2$) degrees of freedom, location $A^{-1}B = E(w|s)$ and precision $P(w|s) = (n-2)A \cdot (C - B'A^{-1}B)^{-1}$, where A, B, and C are now given by (1.41), (1.42), and (1.43), respectively, and a $1 - \Delta$ ($0 \leq \Delta \leq 1$) HPD interval for a future observation w (when $x = x_{n+1}$) is given by

$$E(w|s) \pm t_{\Delta/2, n-2} \sqrt{A^{-1}(C - B'A^{-1}B)(n-2)^{-1}}, \quad n > 3.$$

(1.44)

How does this interval compare to the "usual" prediction interval of a future observation? See Draper and Smith (1966) for details about the conventional simple linear regression analysis.

An Example of Simple Linear Regression

Consider the model

$$y_i = \theta_1 + \theta_2 x_i + e_i, \quad i = 1, 2, \ldots, 30, \tag{3.41}$$

$x_i = i$, the e_i are n.i.d. (0,1), $\theta_1 = 1$, and $\theta_2 = 2$. Thirty y_i values were generated with this model and the parameters of the normal-gamma prior density were determined empirically from the first six (x_i, y_i) pairs. The values of the hyperparameters were

LINEAR REGRESSION MODELS

$$\mu = (.03958, \quad 27426),$$

$$P = \begin{pmatrix} 5.58765 & 19.5568 \\ 19.5568 & 84.7461 \end{pmatrix}, \qquad (3.42)$$

$$\alpha = 1,$$

and

$$\beta = .931275,$$

where these values were calculated as

$$\mu = (z'z)^{-1} z'y^*,$$

$$P = z'z/s^2,$$

$$\beta = (s^2)^{-1},$$

$$s^2 = \frac{y^{*\prime} y^* - y^{*\prime} z (z'z)^{-1} z' y^*}{4},$$

and α was set equal to one. Also,

$$y^* = (y_1, y_2, \ldots, y_6)'$$

and

$$z = \begin{pmatrix} 1 & x_1 \\ 1 & x_2 \\ \cdot & \cdot \\ \cdot & \cdot \\ \cdot & \cdot \\ 1 & x_6 \end{pmatrix}.$$

The posterior analysis is based on the normal-gamma prior density, which was constructed from the first six observations and the likelihood function, which is based on the 7th through 29th pairs (i, y_i), $i = 7, 8, \ldots, 29$, selected from (3.41). Combining the prior and likelihood function gave the following values for the parameters of the joint posterior distribution:

$$E(\theta|s) = \begin{pmatrix} .970903 \\ 2.005566 \end{pmatrix} = \begin{pmatrix} E(\theta_1|s) \\ E(\theta_2|s) \end{pmatrix}, \qquad (3.43)$$

$$P(\theta|s) = \begin{pmatrix} 23.1789 & 351.528 \\ 351.528 & 6931.33 \end{pmatrix},$$

$P(\theta_1|s) = 5.3085,$

$P(\theta_2|s) = 1600.1.$

The average value of a future observation y_{30} given $x_{30} = 30$ is

$E(w|s) = 61.1376$

compared to $y_{30} = 61.1289$, and the precision of the predictive distribution of y_{30} is $P(w|s) = .707206$, where $s = \{(i, y_i): i = y, \ldots, 29\}$.

If one uses a Jeffreys' prior density (3.33), the following values of the parameters of the posterior distribution were calculated:

$$\begin{pmatrix} E(\theta_1|s) \\ E(\theta_2|s) \end{pmatrix} = \begin{pmatrix} 1.0334 \\ 2.00196 \end{pmatrix}. \qquad (3.44)$$

$$P(\theta|s) = \begin{pmatrix} 17.3861 & 312.95 \\ 312.95 & 6398.09 \end{pmatrix},$$

$P(\theta_1|s) = 2.07877,$

$P(\theta_2|s) = 764.989,$

and the mean of the predictive distribution of y_{30} is

$E(w|s) = 61.0922$, when $x_{30} = 30$

when actually $y_{30} = 61.1289$, and the precision of w is $P(w|s) = .637491$.

LINEAR REGRESSION MODELS

By using a normal-gamma prior density with parameters fitted to the first six observations (i, y_i), $i = 1,2,3,4,5,6$, one obtains prior estimates (3.42) of θ_1 and θ_2 which are very close to the true values of the intercept $\theta_1 = 1$ and slope $\theta_2 = 2$ of the regression model (3.41), therefore it is not surprising that the posterior estimates (3.43) are even closer to the true values.

On the other hand, when the prior information is vague, the posterior estimates (3.44) of the regression coefficients are not as close as they were when the prior was fitted to the first six observations. Also, we see the posterior precisions of θ_1 and θ_2 are less when Jeffreys' density expresses our prior information.

Suppose θ_1 and θ_2 are estimated by 95% HPD intervals. The marginal posterior distribution of θ_1 is a t with $n + 2\alpha$ degrees of freedom, location $E(\theta_1|s)$ and precision $P(\theta_1|s)$, thus

$$E(\theta_1|s) \pm t_{.05/2, n+2\alpha} \sqrt{P^{-1}(\theta_1/s)}$$

is a 95% HPD region for θ_1 and reduces to ($n = 23$, $\alpha = 1$)

$$1.0334 \pm (2.060)\sqrt{.693580} \quad \text{or} \quad (-0.39537, 2.462175)$$

As for θ_2, the 95% HPD interval is $2.00556 \pm (2.060)(.02499)$ or (1.95407, 2.05709), which is a much shorter interval than the HPD interval for θ_1 because the latter has a much smaller precision.

Now suppose y_{30} is to be predicted with a 95% HPD interval, then

$$E(w|s) \pm t_{.025, n+2\alpha} \sqrt{P^{-1}(w|s)}$$

is the appropriate formula. Substituting $n = 23$ and $\alpha = 1$, $E(w|s) = 61.1376$ and $P(w|s) = .707206$ gives 61.1376 ± 2.4495 as the interval for forecasting the future observation y_{30} when $x_{30} = 30$.

What are the 95% HPD intervals for θ_1, θ_2, and $w|x_{30} = 30$ when vague prior information is used? The following y values were generated from the model (3.41) and were used in the preceding calculations. Beginning with y_1 and continuing through y_{30}, they are:

2.70178, 4.07126, 7.64881, 7.57121, 12.31, 13.6939, 14.7727, 16.6751, 19.1657, 20.4161, 24.6735, 26.1047, 26.4773, 28.2187, 29.5846, 34.3475, 35.337, 36.2295, 40.5608, 40.6826, 42.5413, 46.2414, 47.9887, 47.5131, 50.6753, 55.0602, 54.9478, 54.8281, 60.5314, 60.0476.

We see they are increasing, as they should because so are the x values since $x_i = i$, $i = 1, 2, \ldots, 30$.

MULTIPLE LINEAR REGRESSION

Consider a linear regression model with two independent variables, then

$$y_i = \theta_1 + \theta_2 x_{i1} + \theta_3 x_{i2} + e_i, \qquad (3.45)$$

where the e_i are n.i.d. $(0, \tau^{-1})$. The n triples (y_i, x_{i1}, x_{i2}), $i = 1, 2, \ldots, n$, are such that y_i is the i-th observation on a dependent variable y and x_{i1} and x_{i2} are the i-th observations on two nonstochastic regressor variables x_1 and x_2, respectively. The unknown parameters are $\theta = (\theta_1, \theta_2, \theta_3)$ and τ, where $\theta \in R^3$ and $\tau > 0$, and given the sample $S = \{(y_i, x_{i1}, x_{i2}): i = 1, 2, \ldots, n\}$, what inferences can be made about the parameters and how does one forcast future values of the dependent variable? Of course, these questions are answered in much the same way as one would answer the same questions about the simple linear regression model and the theory concerning the prior, predictive, and posterior analysis is covered in Chapter 1.

With the simple linear regression model, it was shown the prior analysis is based on the normal-gamma conjugate prior density or the Jeffreys' improper prior, the posterior analysis consists of studying the normal-gamma posterior distribution of θ and τ, and the predictive analysis is then done with the aid of the predictive t distribution. All these techniques apply to the analysis of multiple linear regression models; however, because the model contains more than one independent variable (as compared to a simple linear regression model), the posterior analysis becomes more involved since there are now more parameters in the model and the posterior distribution of θ has one more dimension.

The Posterior and Predictive Analysis

Let us briefly consider the posterior analysis if an improper prior density

$$\xi(\theta, \tau) \propto 1/\tau, \qquad \tau > 0, \qquad \theta \in R^3 \qquad (3.46)$$

is used in assessing one's prior information.

MULTIPLE LINEAR REGRESSION

The joint posterior density of the parameter is

$$\xi(\theta,\tau|s) \propto \tau^{(n/2)-1} \exp - \frac{\tau}{2} \sum_{i=1}^{n} [y_i - \theta_1 - \theta_2 x_{i1} - \theta_3 x_{i2}]^2,$$

(3.47)

where $\theta_i \in R$ (i = 1,2,3) and $\tau > 0$, and is a normal-gamma density, thus the marginal posterior distribution of θ is a trivariate t with n − 3 degrees of freedom, location vector

$$E(\theta|s) = A^{-1}B \qquad (3.48)$$

and precision matrix

$$P(\theta|s) = (n - 3)A(C - B'A^{-1}B)^{-1}, \quad n > 3 \qquad (3.49)$$

where

$$A = \begin{pmatrix} n & \Sigma x_{i1} & \Sigma x_{i2} \\ \Sigma x_{i1} & \Sigma x_{i1}^2 & \Sigma x_{i1}x_{i2} \\ \Sigma x_{i2} & \Sigma x_{i1}x_{i2} & \Sigma x_{i2}^2 \end{pmatrix},$$

$$B = \begin{pmatrix} \Sigma y_i \\ \Sigma y_i x_{i1} \\ \Sigma y_i x_{i2} \end{pmatrix},$$

and

$$C = \sum_{1}^{n} y_i^2 .$$

The three marginal posterior univariate distributions are t distributions. For example, the marginal posterior distribution of θ_2 is a t with n − 3 degrees of freedom, location

$$E(\theta_2|s) = (0,1,0)E(\theta|s) \qquad (3.50)$$

and precision

$$P(\theta_2|s) = [(0,1,0)P^{-1}(\theta|s)(0,1,0)']^{-1} \qquad (3.51)$$

and a $1 - \Delta$ $(0 \le \Delta \le 1)$ HPD region for θ_2 is given by

$$E(\theta_2|s) \pm t_{\Delta/2, n-3} \sqrt{P^{-1}(\theta_2|s)} \qquad (3.52)$$

Consider a test of $H_0: \theta_2 = 0$ versus the alternative $H_0: \theta_2 \ne 0$, then if $0 \notin HPD_\Delta(\theta_2)$, H_0 is rejected at the "significance level" Δ. An HPD region for $(\theta_1, \theta_2, \theta_3)$ is found as follows.

Since $\theta|s \sim t_3[n - 3, E(\theta|s), P(\theta|s)]$,

$$F(\theta) = \frac{1}{3}[\theta - E(\theta|s)]'P(\theta|s)(n - 3)^{-1}[\theta - E(\theta|s)] \qquad (3.53)$$

has an F-distribution with 3 and $n - 3$ degrees of freedom, and a $1 - \Delta$ $(0 < \Delta < 1)$ HPD region for θ is

$$HPD_\Delta(\theta) = \{\theta : F(\theta) \le F_{3, n-3, \Delta}\} \qquad (3.54)$$

and one would reject $H_0: \theta = 0$ versus $H_a: \theta \ne 0$ whenever $F(0) > F_{\Delta; 3, n-3}$. One may show this is equivalent to the size Δ likelihood-ratio test of H_0 versus H_a. This is, of course, the so-called test for significance of regression and the reader is referred to Draper and Smith (1966) for an explanation of the "usual" approach to tests of hypotheses in regression analysis.

Since the marginal posterior distribution of τ is gamma with parameters $\alpha' = (n - 3)/2$ and β', where $2\beta' = C - B'A^{-1}B$, point and interval estimators of this parameter are easily found.

In forecasting a future value of w of Y where

$$w = \theta_1 + \theta_2 x_1^* + \theta_3 x_2^* + e$$

and $e \sim n(0, \tau^{-1})$, one would use the Bayesian predictive distribution (1.39) of Chapter 1, where $z = (1, x_1^*, x_2^*)$, $k = 1$, $P = 0$ (3×3), $\beta = 0$, $\alpha = -3/2$, x is $n \times 3$, namely

MULTIPLE LINEAR REGRESSION

$$X = \begin{pmatrix} 1 & x_{11} & x_{21} \\ 1 & x_{12} & x_{22} \\ . & . & . \\ . & . & . \\ . & . & . \\ 1 & x_{1n} & x_{2n} \end{pmatrix}$$

and $Y = (y_1, y_2, \ldots, y_n)'$, and a $1 - \gamma$ ($0 < \gamma < 1$) HPD region for w is given by (1.40).

An important problem in multiple linear regression is choosing the "best" subset of independent variables to put into the model. That is with two variables x_1 and x_2 one could have the models (always including the intercept)

$$y = \theta_1 + e_1 \qquad \text{(neither } x_1 \text{ nor } x_2\text{)}$$

or

$$y = \theta_{11} + \theta_{12} x_1 + e_2 \qquad \text{(only } x_1 \text{ included)}$$

or

$$y = \theta_{21} + \theta_{22} x_2 + e_3 \qquad \text{(only } x_2 \text{ included)}$$

or

$$y = \theta_{31} + \theta_{32} x_1 + \theta_{33} x_2 + e_4 \qquad \text{(both } x_1 \text{ and } x_2 \text{ included)}$$

and the problem is to choose the appropriate model.

Of course, one's choice of a model depends on one's criterion of choosing the model, and some argue, see Lindley (1968), that one's criterion of choosing a model depends, in turn, on one's purpose of the regression analysis. If the main objective is to estimate the parameters, one's criterion of choosing a model will be based on the posterior distribution of the parameters, but on the other hand, if the main goal is to forecast future observations, one's criterion will involve the Bayesian predictive distribution of those future observations. This is Lindley's (1968) way to solve the problem, however, he uses a decision-theory approach, involving loss functions, and his method will not be taken in this book. Instead a more informal way is introduced, namely, if the main objective is forecasting, the appropriate model is that one with the smallest variance of the one-step-ahead predictive distribution, but if the main goal is to estimate the regression parameters, the appropriate model is the one with the "largest" precision matrix of the marginal posterior distribution of the regression parameters θ. The "largest" precision matrix will be explained later, so for now let us consider the predictive method.

The precision of the predictive distribution of one future observation w is calculated for each of the four models, then the most appropriate model is the one with the largest precision. Suppose one wants to predict y when $x = (1, x_1^*, x_2^*)$ and the Jeffreys' prior density is used, then one would substitute the appropriate values of α, β, P, X, Y, and Z into the precision

$$P(w|s) = (n + 2\alpha)A(C - B'A^{-1}B)^{-1},$$

where A, B, and C are given by (1.39). The following substitutions are:

Model 1: $\alpha = -\frac{1}{2}$, $\beta = P = 0$, $x = (1, 1, \ldots, 1)'$,

$y = (y_1, y_2, \ldots, y_n)$, $z = 1$.

Model 2: $\alpha = -\frac{2}{2}$, $\beta = 0$, $P = 0$ (2×2), $X = \begin{pmatrix} 1 & x_{11} \\ 1 & x_{12} \\ & \vdots \\ 1 & x_{1n} \end{pmatrix}$,

$z = (1, x_1^*)$.

Model 3: $\alpha = -\frac{2}{2}$, $\beta = 0$, $P = 0$ (2×2), $X = \begin{pmatrix} 1 & x_{21} \\ 1 & x_{22} \\ & \vdots \\ 1 & x_{2n} \end{pmatrix}$,

$z = (1, x_2^*)$.

MULTIPLE LINEAR REGRESSION

Model 4: $\alpha = -\frac{3}{2}$, $\beta = 0$, $P = 0$ (3×3), $x = \begin{pmatrix} 1 & x_{11} & x_{21} \\ 1 & x_{12} & x_{22} \\ \vdots & & \\ 1 & x_{1n} & x_{2n} \end{pmatrix}$,

$z = (1, x_1^*, x_2^*)$.

With this way to select the model, model 4 is usually chosen as the most appropriate model, thus if the precision of model 4 is only slightly larger than the precisions of models 2 or 3, then perhaps the final choice of model should be model 2, if the precision of model 3 is larger than that of 2. If model 3 has a larger precision than that of 2, then perhaps model 3 should be the choice of a final model. Obviously subjectivity is heavily involved in the choice of the model.

To summarize the selection procedure, let $P_k(w|s)$ be the precision of the predictive distribution of one future observation w, predicted by model k, where $k = 1, 2, 3, 4$, then the appropriate model is model L, $L = 1, 2, 3, 4$, where

$$P_L(w|s) > \max_{k \neq L} \{P_k(w|s)\},$$

and $P_k(w|s)$ is calculated by the above substitution scheme.

If estimation is one's main goal in the regression analysis, one might choose the model which has the "largest" precision matrix in the marginal posterior distribution of the regression coefficients θ; however, "largest" precision matrix is vague and one must clarify the meaning of "large." In general the precision matrix of the marginal posterior distribution of θ is, from (1.22),

$$P(\theta|s) = (n - p)(x'x)(y'y - y'x(x'x)^{-1}x'y)^{-1}, \qquad (1.22)$$

where p is the number of parameters in the model which is $p = m + 1$, where m is the number of independent variables in the model, and assuming a Jeffreys' prior density for the parameters. Thus one way to choose a model is that one with the largest $|P(\theta|s)|$.

The two criteria for choosing an appropriate model will not always produce the same model. Also, the two selection procedures were developed assuming a Jeffreys' improper prior density, but the general idea is easily extended to regression models with any number of independent variables and when the prior information of the parameters is expressed by a conjugate prior density.

An Example with Two Regressor Variables

An example of a simple linear regression analysis was given in the previous section and it was shown how to calculate the posterior estimates of the regression coefficients and how to predict a future observation.
Consider the following regression model with two regressors,

$$y_i = \theta_1 + \theta_2 x_{i1} + \theta_3 x_{i2} + e_i, \qquad (3.55)$$

$i = 1, 2, \ldots, 30$, $x_{i1} = i$, and $x_{i2} = 1,3,1,5,0,6,5,10,7,13,4,8,17,2,$
$10,20,15,17,22,23,15,18,30,25,10,19,22,20,25$, and 36. The errors e_i and n.i.d. $(0,1)$ and $\theta_1 = 1$, $\theta_2 = 2$, and $\theta_3 = 3$.

Thirty Y values were generated from this model and are given in Table 3.2 and the first six observations (x_{i1}, x_{i2}, y_i), $i = 1,2,3,4,5,6$, provided a way to fit the parameters μ, P (3×3), α and β to a normal-gamma prior density, where

$$\mu = (x^{*\prime} x^*)^{-1} x^{*\prime} z,$$

$$P = x^{*\prime} x^* / s^2,$$

$$s^2 = \frac{z'z - z'x^*(x^{*\prime}x^*)^{-1} x^{*\prime} z}{n - 3}, \quad n = 6,$$

$$\alpha = 1$$

and

$$\beta = (s^2)^{-1}.$$

Also, $z = (y_1, y_2, y_3, y_4, y_5, y_6)'$ and

$$x^* = \begin{pmatrix} 1 & x_{11} & x_{21} \\ 1 & x_{12} & x_{22} \\ \cdot & \cdot & \cdot \\ \cdot & \cdot & \cdot \\ \cdot & \cdot & \cdot \\ 1 & x_{16} & x_{26} \end{pmatrix}$$

and using the formulas the values of the hyperparameters were calculated to be:

MULTIPLE LINEAR REGRESSION

Table 3.2
Y Values for Multiple Linear Regression

y_1 = 5.70178		y_{16} =	80.3337
y_2 = 13.0713		y_{17} =	87.2295
y_3 = 10.6488		y_{18} =	106.561
y_4 = 22.5712		y_{19} =	109.683
y_5 = 12.31		y_{20} =	87.5413
y_6 = 31.6939		y_{21} =	100.251
y_7 = 29.7727		y_{22} =	136.989
y_8 = 46.6751		y_{23} =	122.513
y_9 = 40.1657		y_{24} =	80.6753
y_{10} = 59.4161		y_{25} =	112.06
y_{11} = 36.6735		y_{26} =	120.948
y_{12} = 50.1047		y_{27} =	114.828
y_{13} = 77.4773		y_{28} =	135.531
y_{14} = 34.2187		y_{29} =	138.048
y_{15} = 94.3475		y_{30} =	138.903

$$\mathcal{P} = \begin{pmatrix} 12.1431 & 42.5008 & 32.3815 \\ & 184.17 & 133.574 \\ " & & 145.717 \end{pmatrix}, \qquad (3.56)$$

$\alpha = 1$,

$$\mu = \begin{pmatrix} .26964 \\ 2.47146 \\ 2.6549 \end{pmatrix}$$

and

$\beta = .722657.$

Combining the prior density with parameters (3.56) and the likelihood function for μ, P, α and β given $S = \{(x_{i1}, x_{i2}, y_i): i = 7, \ldots, 29\}$ yields the joint posterior density of the parameters and the marginal posterior distribution of $\theta = (\theta_1, \theta_2, \theta_3)'$ is a t with $n + 2\alpha$ (=25) degrees of freedom, location vector

$$E(\theta|s) = \begin{pmatrix} .951843 \\ 2.02176 \\ 2.98288 \end{pmatrix} \qquad (3.57)$$

and precision matrix

$$P(\theta|s) = \begin{pmatrix} 27.3105 & 354.758 & 302.598 \\ & 6720.7 & 5722.41 \\ & & 5387.58 \end{pmatrix}. \qquad (3.58)$$

It follows that the marginal posterior distribution of θ_1 is also a t with 25 degrees of freedom, location

$$E(\theta_1|s) = .951843 \qquad (3.59)$$

and precision

$$P(\theta_1|s) = 8.5838 \qquad (3.60)$$

As for θ_2, it has mean

$$E(\theta_2|s) = 2.02176 \qquad (3.61)$$

precision

$$P(\theta_2|s) = 534.799 \qquad (3.62)$$

and 25 degrees of freedom.

The parameters of the posterior distribution of θ_3 are

$$E(\theta_3|s) = 2.98288 \qquad (3.63)$$

MULTIPLE LINEAR REGRESSION

and

$$P(\theta_3|s) = 515.142 \tag{3.64}$$

In order to forecast Y_{30} when $x_{1,30} = 30$ and $x_{2,30} = 36$, the predictive distribution of Y_{30} is a univariate t with 25 degrees of freedom, location (1.39)

$$E(w|s) = 168.988 \tag{3.65}$$

and precision

$$P(w|s) = .597336 \tag{3.66}$$

The prior estimate μ, (3.56), of θ is close to θ_2 and θ_3 but not as close to θ_1 (=1); however, as expected, the posterior mean (3.57) of θ is quite close to the "true" value of θ; however, the posterior mean of θ_1 is not as close to $\theta_1 = 1$ as are the posterior means of θ_2 and θ_3 to $\theta_2 = 2$ and $\theta_3 = 3$ respectively.

HPD intervals for these parameters are easily found. For example, consider a 90% interval estimate of θ_1. The general formula is

$$E(\theta_1|s) \pm t_{n+2\alpha,.05} \sqrt{P^{-1}(\theta_1|s)} \tag{3.67}$$

or

.951843 ± .58297

and a 90% HPD interval for θ_2 is 2.02176 ± .07385. What is the 90% HPD interval for θ_3 and how would one test for significance of regression ($\theta_1 = \theta_2 = \theta_3 = 0$)?

Suppose one desires to forecast a future value of Y_{30} (=138.903) given $x_{1,30} = 30$ and $x_{2,30} = 36$. We have seen a point estimate of Y_{30} is given by $E(w|s) = 168.988$ with a predictive precision of .597336 thus a 90% prediction interval for Y_{30} is

168.988 ± 2.20993

and the actual observation of 138.048 lies quite far outside the prediction interval, because the predictive precision is very small.

This example was constructed so that θ_1 and θ_2 have the same "true" values as θ_1 and θ_2, the slope and intercept, respectively, of the simple linear regression model (3.41) and the first regressor in both assume the same values, namely $x_{i1} = i$, $i = 1, 2, \ldots, 30$. The addition of the second regressor affects the marginal posterior distributions of θ_1 and θ_2. For instance $P(\theta_1|s) = 5.35085$ with the simple linear regression model and a data-based prior, while $P(\theta_1|s) = 8.5838$ with the regression containing two regressions and a data-based prior.

NONLINEAR REGRESSION ANALYSIS

Suppose

$$y_i = f(x_i, \theta) + e_i, \quad i = 1, 2, \ldots, n \tag{3.68}$$

where x_i is a $m \times 1$ vector of independent variables, θ is a $p \times 1$ real unknown parameter vector, $x_i \in R^m$, $\theta \in R^p$, $f: R^m \times R^p \to R$ is a known function with domain $R^m \times R^p$ and range contained in R, and the e_i are n.i.d. $(0, \tau^{-1})$, where $\tau > 0$ and unknown. If

$$f(x, \theta) = x\theta, \quad x \in R^m, \quad \theta \in R^p \tag{3.69}$$

where x is a $1 \times m$ vector, then the model is the general linear model, but if $f(x,\theta)$ is not linear in θ, the model (3.68) is called a nonlinear regression model, and the mean of the dependent variable y is a nonlinear function of θ.

It has been shown that if f is linear in θ, then the Bayesian approach is well developed, however when f is nonlinear in θ, Bayesian inferences are more difficult to develop. One problem is the nonlinearity of f makes a conjugate family difficult to identify (except in some special cases). This is obvious from the likelihood function

$$L(\theta, \tau|s) \propto \tau^{n/2} \exp -\frac{\tau}{2} \sum_{i=1}^{n} [y_i - f(x_i, \theta)]^2, \quad \theta \in R^p,$$

$$\tau > 0, \tag{3.70}$$

where the sample is denoted by

NONLINEAR REGRESSION ANALYSIS

$$s = \{(x_{1i}, x_{2i}, \ldots, x_{mi}, y_i), \quad i = 1, 2, \ldots, m\}$$

and

$$x_i = (x_{1i}, x_{2i}, \ldots, x_{mi}).$$

If $p = 1$, there is an exact small-sample Bayesian analysis as follows. Suppose the prior density of θ and τ is

$$\xi(\theta, \tau) \propto \tau^{\alpha-1} e^{-\tau\beta} \xi(\theta), \quad \theta \in R, \quad \tau > 0$$

then the posterior density of θ and τ is

$$\xi(\theta, \tau | s) \propto \xi(\theta) \tau^{(n+2\alpha)/2 - 1} \exp - \frac{\tau}{2} \left\{ \sum_{i=1}^{n} [y_i - f(x_i, \theta)]^2 + 2\beta \right\},$$

$$\theta \in R, \quad \tau > 0, \quad (3.71)$$

where $\xi(\theta)$ is a proper marginal prior density of θ and the marginal prior density of τ is gamma with parameters $\alpha > 0$ and $\beta > 0$.

By integrating (3.71) with respect to τ,

$$\xi(\theta|s) \propto \frac{\xi(\theta)}{\left\{ 2\beta + \sum_{i=1}^{n} [y_i - f(x_i, \theta)]^2 \right\}^{(n+2\alpha)/2}}, \quad \theta \in R \quad (3.72)$$

is the proper marginal posterior density of θ. Since θ is scalar, $\xi(\theta|s)$ is easily normalized, plotted, and the posterior moments, provided they exist, computed. It is assumed f is such a function that $\xi(\theta|s)$ is integrable with respect to θ over R. The marginal posterior density of θ is difficult to determine (except in special cases) in terms of simple functions, however the conditional posterior distribution of τ is gamma with parameters

$$\alpha' = \frac{n + 2\alpha}{2} \tag{3.73}$$

and $\beta'(\theta)$, where

$$2\beta'(\theta) = 2\beta + \sum_{i=1}^{n} [y_i - f(x_i, \theta)]^2, \tag{3.74}$$

106   THE TRADITIONAL LINEAR MODELS

and the marginal posterior mean of τ is

$$E(\tau|s) = \underset{\theta|s}{E}[\tau'|\beta'(\theta)],$$

where the expectation is with respect to the posterior density of θ and one must assume $\xi(\theta|s)/\beta'(\theta)$ is integrable with respect to θ over R. Provided the other moments of τ exist, they may be computed in a similar fashion, in terms of the conditional posterior moments of τ. Of course, $\xi(\tau|s)$ may be determined by a univariate numerical integration of the joint density $\xi(\theta,\tau|s)$, (3.71), and we see if θ is scalar a complete Bayesian analysis is possible.

What is the Bayesian predictive density of a future observation w of the dependent variable y when x = x*? Assuming w is independent of s, the conditional density of w given θ and τ is

$$g(w|\theta,\tau) \propto \tau^{1/2} e^{-\tau[w-f(x^*,\theta)]^2/2}, \quad w \in R \qquad (3.75)$$

and upon forming the product of $g(w|\theta,\tau)$ and $\xi(\theta,\tau|s)$, the joint density of w, θ, and τ given s is

$$h(w,\theta,\tau|s) \propto \xi(\theta)\tau^{(n+2\alpha+1)/2-1} \exp -\frac{\tau}{2}\left\{2\beta + \sum_{i=1}^{n}[y_i - f(x_i,\theta)]^2 + [w - f(x^*,\theta)]^2\right\}, \qquad (3.76)$$

where $w \in R$, $\theta \in R$, and $\tau > 0$.

Eliminating τ from (3.76) gives

$$h(w,\theta|s) \propto \frac{\xi(\theta)}{\left\{2\beta + \sum_{1}^{n}[y_i - f(x_i,\theta)]^2 + [w - f(x^*,\theta)]^2\right\}^{(n+2\alpha+1)/2}},$$

$$w \in R \qquad (3.77)$$

for the joint density of w and θ, given s. But from (3.77), one can identify the conditional distribution of w given θ and s as a t with $(n + 2\alpha)$ degrees of freedom, location

$$E(w|\theta,s) = f(x^*,\theta) \qquad (3.78)$$

and precision

NONLINEAR REGRESSION ANALYSIS

$$P(w|\theta,s) = (n + 2\alpha)\left\{2\beta + \sum_{1}^{n} [y_i - f(x_i,\theta)]^2\right\}^{-1}. \qquad (3.79)$$

How does one forecast a future value of w? If one knew θ, one would use $f(x^*,\theta)$; however, since θ is unknown, one could use the predictive mean of w, namely

$$E(w|s) = \underset{\theta|s}{E}\ E(w|\theta,s)$$

which is calculated by averaging the conditional mean with respect to θ.

By numerical univariate integrations, the Bayesian predictive density of w can be computed by the integral (if it exists)

$$h(w|s) = \int_\theta \xi(\theta|s)h(w|\theta,s)\,d\theta, \qquad w \in R \qquad (3.80)$$

where $\xi(\theta|s)$ is the marginal posterior density of θ and $h(w|\theta,s)$ is the conditional predictive density of w, given θ, which is a t with $n + 2\alpha$ degrees of freedom, location (3.78), and precision (3.79).

One may sum up the Bayesian analysis of nonlinear regression, when θ is scalar, by saying a complete analysis is possible if f is sufficiently well-behaved to insure the existence of certain integrals; however, if θ is of dimension greater than or equal to two, a Bayesian analysis becomes more difficult.

If $\theta \in R^p$, the marginal density of θ is (3.72) but revised to be

$$\xi(\theta|s) \propto \frac{\xi(\theta)}{\left\{2\beta + \sum_{1}^{n} [f(x_i,\theta) - y_i]^2\right\}^{(n+2\alpha)/2}}, \qquad \theta \in R^p \qquad (3.81)$$

where $\xi(\theta)$ is the prior marginal density of θ on R^p, f is chosen so that $\xi(\theta|s)$ is integrable over R^p and the marginal prior density of τ is gamma with parameters α and β.

If $p = 2$, one, by numerical bivariate integration, could develop the normalizing constant, the contours of constant density, and some of the bivariate moments, and by a series of univariate numerical integrations, determine the marginal posterior densities of the components of θ. Of course when $p \geq 3$, the numerical integration problems become more impractical.

When $p \geq 2$, there are some special cases, where an exact and complete Bayesian analysis is possible. For example, let $p = 2$ and consider

$$f(x,\theta) = \theta_2 f_1(x,\theta_1), \quad \theta_1 \in R, \quad \theta_2 \in R \qquad (3.82)$$

where $f_1: R \times R \to R$ and f_1 is such that $\xi(\theta|s)$, (3.81), is integrable over R^p, and if the prior density of θ is

$$\xi(\theta) = \xi_1(\theta_1)\xi_2(\theta_2), \quad \theta \in R^2$$

where $\xi_2(\theta_2)$ is constant over R, then the posterior marginal density of θ is

$$\xi(\theta|s) \propto \frac{\xi_1(\theta_1)}{\left\{2\beta + \sum_1^n [y_i - \theta_2 f_1(x_i,\theta_1)]^2\right\}^{(n+2\alpha)/2}}, \quad \theta \in R$$

(3.83)

hence the conditional posterior density of θ_2 given θ_1 is

$$\xi(\theta_2|\theta_1,s) \propto \frac{1}{\{[\theta_2 - A^{-1}(\theta_1)B(\theta_1)]^2 A(\theta_1) + C(s) - B^2(\theta_1)A^{-1}(\theta_1)\}^{(n+2\alpha)/2}},$$

$$\theta_2 \in R \quad (3.84)$$

where

$$A(\theta_1) = \sum_1^n f_1^2(x_i,\theta_1),$$

$$B(\theta_1) = \sum y_i f_1(x_i,\theta_1),$$

and

$$C(s) = 2\beta + \sum_1^n y_i^2.$$

We see $\theta_2 | \theta_1$ has a univariate t distribution with $n + 2\alpha$ degrees of freedom, location

$$E(\theta_2 | \theta_1, s) = A^{-1}(\theta_1) B(\theta_1),$$

and precision

$$P(\theta_2 | \theta_1, s) = \frac{(n + 2\alpha) A(\theta_1)}{C(s) - B^2(\theta_1) A^{-1}(\theta_1)}.$$

Also, the marginal posterior density of θ_1 is

$$\xi(\theta_1 | s) \propto \frac{\xi_1(\theta_1)}{A^{1/2}(\theta_1) \{C(s) - B^2(\theta_1) A^{-1}(\theta_1)\}^{(n+2\alpha-1)/2}},$$

$$\theta_1 \in R \quad (3.85)$$

and the marginal posterior mean of θ_2 (if it exists) is

$$E(\theta_2 | s) = \underset{\theta_1 | s}{E} E(\theta_2 | \theta_1, s),$$

where $\underset{\theta_1 | s}{E}$ denotes expectation with respect to (3.85), thus $A^{-1}(\theta_1) \cdot B(\theta_1) \xi(\theta_1 | s)$ must be integrable over R. Since $\xi(\theta_1 | s)$ is a scalar function, its properties can be found numerically. Is it possible to find the Bayesian predictive density of a future value w of the independent variable y?

The marginal density of θ_1 is explicitly determined and may be graphed and its moments computed and then it may be used to compute some of the moments of θ_2.

Can one make inferences about τ? The conditional posterior density of τ given θ_1

$$\xi(\tau|\theta_1,s) \propto \tau^{(n+2\alpha-1)/2-1} \exp-\frac{\tau}{2}[C(s) - B^2(\theta_1)A^{-1}(\theta_1)],$$

$$\tau > 0 \qquad (3.86)$$

and is a gamma density with parameters $\alpha' = (n + 2\alpha - 1)/2$ and $\beta'(\theta_1)$, where $2\beta'(\theta_1) = C(s) - B^2(\theta_1)A^{-1}(\theta_1)$. The conditional mean of $\tau|\theta_1$ is

$$E(\tau|\theta_1,s) = \frac{n + 2\alpha - 1}{C(s) - B^2(\theta_1)A^{-1}(\theta_1)}$$

and the marginal posterior mean of τ is computed from

$$E(\tau|s) = \mathop{E}_{\theta_1|s} E(\tau|\theta_1,s),$$

thus $\xi(\theta_1|s)\alpha'/\beta'(\theta_1)$ must be integrable over R. Note, this type of analysis is possible because f was of the special form (3.82).

Let us return to the general case where $\theta \in R^p$ and the joint posterior density of θ and τ is

$$\xi(\theta,\tau|s) \propto \xi(\theta)\tau^{(n+2\alpha)/2-1} \exp-\frac{\tau}{2}\left\{2\beta + \sum_1^n [y_i - f(x_i,\theta)]^2\right\},$$

$$\tau > 0 \qquad (3.87)$$

where, a priori, θ and τ are independent, the marginal prior distribution of τ is gamma with parameters α and β, and $\xi(\theta)$ is the marginal prior density of θ and is such that $\xi(\theta,\tau|s)$ is a proper density.

The marginal posterior density of θ is given by (3.81), thus what inferences are possible about θ?

Partition θ as $\theta = [\theta^{(1)},\theta^{(2)}]$, where $\theta^{(1)} \in R^{p_1}$, $\theta^{(2)} \in R^{p-p_1}$ and $1 \leq p_1 < p$, then the conditional posterior density of $\theta^{(1)}$ given $\theta^{(2)} = \phi \in R^{p-p_1}$

$$\xi_1[\theta^{(1)}|\phi,s] \propto \frac{\xi[\theta^{(1)},\phi]}{\left\{2\beta + \sum_1^n [y_i - f[x_i,\theta^{(1)},\phi]]^2\right\}^{(n+2\alpha)/2}},$$

NONLINEAR REGRESSION ANALYSIS   111

$$\theta^{(1)} \in R^{p_1} \qquad (3.88)$$

hence if $p_1 = 1$, ξ_1 is a scalar function of $\theta^{(1)} \in R$, and ξ_1 versus $\theta^{(1)}$ may be plotted, and each component of θ has an easily determined conditional posterior distribution. If ϕ is "close" to the "true" value of $\theta^{(2)}$, then ξ_1 will not provide inaccurate inferences for $\theta^{(1)}$.

The classical approach to nonlinear regression, see Draper and Smith (1966), expands f in a Taylor's series about known successive values of θ, then by an iterative procedure, attempts to maximize the likelihood function with respect to θ and τ. A similar approach could be taken here, from a Bayesian viewpoint.

One other way to analyze a nonlinear regression problem is to first find the maximum-likelihood estimate of θ, say $\hat{\theta} = [\hat{\theta}(1), \hat{\theta}^{(2)}]$, then substitute $\hat{\theta}^{(2)}$ for ϕ into the conditional posterior density of $\theta^{(1)}$ given $\theta^{(2)} = \phi$. Then posterior inferences for each component of θ are possible, and one may obtain additional information about these parameters beyond what is given by the usual analysis, which consists of the maximum-likelihood estimate $\hat{\theta}$ and its large sample dispersion matrix.

Consider this simple nonlinear regression model

$$y_i = e^{-\theta x_i} + \varepsilon_i, \qquad i = 1, 2, \ldots, n \qquad (3.89)$$

where $\theta \in R$, (x_i, y_i) is the i-th observation, where y_i is the i-th observation of the random variable y, x_i the i-th observation of the regression variable x and the e_i are n.i.d. $(0, \tau^{-1})$, where θ and $\tau > 0$ are unknown parameters. Suppose $n = 30$ (x_i, y_i) pairs are generated from (3.89), where $\theta = 2$ and $\tau = 1$ and that the prior density for θ and τ is

$$\xi(\theta, \tau) \propto \tau^{\alpha-1} e^{-\tau\beta} \tau^{1/2} e^{-(\tau/2)(\theta-\mu)^2 P}, \qquad \theta \in R \text{ and } \tau > 0,$$

$$(3.90)$$

where $\alpha = 2$, $\beta = 1$, $\mu = 3$, and $P = 1$, then the marginal posterior density of θ is

$$\xi(\theta|s) \propto \left\{ 2\beta + \sum_1^n (y_i - e^{-\theta x_i})^2 + (\theta - 3)^2 \right\}^{-(n+2\alpha+1)/2},$$

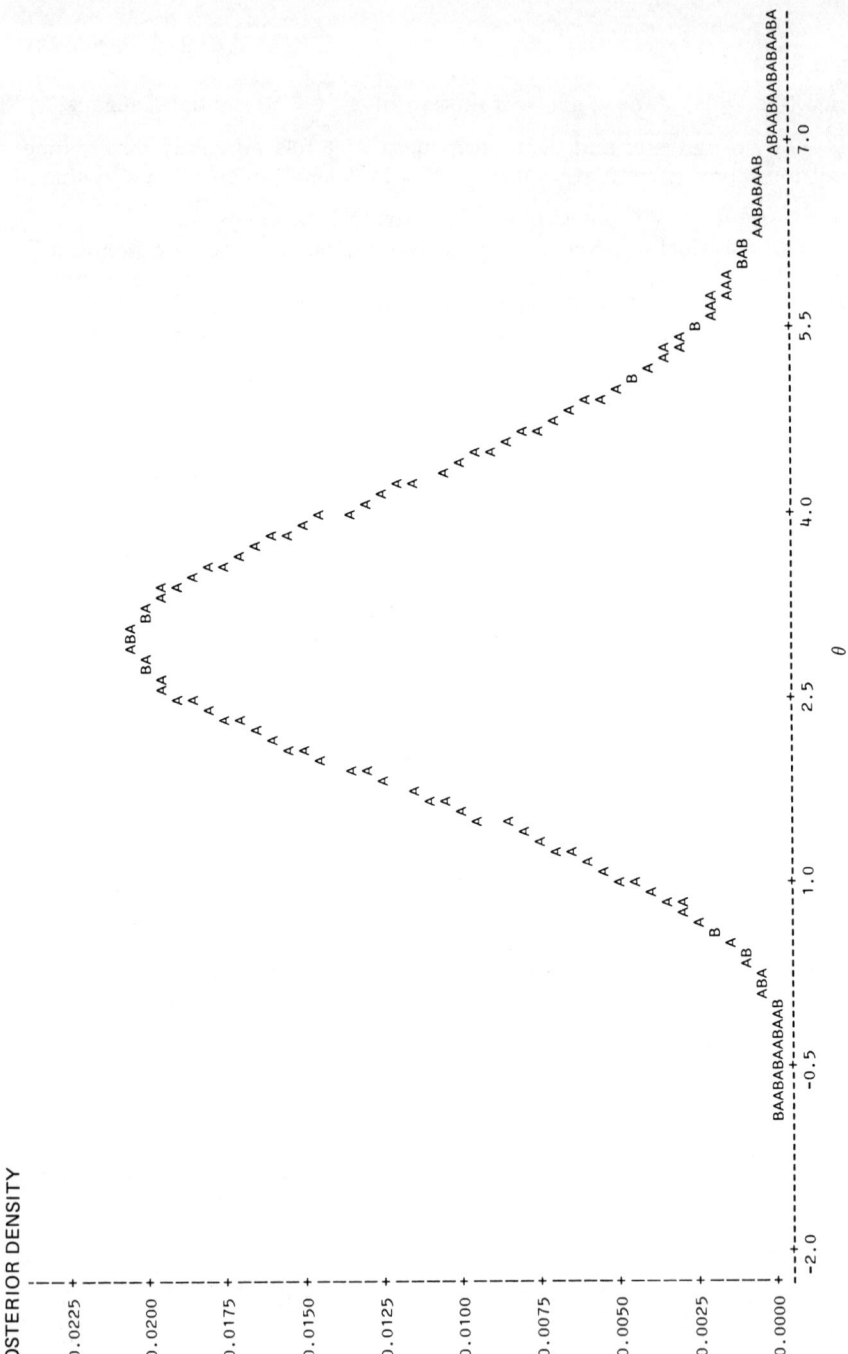

Figure 3.1. Nonlinear regression—Bayesian posterior density.

$$\theta \in R \qquad (3.91)$$

which is plotted in Figure 3.1.

Table 3.3 gives the 30 observations, which were generated from the nonlinear regression model (3.89), and the posterior moments of θ, namely the posterior mean

$$E(\theta|s) = 3.08133,$$

the second posterior moment,

$$E(\theta^2|s) = 10.8659,$$

and the posterior variance

$$Var(\theta|s) = 1.37126.$$

Table 3.3
Observations from Nonlinear Model

N	Y	N	Y
Row 1	0.432834	Row 16	0.19206
Row 2	-1.33177	Row 17	0.636969
Row 3	-0.702886	Row 18	-0.409463
Row 4	-1.64393	Row 19	2.15701
Row 5	-0.828128	Row 20	0.120331
Row 6	-0.738765	Row 21	-0.305884
Row 7	1.07194	Row 22	-0.728323
Row 8	0.731636	Row 23	0.176013
Row 9	0.797699	Row 24	0.409809
Row 10	0.0071879	Row 25	-0.594562
Row 11	0.0268496	Row 26	1.77496
Row 12	1.20312	Row 27	1.06013
Row 13	1.31212	Row 28	0.482367
Row 14	-0.690334	Row 29	0.692809
Row 15	1.08125	Row 30	0.522263

Table 3.4
Posterior Density of θ

Obs.	θ	Posterior density	Obs.	θ	Posterior density
1	-0.94	0.0000000	29	0.74	0.0027856
2	-0.88	0.0000000	30	0.80	0.0031682
3	-0.82	0.0000000	31	0.86	0.0035714
4	-0.76	0.0000000	32	0.92	0.0039949
5	-0.70	0.0000000	33	0.98	0.0044384
6	-0.64	0.0000000	34	1.04	0.0049016
7	-0.58	0.0000000	35	1.10	0.0053843
8	-0.52	0.0000000	36	1.16	0.0058860
9	-0.46	0.0000000	37	1.22	0.0064063
10	-0.40	0.0000000	38	1.28	0.0069445
11	-0.34	0.0000000	39	1.34	0.0074999
12	-0.28	0.0000000	40	1.40	0.0080713
13	-0.22	0.0000000	41	1.46	0.0086575
14	-0.16	0.0000000	42	1.52	0.0092570
15	-0.10	0.0000000	43	1.58	0.0098680
16	-0.04	0.0000000	44	1.64	0.0104886
17	0.02	0.0001189	45	1.70	0.0111164
18	0.08	0.0002839	46	1.76	0.0117491
19	0.14	0.0003522	47	1.82	0.0123838
20	0.20	0.0004532	48	1.88	0.0130176
21	0.26	0.0005919	49	1.94	0.0136474
22	0.32	0.0007659	50	2.00	0.0142698
23	0.38	0.0009728	51	2.06	0.0148813
24	0.44	0.0012104	52	2.12	0.0154785
25	0.50	0.0014762	53	2.18	0.0160575
26	0.56	0.0017683	54	2.24	0.0166148
27	0.62	0.0020847	55	2.30	0.0171465
28	0.68	0.0024242	56	2.36	0.0176490

NONLINEAR REGRESSION ANALYSIS

Table 3.4 (Continued)

Obs.	θ	Posterior density	Obs.	θ	Posterior density
57	2.42	0.0181187	85	4.10	0.0131022
58	2.48	0.0185522	86	4.16	0.0124903
59	2.54	0.0189463	87	4.22	0.0118800
60	2.60	0.0192979	88	4.28	0.0112744
61	2.66	0.0196044	89	4.34	0.0106761
62	2.72	0.0198632	90	4.40	0.0100878
63	2.78	0.0200724	91	4.46	0.0095117
64	2.84	0.0202303	92	4.52	0.0089498
65	2.90	0.0203355	93	4.58	0.0084039
66	2.96	0.0203874	94	4.64	0.0078755
67	3.02	0.0203854	95	4.70	0.0073658
68	3.08	0.0203297	96	4.76	0.0068759
69	3.14	0.0202209	97	4.82	0.0064066
70	3.20	0.0200598	98	4.88	0.0059583
71	3.26	0.0198479	99	4.94	0.0055314
72	3.32	0.0195871	100	5.00	0.0051262
73	3.38	0.0192795	101	5.06	0.0047425
74	3.44	0.0189277	102	5.12	0.0043803
75	3.50	0.0185347	103	5.18	0.0040392
76	3.56	0.0181034	104	5.24	0.0037187
77	3.62	0.0176374	105	5.30	0.0034185
78	3.68	0.0171402	106	5.36	0.0031378
79	3.74	0.0166154	107	5.42	0.0028760
80	3.80	0.0160670	108	5.48	0.0026324
81	3.86	0.0154987	109	5.54	0.0024061
82	3.92	0.0149143	110	5.60	0.0021964
83	3.98	0.0143177	111	5.66	0.0020025
84	4.04	0.0137124	112	5.72	0.0018234

Table 3.4 (Continued)

Obs.	θ	Posterior density	Obs.	θ	Posterior density
113	5.78	0.00165834	132	6.92	0.00022984
114	5.84	0.00150651	133	6.98	0.00020576
115	5.90	0.00136707	134	7.04	0.00018413
116	5.96	0.00123922	135	7.10	0.00016470
117	6.02	0.00112217	136	7.16	0.00014726
118	6.08	0.00101518	137	7.22	0.00013163
119	6.14	0.00091753	138	7.28	0.00011761
120	6.20	0.00082851	139	7.34	0.00010505
121	6.26	0.00074749	140	7.40	0.00009381
122	6.32	0.00067382	141	7.46	0.00008375
123	6.38	0.00060693	142	7.52	0.00007475
124	6.44	0.00054627	143	7.58	0.00006671
125	6.50	0.00049131	144	7.64	0.00005951
126	6.56	0.00044158	145	7.70	0.00005309
127	6.62	0.00039662	146	7.76	0.00004735
128	6.68	0.00035602	147	7.82	0.00004222
129	6.74	0.00031938	148	7.88	0.00003765
130	6.80	0.00028636	149	7.94	0.00003357
131	6.86	0.00025661	150	8.00	0.00002993

These last three moments were computed via numerical integration of the marginal posterior density (3.91).

Table 3.4 is a tabulation of the posterior density of θ, where the domain of the function is from -0.94 to 8.00 to form a partition of 150 θ-values with an interval width of .06. At the end points, $\xi(\theta|s)$ is zero to at least four decimal places and the mode is at $m(\theta|s) = 3.08$.

Upon inspection of the posterior marginal density of θ, one sees the heavy influence of the prior density of θ, which has a mean at $\theta = 3$, while the true value of θ is located at $\theta = 2$, thus the information about θ from the sample does not overcome the highly confident prior information that θ is located at three.

SOME MODELS FOR DESIGNED EXPERIMENTS

The analysis of designed experiments usually consists, among other things, of an analysis of variance, which is a way to determine the effect of various factors on a response variable y. For example, in a completely randomized design t treatments are assigned at random to n experimental units. Suppose that n_i is the number of units receiving the i-th treatment. If y_{ij} denotes the response measured from the j-th unit receiving treatment i, a one-way analysis of variance is performed in order to determine if each treatment produces the same average response.

The analysis of variance partitions the total sum of squares into "between" treatment and "within" treatment components, and if the ratio of their corresponding mean squares is "large," significant differences between the treatments are said to exist, that is, the average responses produced by the treatments are not the same. Fisher (1953), Kempthorne (1952), Cochran and Cox (1957), and Cox (1958) are standard references for the design and analysis of experiments and Graybill (1961) and Searle (1971) give the theory for those models which are appropriate for the analysis of designed experiments.

This section will give the Bayesian analysis of designed experiments and is the analog of the classical approach given by Graybill (1961) and Searle (1971). The so-called one-way model

$$y_{ij} = \theta_i + e_{ij}, \quad i = 1, 2, \ldots, t, \quad j = 1, 2, \ldots, n_i \qquad (3.92)$$

models the completely randomized design, where y_{ij} is the response measured from the j-th unit receiving treatment i, θ_i is the mean of the i-th treatment, and the e_{ij}'s are n.i.d. $(0, \tau^{-1})$, where $\tau > 0$, thus the average response from treatment i is θ_i, which is considered a real unknown parameter.

The analysis of variance is a way to test $H_0: \theta_1 = \theta_2 = \ldots = \theta_t$ versus the alternative that H_0 is not true, and it is equivalent to the likelihood-ratio test. This is shown by Graybill (1961) and Searle (1971). The AOV test is to reject H_0 at level α whenever

$$F = BMS/WMS > F_{\alpha/2, t-1, n-t}, \qquad (3.93)$$

where

$$BMS = \sum_{i=1}^{t} (\bar{y}_{i\cdot} - \bar{y}_{\cdot\cdot})^2 / (t-1)$$

and

$$\text{WMS} = \sum_{1}^{t}\sum_{1}^{n_i} (y_{ij} - \bar{y}_{i.})^2/(n-t).$$

The Bayesian approach to the analysis of the completely randomized design is based on an HPD region for the vector of treatment effects $\theta = (\theta_1, \theta_2, \ldots, \theta_t)'$.

The Completely Randomized Design

The one-way model (3.92) is a special case of the general linear model of Chapter 1, where y is $n \times 1$ ($n = \Sigma_1^t n_i$) and

$$Y = \begin{pmatrix} y^{(1)} \\ y^{(2)} \\ \vdots \\ y^{(t)} \end{pmatrix},$$

where $y^{(i)}$ is $n_i \times 1$ with components $y_{i1}, y_{i2}, \ldots, y_{in_i}$, and the design matrix X is $n \times t$, where

$$X = \begin{pmatrix} x_1 \\ x_2 \\ \vdots \\ x_t \end{pmatrix},$$

where x_i is a $n_i \times t$ matrix with i-th column consisting of ones, and the remaining columns are zero. The error vector ε is $n \times 1$ and contains n.i.d. $(0, \tau)$ random variables.

The likelihood function for θ and τ is

SOME MODELS FOR DESIGNED EXPERIMENTS

$$L(\theta,\tau|y,x) \propto \tau^{n/2} \exp - \frac{\tau}{2} \sum_{i=1}^{t}\sum_{j=1}^{n_i} (y_{ij} - \theta_i)^2,$$

$$\theta \in R^t, \quad \tau > 0 \quad (3.94)$$

and suppose the prior information of the parameters is a normal-gamma distribution with parameters $\mu \in R^p$, P, $\alpha > 0$, and $\beta > 0$, then by Bayes theorem

$$\xi(\theta,\tau|y,x) \propto \tau^{(n+2\alpha+1)/2-1} \exp -\frac{\tau}{2}\left\{2\beta + \sum_{i=1}^{t}\sum_{j=1}^{n_i}(y_{ij}-\theta_i)^2 + (\theta-\mu)'P(\theta-\mu)\right\}, \quad (3.95)$$

where $\theta \in R^p$ and $\tau > 0$, is the joint posterior density of θ and τ, which may be expressed as

$$\xi(\theta,\tau|y,x) \propto \tau^{(n+2\alpha+1)/2-1} \exp -\frac{\tau}{2}\{[\theta - A^{-1}B]'A[\theta - A^{-1}B] + C - B'A^{-1}B\}, \quad (3.96)$$

where

$$A = \text{Diag}(n_1, n_2, \ldots, n_t) + P = D + P, \quad (3.97)$$

$$B = \begin{pmatrix} y_{1\cdot} \\ y_{2\cdot} \\ \cdot \\ \cdot \\ \cdot \\ y_{t\cdot} \end{pmatrix} + P\mu = T + P\mu \quad (3.98)$$

and

$$C = y'y + 2\beta + \mu'P\mu, \quad (3.99)$$

where $y'y = \sum_{i=1}^{t}\sum_{j=1}^{n_i} y_{ij}^2$ and $y_{i\cdot} = \sum_{j=1}^{n_i} y_{ij}$ for $i = 1, 2, \ldots, t$.

Since the treatment effects are the parameters of interest, the marginal posterior distribution of θ is a t distribution with mean

$$E[\theta|x,y] = A^{-1}B = (D + \mu)^{-1}(T + \mathcal{P}), \qquad (3.100)$$

precision matrix

$$P[\theta|x,y] = \frac{(n + 2\alpha)A}{C - B'A^{-1}B}, \qquad (3.101)$$

and $n + 2\alpha$ degrees of freedom. The posterior dispersion matrix of θ is

$$D[\theta|x,y] = (C - B'A^{-1}B)A^{-1}(n + 2\alpha - 2)(n + 2\alpha)^{-1} \qquad (3.102)$$

Now how does one test $H_0: \theta_1 = \theta_2 = \ldots = \theta_t$? Following Box and Tiao (1973), one finds an HPD region for γ, where $\gamma = G\theta$ such that H_0 is true if and only if $\gamma = 0$. For example, γ might be a $(t - 1) \times 1$ vector of contrasts γ_i of the t treatment effects, say

$$\gamma_i = \theta_i - \theta_{i+1}, \quad i = 1, 2, \ldots, t - 1,$$

hence G is a $(t - 1) \times t$ matrix

$$G = \begin{pmatrix} 1 & -1 & 0 & \cdots & 0 & 0 \\ 0 & 1 & -1 & \cdots & 0 & 0 \\ & & \vdots & & & \\ 0 & 0 & 0 & \cdots & 1 & -1 \end{pmatrix}$$

Transforming from θ to γ gives a t-distribution for γ with mean $GA^{-1}B$ and precision matrix $[GP^{-1}[\theta|x,y]G']^{-1}$ with $n + 2\alpha$ degrees of freedom and a $1 - \Delta$ HPD region for γ is

$$HPD_\Delta(\gamma) = \{\gamma: F(\gamma) \leq F_{\Delta;t-1,n+2\alpha}\} \qquad (3.103)$$

where

$$F(\gamma) = [\gamma - GA^{-1}B]'[GP^{-1}[\theta|x,y]G']^{-1}[\gamma - GA^{-1}B](t - 1)^{-1} \qquad (3.104)$$

SOME MODELS FOR DESIGNED EXPERIMENTS   121

has an F distribution with $t - 1$ and $n + 2\alpha$ degrees of freedom.
Thus, if $F(0) > F_{\Delta;t-1,n+2\alpha}$, then H_0 is rejected at level Δ, $0 < \Delta < 1$. One important problem remains to be solved, namely how does one assign values to the hyperparameters μ, P, α and β? First, let us see what happens when one assigns a Jeffreys' prior

$$\xi(\theta,\tau) \propto \tau^{-1}, \quad \theta \in R^t, \quad \tau > 0, \quad (3.105)$$

which is equivalent to letting $\beta \to 0$, $P \to 0$ ($t \times t$), and $\alpha \to -t/2$ in the posterior density of θ. This yields

$$A^{-1}B = \begin{pmatrix} \bar{y}_{1\cdot} \\ \bar{y}_{2\cdot} \\ \cdot \\ \cdot \\ \cdot \\ \bar{y}_{t\cdot} \end{pmatrix} \quad (3.106)$$

for the posterior mean of θ, which is the vector of treatment means, while the precision matrix is

$$P[\theta|x,y] = \frac{D}{s^2}, \quad (3.107)$$

where $(n - t)s^2 = \sum_{i=1}^{t} \sum_{j=1}^{n_i} y_{ij}^2 - \sum_{i=1}^{t} y_{i\cdot}^2/n_i$, and s^2 is called the within mean square of the AOV. Substituting these into $F(\gamma)$, then $F(0) > F_{\Delta;t-1,n-t}$ if and only if $BMS/WMS > F_{\Delta,t-1,n-t}$, which is the AOV test of H_0 the equality of treatment effects. The HPD method of testing H_0, using an improper prior density, is equivalent to the AOV; however, the HPD region for γ, given by (3.103), is more general than the AOV because the former allows prior information in the form of a normal-gamma density with parameters μ, P, α and β. But this poses another problem, since the hyperparameters must be assigned legitimate values that actually express one's prior information. How should values for μ, P, α and β be assigned?

By using the improper prior density one is saying the treatment effects and error precision are independent, that the treatments all have the same effect on the average response, and finally that error precisions closer to zero are more likely than those far away from zero, thus a Jeffreys' improper prior density weakly supports the null hypothesis. The same support for the null hypothesis can be achieved with the conjugate prior density if the components of μ are all equal and if P is diagonal. Still another way to express prior knowledge of the treatment effects θ_i is to assume them to be exchangeable, i.e., one's prior opinion about θ_1 is the same as that of θ_3, or any other θ_i, and similarly for pairs and triplets. In particular, one could assume the treatment effects are i.i.d. $n(0,\eta)$, and the θ_i $(i = 1, 2, \ldots, t)$ would be exchangeable, and the null hypothesis would be supported, a priori. Exchangeability is formulated by de Finetti (1972) and applied by Lindley and Smith (1972) to the Bayesian analysis of linear models, but will not be emphasized here.

It is relatively easy to support the null hypothesis with a normal-gamma prior, however it remains to assign values to the hyperparameters. If one has a past experiment

$$z = x^*\theta + \varepsilon, \qquad (3.108)$$

with the same treatments, one may use the prior predictive distribution of z to assign values to the hyperparameters, and this method was explained in the second section of the chapter. If (3.108) represents a future experiment, the experimenter is asked to predict z; the future observations y, when the design is completely randomized; that is, the experimenter must assign at least one response for each of the t treatments, then the values of the hyperparameters are assigned by fitting them to the prior predictive density (3.3). Of course, past experiments of the same form as the one which will be performed give less subjective values for the hyperparameters than hypothetical future experiments.

Returning to the analysis of a completely randomized design, the AOV or the HPD method of testing the equality of treatment effects is only one step in the analysis of such a design. It should be remembered, the marginal posterior distribution of θ tells the complete story about the treatment effects in the sense that once the data y are observed, all inferences are based on this distribution of θ.

If H_0 is rejected, a careful inspection of the marginal posterior distribution of θ is necessary in order to see in what way the treatments are affecting the average response. For example, are the treatment effects partitioned into various groups or clusters?

Consider three treatments and their effects θ_1, θ_2, and θ_3. Suppose for every $\varepsilon > 0$, $P[|\theta_1 - \theta_2| < \varepsilon|y] > P[|\theta_2 - \theta_3| < \varepsilon|y]$, then the data suggest θ_2 is "closer" to θ_1 than it is to θ_3, and thus $\{\theta_2, \theta_1\}$ forms one cluster, while $\{\theta_3\}$ is by itself, so to speak. These probabilities are easily evaluated, because $\theta_i - \theta_j$ has a known univariate t distribution and one may compute $P[|\theta_i - \theta_j| < \varepsilon|y]$ by Student's t-tables.

THE ANALYSIS OF TWO-FACTOR EXPERIMENTS

In the previous section, the completely randomized experiment was analyzed. Such a one-way layout is an example of a one-factor experiment, where the various levels of that factor may have distinct influences on the level of the average response. The factor was the treatment and the treatment effects were the levels of the treatment factor.

Such an experiment was analyzed as if the t treatments were fixed, that is, one is only interested in making inferences about the set of treatments which were included in the experiment and not to a population of treatments, from which the treatments were selected. In the latter case, the treatment factor would have been considered random, that is the different treatments would have been selected at random from a population of treatments, and the experiment would have been examined by the techniques of Chapter 4. This chapter deals only with fixed factors.

With a two-factor experiment, the average response s_{ij} may be influenced by the levels of the two factors, namely the i-th level of the first and the j-th level of the second factor, where $i = 1, 2, \ldots, t$ and $j = 1, 2, \ldots, b$. Suppose for each of the bt combinations of factor levels, one has n_{ij} measurements on the response and suppose the $n = \sum_1^t \sum_1^b n_{ij}$ observations are independent, then if y_{ijk} is the k-th measurement, when level i of the first factor and level j of the second are present,

$$y_{ijk} = s_{ij} + e_{ijk}$$

where the e_{ijk}'s are independent random variables with mean zero and precision τ, i.e., the average response is $E(y_{ijk}) = s_{ij}$ for all i, j, and k.

If the two factors are additive, i.e., $s_{ij} = m + a_i + b_j$, then a_i is referred to as the i-th effect of the first factor and b_j the effect of the j-th level of the second factor, where the a_i, the b_j and m are unknown real parameters. If the two factors are not additive $s_{ij} = m + a_i + b_j + (ab)_{ij}$, $(ab)_{ij} \neq 0$, for some i and j, and $(ab)_{ij}$ is called the interaction between the i-th and j-th levels of the two factors, where the tb interaction effects are unknown real parameters, and in this case there are $tb + t + b + 1$ parameters, excluding the common precision parameter τ of the errors.

Additive Models

Let us first consider an additive model,

$$y_{ijk} = m + a_i + b_j + e_{ijk} \qquad (3.109)$$

where $i = 1, 2, \ldots, t$, $j = 1, 2, \ldots, b$, $k = 1, 2, \ldots, r$, and $e_{ijk} \sim n(0, \tau^{-1})$, $\tau > 0$, $a_i \in R$, $b_j \in R$, and $m \in R$. If $\theta = (\theta_0, \theta_1', \theta_2')'$ where $\theta_0 = m$,

$$\theta_1 = \begin{pmatrix} a_1 \\ a_2 \\ \cdot \\ \cdot \\ \cdot \\ a_t \end{pmatrix},$$

and

$$\theta_2 = \begin{pmatrix} b_1 \\ b_2 \\ \cdot \\ \cdot \\ \cdot \\ b_b \end{pmatrix},$$

then the model is written as

$$y = x\theta + e, \qquad (3.110)$$

where x is a $n \times p$ ($p = 1 + b + t$) design matrix, $y = (y_{111}, y_{112}, \ldots, y_{11n_{11}}, \ldots, y_{tb1}, \ldots, y_{tbn_{tb}})'$, and e is the $n \times 1$ vectors of error terms e_{ijk}, thus θ_1 is the $t \times 1$ vector of the effects of the first factor, and θ_2 the $b \times 1$ vector of effects of the second factor. The design matrix x is such that the first n_{11} rows consist of the $(1 + t + b)$ vector $(1; 1, 0, \ldots, 0; 1, 0, \ldots, 0)$, the next n_{12} rows consist of the vector $(1; 1, 0, \ldots, 0; 0, 1, 0, \ldots, 0)$, and so on, hence x is of less than full rank p, namely of rank $p - 2$.

THE ANALYSIS OF TWO-FACTOR EXPERIMENTS   125

Since x is less than full rank, if one combines the likelihood function of θ and τ induced by (3.110) with Jeffreys' improper prior density

$$\xi(\theta,\tau) \propto 1/\tau, \quad \tau > 0, \quad \theta \in R^p \tag{3.111}$$

the posterior density of the parameters will not be proper. Of course this difficulty can be avoided by employing a normal-gamma prior density in lieu of (3.111); however, some may object to this alternative because it may be very difficult to assign a proper prior distribution.

Fortunately, one may transform a less than full-rank model (3.110) to full rank

$$y = z\alpha + e, \tag{3.112}$$

where z is $n \times k$ ($k < p$) of full rank, α is $k \times 1$, namely,

$$\alpha = u\theta$$

where u is a $k \times p$ known matrix, and as before, $e \sim n(0, \tau^{-1} I_n)$.

According to Graybill (1961, pages 235-236), z and u are constructed as follows. Since x'x is a $p \times p$ and symmetric positive semi-definite matrix, there exists a non-singular w^* ($p \times p$) matrix such that

$$(w^*)'x'xw^* = \begin{pmatrix} B & 0 \\ 0 & 0 \end{pmatrix}$$

where B is $k \times k$ of rank k. If $w^* = (w, w_1)$, where w is $p \times k$ and $(w^*)^{-1} = u^* = (u', u_1')'$, where u is $k \times p$, then

$$z = xw$$

and

$$\alpha = u\theta.$$

Consider the two-way additive model (3.109) with $p = b + t + 1$ parameters where x is of rank $k = p - 2 = b + t - 1$, then obviously, one may use the above reparametrization to arrive at a full-rank representation. How should one choose α, u, and w? Since the primary purpose of the analysis is to investigate the effect of the two factors on the average response $m + a_i + b_j$, one way to choose α is to let

$$\alpha = \begin{pmatrix} m + a_1 + b_1 \\ a_1 - a_2 \\ a_2 - a_3 \\ \vdots \\ a_t - a_{t-1} \\ b_1 - b_2 \\ b_2 - b_3 \\ \vdots \\ b_b - b_{t-1} \end{pmatrix} = \begin{pmatrix} \alpha_1 \\ \alpha_2 \\ \alpha_3 \end{pmatrix}, \qquad (3.113)$$

where $\alpha_1 = m + a_1 + b_1$. This means u must be

$$u = \begin{pmatrix} 1; & 1, & 0, & \ldots, & 0; & 1, & 0, & \ldots, & 0 \\ 0; & 1, & -1, & 0, & \ldots, & 0; & 0, & 0, & \ldots, & 0 \\ 0; & 0, & 1, & -1, & \ldots, & 0; & 0, & 0, & \ldots, & 0 \\ & & & \vdots & & & & & & \\ 0; & 0, & \ldots, & & 1, & -1; & 0, & 0, & \ldots, & 0 \\ 0; & 0, & \ldots, & & 0; & 1, & -1, & 0, & \ldots, & 0 \\ & & & \vdots & & & & & & \\ 0; & 0, & \ldots, & & 0; & 0, & 0, & \ldots, & 1, & -1 \end{pmatrix},$$

(3.114)

which is $k \times p$. Thus let

THE ANALYSIS OF TWO-FACTOR EXPERIMENTS

$$u^* = \begin{pmatrix} u \\ u_1 \end{pmatrix}$$

where u_1 is $(p - k) \times p$ so that u^* is of full rank. Now let $w^* = (u^*)^{-1}$, where $w^* = (w, w_1)$, then $z = xw$, and the reparametrization is complete. The likelihood function for α and τ is

$$L(\alpha, \tau | y) \propto \tau^{rbt/2} \exp - \frac{\tau}{2}(y - z\alpha)'(y - z\alpha), \quad \alpha \in R^k, \quad \tau > 0,$$

where α is of order $b + t - 1$ and when combined with the Jeffreys' improper prior density

$$\xi(\alpha, \tau) \propto 1/\tau, \quad \tau > 0, \quad \alpha \in R^k,$$

gives

$$\xi(\alpha, \tau | y) \propto \tau^{rbt/2-1} \exp - \frac{\tau}{2}[(\alpha - \hat{\alpha})'z'z(\alpha - \hat{\alpha})$$

$$+ (y - z\hat{\alpha})'(y - z\hat{\alpha})], \quad (3.115)$$

for the joint posterior distribution of α and τ, where $\hat{\alpha} = (z'z)^{-1}z'y$, $\alpha \in R^k$, and $\tau > 0$, and the usual Bayesian analysis follows. It can be shown that $\hat{\alpha}$ is such that its first component is $\bar{y}_{...} + (\bar{y}_{1..} - \bar{y}_{...}) + (\bar{y}_{.2.} - \bar{y}_{...})$, the next $a - 1$ components are $\bar{y}_{i..} - \bar{y}_{i+1..}$, $i = 1, 2, \ldots, a - 1$, and the last $b - 1$ components are $\bar{y}_{.j.} - \bar{y}_{.j+1.}$, where $j = 1, 2, \ldots, b - 1$, where

$$\bar{y}_{...} = \sum_i \sum_j \sum_k y_{ijk}/btr$$

$$\bar{y}_{i..} = \sum_j \sum_k y_{ijk}/rb$$

and

$$\bar{y}_{.j.} = \sum_i \sum_k y_{ijk}/rt.$$

Suppose one wants to make inferences about the first factor $(a_1, a_2, \ldots, a_t)$, then one would want to know the marginal posterior distribution of $\alpha_2 = (a_1 - a_2, a_2 - a_3, \ldots, a_{t-1} - a_t)$, which is known once one knows the posterior distribution of α, which, from (3.115), is a t-distribution with $rbt - (t + b + 1)$ degrees of freedom, location

$$E(\alpha|y) = \hat{\alpha}, \qquad (3.116)$$

and precision

$$P(\alpha|y) = \frac{[rbt - (t + b + 1)]z'z}{(y - z\hat{\alpha})'(y - z\hat{\alpha})} \qquad (3.117)$$

thus α_2 also has a t-distribution with $rbt - (t + b + 1)$ degrees of freedom, location

$$E(\alpha_2|y) = (\phi_1, I_{t-1}, \phi_2)\hat{\alpha}, \qquad (3.118)$$

where ϕ_1 is $(t - 1) \times 1$ matrix of zeros, I_{t-1} is the identity matrix of order $t - 1$, and ϕ_2 is a zero matrix of order $(t - 1) \times (b - 1)$. The precision matrix of α_2 is

$$P(\alpha_2|y) = [(\phi_1, I_{t-1}, \phi_2)P^{-1}(\alpha|y)(\phi_1, I_{t-1}, \phi_2)']^{-1}. \qquad (3.119)$$

As for $\alpha_3 = (b_1 - b_2, b_2 - b_3, \ldots, b_b - b_{b-1})'$, the other factor in the model, its posterior distribution is also a t with $rbt - (t + b + 1)$ degrees of freedom, location vector

$$E(\alpha_3|y) = (\phi_1^*, \phi_2^*, I_{b-1})E(\alpha|y), \qquad (3.120)$$

where ϕ_1^* is a $(b - 1) \times 1$ zero vector, and ϕ_2^* is a $(b - 1) \times (t - 1)$ matrix of zeros, thus the precision matrix of the posterior distribution of α_3 is

$$P(\alpha_3|y) = [(\phi_1^*, \phi_2^*, I_{b-1})P^{-1}(\alpha|y)(\phi_1^*, \phi_2^*, I_{b-1})']^{-1}. \qquad (3.121)$$

Of course, the marginal posterior distribution of τ is gamma with parameters $[rbt - (t + b + 1)]/2$ and $(y - z\hat{\alpha})'(y - z\hat{\alpha})/2$.

In problems like this, one is usually interested in the effect of the levels of the two factors on the average response and in particular if

THE ANALYSIS OF TWO-FACTOR EXPERIMENTS

all the levels of one factor have the same effect. For example, do all the t levels of the first factor have the same effect, that is, is H_0: $a_1 = a_2 = \ldots = a_t$, which is true whenever $\alpha_2 = 0$? The usual approach is to do an analysis of variance and the Bayesian approach is to find an HPD region for α_2.

Since the random variable

$$F(\alpha_2|y) = (t-1)^{-1}(\alpha_2 - E(\alpha_2|y)'P(\alpha_2|y)(\alpha_2 - \hat{\alpha}) \qquad (3.122)$$

has an F distribution with $(t-1)$ and $rbt - (t+b+1)$ degrees of freedom, a $1 - \Delta$ $(0 < \Delta < 1)$ HPD region for α_2 is

$$HPD_\Delta(\alpha_2) = \{\alpha_2: F(\alpha_2|y) \leq F_{\Delta; t-1, rbt-t-b-1}\} \qquad (3.123)$$

and

$$H_0: a_1 = a_2 = \ldots = a_t$$

is rejected when $\alpha_2 = 0[(t-1) \times 1]$ is not contained in the region or equivalently whenever $F(0|y) > F_{\Delta; t-1, rbt-t-b-1}$.

One may test the hypothesis

$$H_0^*: b_1 = b_2 = \ldots = b_b \quad (\alpha_3 = 0)$$

in a similar way by constructing an HPD region for α_3.

The Randomized Block Design

The two-way model (3.109) may be used to analyze the experimental results of a randomized block design. In such a layout, the n experimental units are arranged into b blocks such that each block has t units, where t treatments are assigned at random to the t units of each block assuming no interaction between the treatments and experimental material, a randomized block design is modeled by (3.109), where $r = 1$, and the levels of the first factor $a_1, a_2, \ldots, a_t$ are called treatment effects, while the block effects are $b_1, b_2, \ldots, b_b$, the levels of the second factor. The model is

$$y_{ij} = m + a_i + b_j + e_{ij} \qquad (3.124)$$

where the e_{ij}'s are n.i.d. $(0, \tau^{-1})$, $m \in R$, $a_i \in R$, and $b_j \in R$.

Obviously, one may use the results of the preceding section to analyze a randomized block design. First, one would reparameterize the model to full rank and test for equality of treatment effects by using the HPD region for α_2 given by (3.123).

Experiments with Interaction

The two-way classification model with interaction is given by

$$y_{ijk} = m + a_i + b_j + (ab)_{ij} + e_{ijk}, \qquad (3.125)$$

where $i = 1, 2, \ldots, t$, $j = 1, 2, \ldots, b$, and $k = 1, 2, \ldots, r$. Also, y_{ijk} is the k-th observation when the i-th level of the first factor and the j-th level of the second are present. The a_i, $i = 1, 2, \ldots, t$, are the effects of the first factor while the effects of the second are the b_j, $j = 1, 2, \ldots, b$, and $(ab)_{ij}$ is called the interaction between the i-th and the j-th levels of the two factors. As before, assume the e_{ijk}'s are n.i.d. $(0, \tau^{-1})$, $\tau > 0$, then one may express (3.125) as $y = x\theta + e$, where y is the $n(=rbt) \times 1$ vector of observations, x is the $n \times p(= bt + b + t + 1)$ design matrix, θ is the $p \times 1$ vector

$$\theta = \begin{pmatrix} \theta_1 \\ \theta_2 \\ \theta_3 \\ \theta_4 \end{pmatrix}, \qquad (3.126)$$

where $\theta_1 = m$, θ_2 is $t \times 1$, namely

$$\theta_2 = \begin{pmatrix} a_1 \\ a_2 \\ \cdot \\ \cdot \\ \cdot \\ a_t \end{pmatrix}, \qquad (3.127)$$

THE ANALYSIS OF TWO-FACTOR EXPERIMENTS

$\theta_3 = (b_1, b_2, \ldots, b_b)'$, and θ_4 is tb $\times$ 1, namely

$$\theta_4 = \begin{pmatrix} (ab)_{11} \\ \cdot \\ \cdot \\ \cdot \\ (ab)_{tb} \end{pmatrix}. \tag{3.128}$$

Finally, e is an $n \times 1$ normal random vector distributed as $n(0, \tau^{-1}I_n)$ and $e = (e_{111}, e_{112}, \ldots, e_{tbr})'$.

Knowing θ allows one to construct the design matrix x which is $n \times p$ and of rank $k^* = bt$. Thus since x is of less than full rank, the model $y = x\theta + e$ cannot be analyzed with Jeffreys' improper prior density. But, one may transform to a full rank model

$$y = z\alpha + e, \tag{3.129}$$

where α is $k^* \times 1$ and z is $n \times k^*$ and is of full rank.

One way to choose α is to let α be (bt) $\times$ 1, where

$$\alpha = \begin{pmatrix} m + a_1 + b_1 + (ab)_{11} \\ m + a_1 + b_2 + (ab)_{12} \\ \cdot \\ \cdot \\ \cdot \\ m + a_t + b_b + (ab)_{tb} \end{pmatrix} = u\theta, \tag{3.130}$$

thus u is bt $\times$ (1 + t + b + bt) and known, since one knows θ from (3.126). Now let $u^* = (u', u_1')'$, where u_1 is any matrix such that u^* is non-singular. Now let $w^* = (u^*)^{-1}$, where $w^* = (w, w_1)$ and w is $p \times k^*$, then $z = xw$ and the full rank representation (3.129) of the original model is complete.

One now analyzes this model in much the same way the additive model was analyzed in the previous section. We see from (3.129) that y is normally distributed with mean of $z\alpha$ and precision matrix $\tau^{-1}I_n$, thus if α and τ have a Jeffreys' prior density

$$\xi(\alpha, \tau) \propto 1/\tau, \quad \tau > 0, \quad \alpha \in R^{k^*}$$

the joint posterior density of α and τ is

$$\xi(\alpha,\tau|y) \propto \tau^{rbt/2-1} \exp -\frac{\tau}{2}[(\alpha-\hat{\alpha})'z'z(\alpha-\hat{\alpha})$$

$$+ (y - z\hat{\alpha})'(y - z\hat{\alpha})], \qquad (3.131)$$

where $\alpha \in r^{k*}$, $\tau > 0$, and $\hat{\alpha} = (z'z)^{-1}z'y$. The marginal posterior density of α is

$$\xi(\alpha|y) \propto [(y - z\hat{\alpha})'(y - z\hat{\alpha}) + (\alpha-\hat{\alpha})'z'z(\alpha-\hat{\alpha})]^{rbt/2},$$

$$\alpha \in R^{bt} \qquad (3.132)$$

hence α has a t distribution with $rbt - bt$ degrees of freedom, location vector

$$E(\alpha|y) = \hat{\alpha}, \qquad (3.133)$$

and precision matrix

$$P(\alpha|y) = \frac{bt(r-1)z'z}{(y - z\hat{\alpha})'(y - z\hat{\alpha})}. \qquad (3.134)$$

It is interesting to observe that if $r = 1$, there are zero degrees of freedom and the marginal posterior distribution is improper, which is also a problem when one analyzes the experiment the conventional way, because then the error sum of squares is zero and one is unable to estimate τ^{-1}, the error variance. One way for the Bayesian to avoid this is to use a normal-gamma prior density for α and τ, but it perhaps would be a difficult task to assign values to the hyperparameters. Let us, for the time being, assume $r > 1$ and avoid the difficulty.

How does one analyze such a model? Remember, one is interested in the way the levels of the two factors effect the average response

$$S_{ij} = m + a_i + b_j + (ab)_{ij},$$

for $i = 1, 2, \ldots, a$ and $j = 1, 2, \ldots, b$, and the model is said to have no interaction if $(S_{ij} - S_{i'j}) - (S_{ij'} - S_{i'j'}) = 0$ for all i, i', j and j', which is equivalent to saying $(ab)_{ij} - (ab)_{i'j} - (ab)_{ij'} + (ab)_{i'j'} = 0$ for all i, i', j, and j', otherwise the model is said to be nonadditive or is a model with interaction. In the analysis of such models, one first checks for the presence of interaction and if there is none, one examines the levels $a_1, a_2, \ldots, a_t$ of the first factor and then the b levels of the second factor.

THE ANALYSIS OF TWO-FACTOR EXPERIMENTS

Thus, the first task is to develop a test for no interaction, namely, test the hypothesis

H_0: $(ab)^*_{ij} = 0$, where $i = 1, 2, \ldots, a$ and $j = 1, 2, \ldots, b$, where

$(ab)^*_{ij} = (ab)_{ij} - \overline{(ab)}_{i\cdot} - \overline{(ab)}_{\cdot j} + \overline{(ab)}_{\cdot\cdot}$, and

$\overline{(ab)}_{i\cdot} = \sum_{j=1}^{b} (ab)_{ij}/b$, $\overline{(ab)}_{\cdot j} = \sum_{i=1}^{t} (ab)_{ij}/t$, and

$\overline{(ab)}_{\cdot\cdot} = \sum_{1}^{t}\sum_{1}^{b} (ab)_{ij}/tb$.

One may show there is no interaction if and only if $(ab)^*_{ij} = 0$ for all i and j.

For example, if $t = b = 2$, H_0 implies $T\alpha = 0$ (1×1), where T is the matrix $T = (1, -1, -1, 1)'$, thus to test for no interaction, one must find an HPD region for $T\alpha$.

In general, to test for no interaction, one may find a $(t-1) \cdot (b-1) \times bt$ matrix T, such that no interaction H_0 is equivalent to

H_0: $\gamma = T\alpha = 0[(t-1)(b-1) \times 1]$. (3.135)

Since $\alpha|\gamma \sim t[rbt - bt, E(\alpha|y), P(\alpha|y)]$, where the location and precision are given by (3.133) and (3.134), respectively, $\gamma = T\alpha$ also has a t distribution with $bt(r-1)$ degrees of freedom, location vector

$E(\gamma|y) = TE(\alpha|y)$ (3.136)

and precision matrix

$P(\gamma|y) = [TP^{-1}(\alpha|y)T']^{-1}$, (3.137)

and a $1 - \Delta$ $(0 < \Delta < 1)$ HPD region for γ is

$HPD_\Delta(\gamma) = \{F(\gamma|y) \leq F_{\Delta, (t-1)(b-1), (r-1)bt}\}$ (3.138)

where

$F(\gamma|y) = [\gamma - E(\gamma|y)]'P(\gamma|y)[\gamma - E(\gamma|y)](t-1)^{-1}(b-1)^{-1}$.
(3.139)

If no interaction is indicated, that is, if $F(0|\gamma) > F_{\Delta,(t-1)(b-1),bt(r-1)}$, one may analyze the additive model

$$y_{ijk} = m + a_i + b_j + e_{ijk}$$

according to the methods of the previous section.

When $r = 1$, there is no Bayesian way to test for interaction, because Jeffreys' density was used to express prior information; however, one may develop a complete Bayesian analysis if one may express prior information by a proper prior density, such as the normal-gamma density for α and τ.

The above model can be used to analyze several experimental layouts. For example, a two-factor factorial arrangement of treatments in a completely randomized design could be examined by the methods presented in this section.

THE ANALYSIS OF COVARIANCE

Analysis of covariance models combine the regression and the models for designed experiments into one model. For example, consider a completely randomized design consisting of t treatments, where the i-th treatment is assigned to n_i experimental units, and the response y_{ij} and a regressor or covariable x_{ij} are measured on the j-th unit receiving the i-th treatment. A model for such a situation is given by

$$y_{ij} = a_i + x_{ij}\gamma + e_{ij}, \qquad (3.140)$$

where $a_i \in R$, $x_{ij} \in R$, $\gamma \in R$, and the e_{ij} are n.i.d. $(0, \tau^{-1})$, where $\tau > 0$ and $i = 1, 2, \ldots, t$ and $j = 1, 2, \ldots, n_i$. The unknown parameters are the treatment effects $a_1, a_2, \ldots, a_t$, γ, and τ, and one is usually interested in the effect of treatments on the average responses $E(y_{ij}) = a_i + x_{ij}\gamma$.

The covariable or concomitant variable x is introduced to increase the precision estimating the treatment effects. The covariable should be highly related to the main response y and is not to be affected by the treatments.

Our objective is to examine the effect of the treatments on the average response after taking into account or "adjusting" for the effect of the covariable. One is referred to Cox (1958) for a discussion

THE ANALYSIS OF COVARIANCE

of experimental design considerations when one uses concomitant observations and to Graybill (1961) for the conventional analysis of variance of such designs.

From a Bayesian viewpoint, in order to study the treatment effects, adjusted for the concomitant variable, one must isolate the marginal posterior distribution of θ_1 where $\theta_1 = (a_1, a_2, \ldots, a_t)'$ is the vector of treatment effects, and it is necessary to assign a prior distribution to the parameters θ and τ where $\theta = (\theta_1', \theta_2)'$ and $\theta_2 = \gamma$. The Bayesian analysis will produce results similar to the analysis of variance if one uses

$$\xi(\theta, \tau) \propto 1/\tau, \quad \tau > 0, \quad \theta \in R^{t+1} \tag{3.141}$$

for the prior density; but it is no trouble if one uses a normal-gamma prior density.

The likelihood function for θ and τ is

$$L(\theta, \tau | y) \propto \tau^{n/2} \exp -\frac{\tau}{2}[(\theta - A^{-1}B)'A(\theta - A^{-1}B) + C - B'A^{-1}B],$$

$$\tau > 0, \quad \theta \in R^{t+1} \tag{3.142}$$

where A is a $(t + 1) \times (t + 1)$ matrix

$$A = \begin{pmatrix} n_1 & & & & x_{1\cdot} \\ & n_2 & & \emptyset & x_{2\cdot} \\ & & & & \vdots \\ & \emptyset & & n_t & x_{t\cdot} \\ x_{1\cdot} & x_{2\cdot} & \cdots & x_{t\cdot} & \Sigma\Sigma x^2 y \end{pmatrix},$$

B is the $(t + 1) \times 1$ vector

$$B = \begin{pmatrix} y_{1\cdot} \\ y_{2\cdot} \\ \vdots \\ y_{t\cdot} \\ \Sigma\Sigma x_{ij} y_{ij} \end{pmatrix}$$

and

$$C = \sum\sum y_{ij}^2.$$

The marginal posterior density of θ is a t with $n - (t + 1)$ degrees of freedom, location vector

$$E(\theta|y) = A^{-1}B$$

and precision matrix

$$P(\theta|y) = \frac{(n - t - 1)A}{C - B'A^{-1}B}.$$

Inferences for the treatment effects are to be based on the marginal posterior distribution of θ_1, which is also a t with $n - t - 1$ degrees of freedom, location vector

$$E(\theta_1|y) = (I_t, \phi)A^{-1}B,$$

where ϕ is a t × 1 zero vector, and precision matrix

$$P(\theta_1|y) = [(I_t, \phi)P^{-1}(\theta|y)(I_t, \phi)']^{-1}.$$

How does one test

$$H_0: a_1 = a_2 \ldots = a_t$$

for equality of treatment effects? Since H_0 is equivalent to $H_0: a_i - a_{(i+1)} = 0$, $i = 1, 2, \ldots, t - 1$, consider

$$S = T\theta_1,$$

where T is a (t − 1) × t vector

$$T = \begin{pmatrix} 1 & -1 & 0 & \ldots & & 0 \\ 0 & 1 & -1 & 0 & \ldots & 0 \\ \vdots & \vdots & & & & \vdots \\ \vdots & & \ddots & & & \vdots \\ 0 & 0 & \ldots & 0 & 1 & -1 \end{pmatrix},$$

then $S = 0[(t - 1) \times 1]$, if and only if H_0 is true and S has a t distribution with $n - t - 1$ degrees of freedom, location vector $TE(\theta_1|y)$ and precision matrix

$$P(S|y) = [TP^{-1}(\theta_1|y)T']^{-1}.$$

It is known that

$$F(S|y) = [S - E(S|y)]'P(S|y)[S - E(S|y)](t - 1)^{-1}$$

has an F distribution with $t - 1$ and $n - t - 1$ degrees of freedom, thus H_0 is rejected at significance level Δ ($0 < \Delta < 1$) if $F(0|y) > F_{\Delta;t-1,n-t-1}$ and it can be shown that this test is equivalent to the analysis of covariance procedure.

Of course, one perhaps should first test to see if the concomitant variable x is actually needed in the model, i.e., is $\gamma = 0$? This can be done in the usual way by finding the marginal posterior distribution of γ from the marginal posterior density of θ and constructing an HPD region for γ.

The above analysis of the completely randomized design with one covariable is easily extended to other designs, for example to a randomized block design with one or more concomitant variables.

COMMENTS AND CONCLUSIONS

The Bayesian methodology for the standard statistical models of regression and for designed experiments is given in this chapter. The methodology is based on the theory which was developed in Chapter 1, and the theory gave the prior, posterior, and predictive analysis for the standard linear model.

Before analyzing the regression models, Bayesian inferences for two normal populations were examined and it was found that when the populations have a common mean but distinct variances, the marginal posterior distribution of the common mean was a poly-t, a distribution which will be again encountered in the chapters on mixed models and linear dynamic systems.

The section on regression analysis was partitioned into subsections on simple linear, multiple linear, and nonlinear regression problems. The simple and multiple linear models were studied on the basis of prior, posterior, and predictive analyses. With regard to nonlinear regression, it was shown that it is indeed difficult to specify the marginal posterior distribution of the parameters when there are several parameters, because the distribution is not well-known as it is when the regression relation is linear.

This chapter gives only a brief introduction to the Bayesian analysis of models used to analyze designed experiments. If the design is completely randomized the Bayesian analysis (with Jeffreys' prior density) is more or less equivalent to the analysis of variance procedure for testing the equality of treatment effects. On the other hand, when examining two-factor experiments, such as randomized block designs, one must reparametrize the less than full rank (design matrix) models to full rank equivalents in order to produce the standard F tests of the analysis of variance (again using Jeffreys' prior density). It was necessary to reparametrize the model to full rank because if one uses Jeffreys' prior density in conjunction with the less than full rank model, the posterior distribution of the parameters is improper.

The chapter is concluded with a Bayesian test of the equality of treatment effects, where the design is completely randomized and one concomitant variable is measured on the experimental units.

One may conclude that it is possible to develop a complete Bayesian methodology for the analysis of designed experiments and that the analysis will be more informative than the traditional analysis of variance.

EXERCISES

1. If X is a Bernoulli random variable with parameter θ, $0 \leq \theta \leq 1$ and θ has a Beta distribution with parameters α and β ($\alpha > 0$, $\beta > 0$), how would one set values for the hyperparameters α and β if one had past data $X_1^*, X_2^*, \ldots, X_m^*$, $X_i^* = 0, 1$, $i = 1, 2, \ldots, m$? Use the prior predictive distribution of X.
2. Suppose X is $n(\theta, \tau)$ and that θ and τ have a prior normal-gamma distribution with hyperparameters μ, P, α and β. If $z_1, z_2, \ldots, z_m$ represent earlier values of X, use them to set values for the hyperparameters.
3. Refer to equation (3.15) and construct: (a) a HPD region for $\theta_1 - \theta_2$, and (b) a HPD region for τ, the common precision.
4. Derive the joint prior density of τ_1 and τ_2, that is verify equation (3.18).
5. Refer to equation (3.28), the joint posterior density of θ_1 and θ_2. Are θ_1 and θ_2 independent? Let $\gamma = \theta_1 - \theta_2$ and develop a normal approximation to the marginal posterior density of γ.

EXERCISES 139

6. In the Behrens-Fisher problem, what is the marginal posterior density of τ_1/τ_2? Refer to equation (3.21).

7. In the example of simple linear regression in this chapter (see pp. 85-94) find the mean of the marginal posterior distribution of τ^{-1}, using the data-based prior.

8. In the section on multiple linear regression, explain why equation (3.52) gives a HPD region for θ_2.

9. Refer to equation (3.103) but use Jeffreys' improper prior density (3.105), then show the analysis of variance test of $H_0: \theta_1 = \theta_2 = \ldots = \theta_t$ is equivalent to the Bayes procedure of using the HPD region of (3.103).

10. Consider the k linear models with common parameter θ

$$y_i = X_i \theta + e_i, \quad i = 1, 2, \ldots, k$$

where the y_i are $n \times 1$, X_i is $n \times p$, θ is $p \times 1$ and $e_1, e_2, \ldots, e_k$ are n.i.d. $(0, \tau_i^{-1} I_n)$, and $\theta \in R^p$ and the $\tau_i > 0$ are unknown parameters.

   (a) What is the conjugate prior density for the parameters θ, $\tau_1, \tau_2, \ldots, \tau_k$?
   (b) What is the marginal posterior density of θ? Hint: See equation (3.28).
   (c) Derive the marginal posterior density of $\rho = (\tau_1, \tau_2, \ldots, \tau_k)$.

   To answer (a) and (b) use the conjugate prior density of the parameters.

11. Consider a linear model

$$Y_{ij} = \theta_i + \beta X_{ij} + e_{ij},$$

where $\theta_i \in R$, $\beta \in R$, the X_{ij} are known values of an independent variable X, Y_{ij} is the j-th observation of a random variable for group i, and the e_{ij} are n.i.d. $(0, \tau^{-1})$, where $i = 1, 2, \ldots, a$, $j = 1, 2, \ldots, b$, and $\tau(> 0)$ is an unknown precision. This model is that corresponding to a groups, on each of which b pairs of observations (X_{ij}, Y_{ij}) are made. The independent variable X represents a covariable which is related to the dependent variable Y. Using the improper prior density

$$\xi(\theta,\tau,\beta) \propto \frac{1}{\tau}, \quad \theta \in R^a, \quad \tau > 0, \quad \beta \in R,$$

where $\theta = (\theta_1, \theta_2, \ldots, \theta_a)'$, do the following:
(a) Find the marginal posterior density of β.
(b) Find an HPD region for β and test $H_1: \beta = 0$.
(c) Find an HPD region for θ and test $H_2: \theta_1 = \theta_2 = \ldots = \theta_a$.

REFERENCES

Box, G. E. P. and G. C. Tiao (1973). *Bayesian Inference in Statistical Analysis*, Addison-Wesley, Reading, Mass.

Cochran, W. G. and G. M. Cox (1957). *Experimental Designs*, 2nd ed., John Wiley and Sons, New York.

Cox, D. R. (1958). *Planning of Experiments*, John Wiley and Sons, New York.

de Finetti, Bruno (1972). *Probability, Induction, and Statistics, the Art of Guessing*, John Wiley and Sons, London.

Draper, N. R. and H. Smith (1966). *Applied Regression Analysis*, John Wiley and Sons, Inc., New York.

Fisher, R. A. (1951). *Design of Experiments*, 6th ed., Oliver and Boyd, Edinburgh.

Graybill, F. A. (1961). *An Introduction to Linear Statistical Models*, McGraw Hill, New York.

Kadane, Joseph B., James M. Dickey, Robert L. Winkler, Wayne S. Smith and Stephen C. Peters (1980). "Interactive elicitation of opinion for a normal linear model," Journal of American Statistical Association, Vol. 75, pp. 845-854.

Kempthorne, O. (1952). *Design and Analysis of Experiments*, John Wiley and Sons, Inc., New York.

Lindley, D. V. (1965). *Introduction to Probability and Statistics from a Bayesian Viewpoint*, University Press, Cambridge.

Lindley, D. V. (1968). "The choice of variables in multiple regression," Journal of the Royal Statistical Society, Series B, Vol. 30, pp. 31-66.

Lindley, D. V. and A. F. M. Smith (1972). "Bayes estimates for the linear model," Journal of the Royal Statistical Society, Series B, Vol. 34, pp. 1-42.

Patil, V. H. (1964). "The Behrens-Fisher problem and its Bayesian solution," Journal of the Indian Statistical Association, Vol. 2, pp. 21-31.

Scheffé, Henry (1943). "On solutions to the Behrens-Fisher problem based on the t-distribution," Annals of Mathematical Statistics, Vol. 14, pp. 35-44.

REFERENCES

Searle, S. R. (1971). *Linear Models*, John Wiley and Sons, Inc., New York.

Sedory, S. A. (1981). Pursuing the Poly-t. Unpublished Masters Report, Department of Statistics, Oklahoma State University.

Winkler, Robert L. (1977). "Prior distributions and model building in regression analysis," in *New Developments in the Applications of Bayesian Methods*, edited by Ahmet Aykac and Carlo Baumat, North-Holland, Amsterdam.

Yusoff, Mat Abdullah (1982). *Bayesian Inference with the Poly-t Distribution*. Ph.D. dissertation, Oklahoma State University.

Zellner, Arnold (1971). *An Introduction to Bayesian Inference in Econometrics*, John Wiley and Sons, Inc., New York.

Zellner, Arnold (1980). "On Bayesian regression analysis with g prior distributions," Technical reprint, H. G. B. Alexander Research Foundation, Graduate School of Business, University of Chicago.

4
THE MIXED MODEL

INTRODUCTION

The mixed model is more general than the fixed because in addition to containing fixed effects the model contains so called random effects. The random factors of the model are unobservable random variables and these have variances called variance components which are the primary parameters of interest. In the traditional analysis, the fixed effects are parameters but are not regarded as random variables.

Fixed, mixed, and random are terms which came from the sampling theory framework and are not applicable to the Bayesian approach; nevertheless we will continue to use them, but within a Bayesian context.

The fixed model is the general linear model $Y = X\theta + e$ of Chapter 1, where one's main interest lies with the posterior distribution of θ and not with the parameters per se of the posterior distribution of θ.

With a mixed model

$$y = X\theta + \sum_{1}^{c} u_i b_i + e, \qquad (4.1)$$

the random factors $b_1, b_2, \ldots, b_c$ are independent random vectors with zero means and dispersion matrices $\sigma_i^2 I_{m_i}$, and one's main interest is with the variance components σ_i^2, $i = 1, 2, \ldots, c$, not the posterior distribution of the random factors. If θ is a vector of fixed effects, one is also interested in the posterior distribution of θ, not with the parameter of the posterior distribution, although these are indispensable for describing the posterior of θ. The usual assumption about the random factors is that they are independent and normally distributed, namely $N[0, \tau_i^{-1} I_{m_i}]$, where $\tau_i = \sigma_1^{-2}$, which is equivalent to a prior assumption about the random factor parameters. Now, since the variance components are parameters of the random factors, one must introduce a prior distribution for their analysis.

Note there are two levels of parameters. The primary level consists of θ, the vector of fixed effects, the random factors $b_1, \ldots, b_c$, and the error precision $\tau = \sigma^{-2}$, where the random error term $e \sim N[0, \tau^{-1} I_n]$, and is independent of the random factors. The secondary level consists of the vector of precision components, $\rho = (\tau_1, \tau_2, \ldots, \tau_c)'$. Thus, the precision components (or variance components) are parameters of some of the primary level parameters of the model.

For a complete description of the model we let X be a full rank $n \times p$ matrix, and let the u_i be $n \times m_i$ known design matrices.

There are $p + m + c + 1$ parameters, with $p + m + 1$ primary level parameters, and c secondary level parameters. The random factors are usually regarded as nuisance parameters, thus there are $p + c + 1$ parameters of interest.

Most of the work concerning random (p = 1) and mixed models has centered on estimating the variance components τ_i^{-1} and until 1967 the methodology was to equate the analysis of variance sum of squares to their expectations and solve the resulting system of linear equations for estimates of the variance components. For example, in a one-way random model, the between and within mean squares are equated to their expectations, giving analysis of variance estimates of the between and within components. See Searle (1971), pp. 385-389 for an example. The original methodology was introduced by Daniels (1939) and Winsor and Clarke (1940), and the sampling properties of the estimators studied by Graybill (1954), Graybill and Wortham (1956), and Graybill and Hultquist (1961).

For unbalanced data, Henderson (1953) introduced three analysis of variance methods of estimating variance components, and the properties of these estimators were examined by Searle (1971). The next new development was maximum likelihood estimation given by Hartley and Rao in 1967 and since then there have appeared many new methodologies including restricted maximum likelihood, minimum norm quadratic unbiased estimation or MINQUE, iterative MINIQUE, MIVQUE,

THE PRIOR ANALYSIS

or minimum variance quadratic unbiased estimation. These and other methods of estimating variance components are described by Searle (1978).

Box and Tiao (1973) is the only Bayesian book dealing with the variance components of random and mixed models. They give a very thorough treatment of the subject and the methodology is based on a numerical determination of the one- and two-dimensional marginal posterior distribution of the variance components. Thus with the one-way random model, which has two variance components, the one two-dimensional and two one-dimensional marginal posterior distributions are given for two data sets of Chapter 4. One interesting feature of their analysis is the choice of prior distribution. They show a Jeffreys' type vague non-informative prior on the variance components is equivalent to a vague prior on the expected mean squares of the analysis of variance.

They continue the study of variance components by examining the two-fold nested random model, the two-way random, and the two-way mixed models. Other studies from a Bayesian approach have been taken by Hill (1965, 1967), who studied the one-way model, and Stone and Springer (1965) who criticize the Box and Tiao choice of prior distribution.

Regardless of one's approach, it is safe to say that to estimate variance components is indeed a difficult task. There are so many sampling methods, it is difficult to declare any one as superior to the others, while with the existing Bayesian techniques one must rely on multi-dimensional numerical integrations. I suspect, variance components, except for the error variance, are difficult to estimate because they are secondary level parameters (at least more difficult to estimate than primary level parameters). This suspicion is motivated by Goel and DeGroot (1980), who assess the amount of information of hyperparameters in hierarchical models.

Inference about parameters of a mixed model will be accomplished by first determining the one-dimensional marginal posterior distribution of the variance components and the marginal posterior distribution of the fixed effects θ, then finding the joint posterior mode of all the parameters in the model.

THE PRIOR ANALYSIS

To recapitulate, consider the model (4.1), where y is a $n \times 1$ vector of observations, X an $n \times p$ known full-rank design matrix, u_i a $n \times m_i$ known design matrix, θ a $p \times 1$ real unknown parameter vector, b_i a $m_i \times 1$ random real parameter vector, and e a $n \times 1$ normal vector with mean 0 and precision τI_n. Furthermore assume $b_1, b_2, \ldots, b_c$, and e are independent.

Thus given θ, the b_i, and $\tau(>0)$, y is normal with mean $X\theta + ub$ and precision τI_n.

The prior information is introduced in two stages: First, the conditional prior distribution of b_i given τ_i is $b_i \sim N(0, \tau_i^{-1} I_{m_i})$, $i = 1, 2, \ldots, c$, which is the usual assumption about the random factors, θ has a marginal prior density which is constant over R^p, and τ is gamma with positive parameters α and β. Also assume θ, b, and τ are independent, a priori. Secondly, assume the secondary parameters $\tau_1, \tau_2, \ldots, \tau_c$ are independent and that τ_i is gamma with positive parameters α_i and β_i.

Why this choice of prior distribution? As will be seen, this form of the prior density, namely,

$$p(\theta, b, \tau | \rho) \propto \prod_{i=1}^{c} \tau_i^{m_i/2} e^{-\tau_i b_i' b_i / 2} \tau^{\alpha-1} e^{-\tau \beta},$$

$$p(\rho) \propto \prod_{i=1}^{c} \tau_i^{\alpha_i - 1} e^{-\tau_i \beta_i}$$

(4.2)

for $\theta \in R^p$, $\tau > 0$, $b_i \in R^{m_i}$, and $\tau_i > 0$, produces an analytically tractable posterior distribution for the parameters. If one believes this prior is not flexible enough to express one's prior opinion about θ, ρ, and τ, one may use mixtures of these distributions which will allow more flexibility. A normal prior distribution could have been used with θ, but this yields messy posterior distributions. Or we could give θ and τ a normal-gamma distribution, b and ρ a normal-gamma, and let θ and b be independent. This would give messy but tractable posterior distributions. Note that (4.2) is an improper prior density.

The likelihood function or probability density of y given θ, b, and τ is

$$L(\theta, b, \tau) \propto \tau^{n/2} \exp -\frac{\tau}{2}(y - X\theta - ub)'(y - X\theta - ub) \quad (4.3)$$

for $\theta \in R^p$, $b \in R^m$, and $\tau > 0$, where $m = \sum_{1}^{c} m_i$.

THE POSTERIOR ANALYSIS

This will be combined with (4.2), by Bayes' theorem, to give the posterior distribution of all the parameters.

THE POSTERIOR ANALYSIS

In this section, joint and marginal posterior distributions for the parameters will be derived, thereby giving us the foundation for making inferences about the parameters of the model.

First, the joint distribution of all the parameters is found by combining the likelihood function with the two-stage prior density, then the conditional posterior distributions of θ, b, τ, and ρ, given the other parameters is found. Second, the conditional distributions of θ given b, τ given b, and ρ given b are derived and lastly, the marginal posterior density of ρ and τ is found.

Combining the likelihood function (4.3) with the prior density (4.2), we have

$$p(\theta, b, \tau, \rho | y) \propto \tau^{(n+2\alpha)/2-1} \exp -\frac{\tau}{2}\{2\beta + y'Ry - \hat{b}'u'Ru\hat{b}$$

$$+ (b - \hat{b})'u'Ru(b - \hat{b})$$

$$+ (\theta - \hat{\theta})'X'X(\theta - \hat{\theta})\} \prod_{1}^{c} \tau_i^{((m_i + 2\alpha_i)/2)-1}$$

$$\times \exp -\frac{\tau_i}{2}(2\beta_i + b_i'b_i), \qquad (4.4)$$

where $\theta \in R^p$, $b \in R^m$, $\tau > 0$, and $\tau_i > 0$, as the joint posterior density of all the parameters. The various quantities in this expression are $\hat{b} = (u'Ru)^{-}u'Ry$, $R = I - X(X'X)^{-1}X'$, and $\hat{\theta} = (X'X)^{-1}X'(y - ub)$, where A^{-} is the unique Moore-Penrose generalized inverse of the matrix A.

An equivalent representation of this density is

$$p(\theta, b, \tau, \rho | y) \propto \tau^{(n+2\alpha)/2-1} \exp -\frac{\tau}{2}[2\beta + (y - X\theta - ub)'$$

$$\times (y - X\theta - ub)] \prod_{1}^{c} \tau_i^{(m_i + 2\alpha_i)/2-1}$$

$$\times \exp -\frac{\tau_i}{2} (b_i'b_i + 2\beta_i) \tag{4.5}$$

where $\theta \in R^p$, $\tau > 0$, $\tau_i > 0$, and $b \in R^m$. This form of the density allows one to readily deduce the many conditional posterior distributions of the parameters. Thus we have

Theorem 4.1. The conditional posterior distribution of θ given b, τ, and ρ is normal with mean $\hat{\theta}$ and precision matrix $\tau X'X$. The conditional posterior distribution of b given θ, τ, and ρ is normal with mean $[\tau u'u + A(\rho)]^{-1} \tau u'(y - X\theta)$ and precision matrix $\tau u'u + A(\rho)$, where $A(\rho)$ is the m × m block diagonal matrix with i-th diagonal matrix $\tau_i I_{m_i}$, $i = 1, 2, \ldots, c$. The conditional posterior distribution of τ given θ, b, and ρ is gamma with parameters $(n + 2\alpha)/2$ and $[2\beta + (y - X\theta - ub)'(y - X\theta - ub)]/2$.

Finally, the conditional distribution of ρ given θ, b, and τ is such that $\tau_1, \tau_2, \ldots, \tau_c$ are independent and τ_i is gamma with parameters $(m_i + 2\alpha_i)/2$ and $(2\beta_i + b_i'b_i)/2$, $i = 1, 2, \ldots, c$.

Thus the conditional distribution of each parameter given the other is a well-known density and, as will be shown, provides a convenient way to estimate the mode of the joint posterior density.

Of course, our goal is to determine the marginal posterior distribution of θ, τ, and ρ, regarding b as a nuisance parameter, but first consider the conditional distributions of θ given b and (τ, ρ) given b.

Theorem 4.2. The conditional posterior distribution of θ conditional on b is a multivariate t with mean vector $\hat{\theta}$ and precision matrix

$$T = \frac{(n - p + 2\alpha)X'X}{2\beta + y'Ry - \hat{b}'u'Ru\hat{b} + (b - \hat{b})'u'Ru(b - \hat{b})}. \tag{4.6}$$

The conditional distribution of τ conditional on b is gamma with parameters $(n - p + 2\alpha)/2$ and $[(2\beta + y'Ry - \hat{b}'u'Ru\hat{b} + (b - \hat{b})'u'Ru(b - \hat{b})]/2$. Also, conditional on b; τ, $\tau_1, \tau_2, \ldots, \tau_c$ are independent and τ_i is gamma with parameters $(m_i + 2\alpha_i)/2$ and $(2\beta_i + b_i'b_i)/2$, $i = 1, 2, \ldots, c$.

We will use the conditional distributions of the previous theorem in order to find the marginal posterior distributions of θ, τ, and ρ, and to do this, we need, also, the marginal posterior distribution of b, which is given by

THE POSTERIOR ANALYSIS

Theorem 4.3. The marginal posterior density of b, the vector of random effects, is

$$p(b|y) \propto [1 + (b - \hat{b})'A(b - \hat{b})]^{-(n-p+2\alpha)/2}$$

$$\times \prod_{1}^{c} (1 + b_i'A_i b_i)^{-(m_i + 2\alpha_i)/2}, \qquad (4.7)$$

where $b \in R^m$,

$$A = \frac{u'Ru}{2\beta + y'Ry - \hat{b}'u'Ru\hat{b}},$$

and

$$A_i = (2\beta_i)^{-1} I_{m_i}, \quad i = 1, 2, \ldots, c.$$

This density is called a multiple t or poly-t density and is discussed at length by Dreze (1977). It arises in many cases of regression analysis and Dreze discusses the various situations it will occur in in econometrics. Except in the trivial case, when c = 1, the usual t density, the poly-t distribution is difficult to work with because the normalizing constant is unknown.

For an exact Bayesian analysis of the variance components, one needs their marginal distributions. Consider the following.

Theorem 4.4. The marginal posterior density of τ and ρ is

$$p(\tau, \rho | y) \propto A(\tau, \rho) |B(\tau, \rho)|^{-1/2} \exp - \frac{1}{2} \hat{b}' B_1(\rho)$$

$$\times [B^{-1}(\tau, \rho) - B_1^{-1}(\rho)] B_1(\rho) \hat{b}, \qquad (4.8)$$

where $\tau > 0$ and $\tau_i > 0$, $i = 1, 2, \ldots, c$.

$$A(\tau, \rho) = \tau^{(n-p+2\alpha)/2 - 1} \exp - \frac{\tau}{2} (2\beta + y'Ry - \hat{b}'u'Ru\hat{b})$$

$$\times \prod_{1}^{c} \tau_i^{(m_i + 2\alpha_i)/2 - 1} \exp - \tau_i \beta_i$$

$$\beta(\tau,\rho) = \beta_1(\rho) + A(\rho),$$

where

$$\beta_1(\rho) = \tau u'Ru.$$

The above joint posterior density of τ and ρ is a very complicated expression and is analytically intractable. For instance it seems impossible to integrate it with respect to the precision components.

Another way of finding the marginal densities of the variance components is to combine the marginal density of b with the conditional densities of the variance components given b, then integrate with respect to b, but this will require an approximation to the density of b.

APPROXIMATIONS

The posterior distribution of b is a key factor in obtaining the conditional distribution of the variance components as well as determining the posterior marginal distribution of these parameters. The distribution of b is very difficult to handle, thus one way of solving the problem is to find an approximation in terms of simple expressions.

Since a multivariate t distribution may be approximated by a multivariate normal with the same first two moments as the multivariate t distribution, $[1 + (1/k)(X - \mu)'A(X - \mu)]^{-(n+k)/2}$ may be approximated by $\exp - 1/2[(k-2)/k](X-\mu)'A(X-\mu)$ for any $X \in R^n$ and any non-negative definite matrix A.

Using this approximation with each of the c factors of (4.7), one may show the density of b may be approximated by

$$p(b|y) \propto \exp - \frac{1}{2}(b - b^*)'A^*(b - b^*), \qquad b \in R^m, \qquad (4.9)$$

where

$$A^* = A_1 + A_2,$$

$$b^* = (A^*)^{-1}A_1\hat{b},$$

$$A_1 = (n - p + 2\alpha - m - 2)u'Ru[2\beta + y'Ry - \hat{b}'u'Ru\hat{b}],$$

APPROXIMATIONS

and

$$A_2 = \{DIAG[(\alpha_i - 1)/\beta_i]I_{m_i}\}$$

The matrix A_2 is m × m block diagonal with i-th diagonal matrix $[(\alpha_i - 1)/\beta_i]I_{m_i}$, where $i = 1, 2, \ldots, c$.

Thus the posterior distribution of b is approximately normal with mean vector b* and precision matrix A*, however we would expect the approximation not to be good unless $\hat{b}$ is close to the zero vector. See Box and Tiao (1973), Chapter 9, where they discuss the poly-t distribution. Later, we will discuss how good the approximation is, using a one-way random model.

The mean of b, namely b*, is the approximate mean of the distribution of b and can be used as the conditioning value of b. For example, if conditional point estimates for the parameters are needed, the mean of these conditional posterior distributions of Theorem 4.2 can be taken as estimates. They are

$$\hat{\theta} = (X'X)^{-1}X'(y - ub^*)$$

for the fixed effects,

$$\hat{\sigma}^2 = (n + 2\alpha - p - 2)^{-1}(2\beta + y'Ry - \hat{b}'u'Ru\hat{b})$$

for the error variance, and

$$\hat{\sigma}_i^2 = (2\beta_i + b_i^{*'}b_i^*)/(m_i + 2\alpha_i - 2),$$

for the variance components. Note, $b^* = (b_1^{*'}, b_2^{*'}, \ldots, b_c^{*'})'$ is the partitioned form of the approximate mean of the posterior distribution of b.

Our objective here is to determine a convenient approximation to the marginal posterior distribution of the variance components. To do this, we need the distribution of quadratic forms in normal variables.

Let Z be a random n-vector having a normal distribution with mean μ and positive definite dispersion matrix D, or symbolically let $Z \sim N(\mu, D)$. Let $Q(Z) = (Z - Z_0)'M(Z - Z_0)$, where Z_0 is a known n vector and M a known non-negative definite matrix, then it is clear the distribution of $Q(Z)$ is the same as that of $(Z^* - Z_1)'M(Z^* - Z_1)$ where $Z_1 = Z_0 - \mu$ and $Z^* \sim N(0, D)$.

Since D is positive definite, there exists a nonsingular lower triangular matrix L such that $D = LL'$. Also, since $L'ML$ is symmetric there exists an orthogonal matrix P such that $P'L'MLP = \Lambda$, the diagonal matrix of eigenvalues of $L'ML$. Using the transformation $Z^* = LPW$, one may show the distribution of $Q(Z)$ is that of

$$\sum_1^n \lambda_i (W_i - w_i)^2,$$

where

$$\Lambda = \text{diag}(\lambda_1, \lambda_2, \ldots, \lambda_n),$$

$$w' = (w_1, w_2, \ldots, w_n),$$

$$W' = (W_1, W_2, \ldots, W_n),$$

$$w = P'L^{-1}Z_1, \text{ and}$$

n = number of rows of M.

If n' = rank of $m < n$, then assuming the last $n - n'$ of the λ's to be zero, one may show the distribution of $Q(Z)$ is the same as that of

$$Q(W) = \sum_1^n \lambda_i (W_i - w_i)^2.$$

Ruben (1960, 1962) has shown that for given Λ, n, and w, the distribution of $Q(W)$ may be expressed as

$$F_n(q/\Lambda, w) = \Pr[Q(z) \leq q]$$

$$= \Pr\left[\sum_{j=1}^n \lambda_j (W_j - w_j)^2 \leq q\right]$$

$$= \sum_{j=0}^\infty e_j \Pr(\chi^2_{n+2j} < q/c), \qquad (4.10)$$

where

$$e_0 = \exp\left(-\frac{1}{2}\sum_{j=1}^{n} w_j^2\right)\left[\prod_{j=1}^{n} (c/\lambda_j)^{1/2}\right],$$

$$e_r = (1/2r)\sum_{j=0}^{n-1} G_{r-j} e_j \quad (r \geq 1),$$

$$G_r = \sum_{j=1}^{n}(1 - c/\lambda_j)^r + rc\sum_{j=1}^{n}(w_j^2/\lambda_j)(1 - c/\lambda_j)^{r-1} \quad (r \geq 1),$$

and c is an arbitrary positive constant.

Ruben has also shown that the series on the right of (4.10) is uniformly convergent over any finite interval of q and the bound for error in the above series is the n-th term. Furthermore it is obvious that the distribution in (4.10) can be an infinite mixture of chi-square densities provided c is chosen to be less than the minimum of $(\lambda_1, \lambda_2, \ldots, \lambda_n)$, to insure each $e_i > 0$.

APPROXIMATION TO THE POSTERIOR DISTRIBUTION OF σ^2

From Theorem 4.2, the conditional distribution of τ given b has a gamma distribution with parameters $\alpha^* = (n - p + 2\alpha)/2$ and $\beta^* = (1/2)(2\beta + y'Ry - \hat{b}'u'Ru\hat{b} + Q)$, where $Q = (b - \hat{b})'u'Ru(b - \hat{b})$.

From Ruben's representation, let b^*, $\hat{b}$, A^{*-1} and u'Ru, replace μ, Z_0, D, and M respectively, and let $s = n = $ rank of u'Ru, then the marginal density of τ is

$$p(\tau|y) = \int_0^\infty \frac{(\beta^*)^{\alpha^*}\tau^{\alpha^*-1}e^{-\tau\beta^*}}{\Gamma(\alpha^*)}$$

$$\times \sum_{j=0}^{\infty} e_j\left[\frac{d}{dq}\Pr(\chi^2_{s+2j} < q/c)\right]dq$$

$$= \sum_{j=0}^{\infty} e_j \int_0^\infty [(d+q)/2]^k \tau^{k-1} \exp -\frac{1}{2}[q/e + (d+q)\tau]$$

$$\times (q/c)^{s/2+j-1}(1/c)dq, \quad \tau > 0 \qquad (4.11)$$

where $2k = n - p + 2\alpha$, and $d = 2\beta + y'Ry - \hat{b}'u'Ru\hat{b}$, and $c > 0$.

Now let k be an integer. To insure this, one needs to make a slight adjustment in the value of α, but the change in α need not be larger than 0.5. Using this assumption, the density of σ^2 is given by

$$p(\sigma^2|y) = ke^{-d/2\sigma^2}(d/2)^k(1/\sigma^2)^{(k+1)}$$

$$\times \sum_{j=0}^{\infty} \frac{e_j}{\Gamma(s/2+j)(1+c/\sigma^2)^{s/2+j}}$$

$$\times \sum_{r=0}^{k} \frac{\Gamma(s/2+r+j)}{\Gamma(r+1)\Gamma(k-r+1)[(d/2)(1/c+1/\sigma^2)]^r}.$$

$$(4.12)$$

APPROXIMATION TO THE POSTERIOR DISTRIBUTION OF THE VARIANCE COMPONENTS

Consider the i-th precision component τ_i, then using Ruben's result, replace μ, z_0, D, and M by b_i^*, 0, A_i^{*-1}, and I_{m_i} respectively and let $s_i = m_i$ = rank of I_{m_i}, where A_i^* is the precision matrix of b_i in the normal approximation to b, given by (4.9).

The posterior marginal density of τ_i is

$$p(\tau_i|y) = \frac{1}{\Gamma(\alpha_i^*)} \int_0^{\infty} (\beta_i^*)^{\alpha_i^*} \tau_i^{\alpha_i^*-1} \exp - \beta_i^*\tau_i$$

$$\times \sum_{j=0}^{\infty} e_{ij}\left[\frac{d}{dq_i} \Pr(\chi^2_{s_i+2j} < q_i/c_i)\right] dq_i,$$

where $\tau_i > 0$, $\alpha_i^* = (m_i + 2\alpha_i)/2$, $\beta_i^* = (2\beta_i + b_i'b_i)/2$, and the e_{ij} are given by (4.10), with $e_{ij} = e_j$, and c_i is an arbitrary positive constant.

POSTERIOR INFERENCES

One may let $k_i = (m_i + 2\alpha_i)/2$ be an integer with a slight adjustment in α_i, then letting $\sigma_i^2 = \tau_i^{-1}$, the above formula reduces to

$$p(\sigma_i^2|y) = k_i(d_i/2)^{k_i}(1/\sigma_i^2)^{(k_i+1)} \exp - d_i/2\sigma_i^2$$

$$\times \sum_{j=0}^{\infty} \frac{e_{ij}}{\Gamma(s_i/2 + j)(1 + c_i/\sigma_i^2)^{(s_i/2+j)}}$$

$$\times \sum_{r=0}^{k_i} \frac{\Gamma(s_i/2 + r + j)}{\Gamma(r+1)\Gamma(k_i - r + 1)[(d_i/2)(1/c_i + 1/\sigma_i^2)]} \quad (4.13)$$

where $\sigma_i^2 > 0$.

Using the normal approximation to b and Ruben's representation of a quadratic form in normal variables, formulas (4.12) and (4.13) are approximate densities for the marginal posterior distribution of the error variance and i-th variance component respectively. Ruben's result is exact, however, the normal distribution of b is an approximation to the true distribution of b, given by (4.9), therefore the adequacy of the approximation to the posterior distribution of the variance components and error variance depends on the adequacy of the normal approximation to b.

Thus, up to now, the exact marginal and conditional posterior distributions of the parameters of the mixed model have been derived. Since the exact marginal density (4.8) of the error variance and variance components is analytically intractable, it is necessary to represent their marginal density in a more convenient form, thus the normal approximation to b and Ruben's representation was employed. These derivations are found in Rajagopalan (1980) and Rajagopalan and Broemeling (1983).

POSTERIOR INFERENCES

Inferences for the parameters of the mixed model are to be based on the appropriate posterior distribution and two approaches will be taken here. First, in regard to estimating the error variance and variance components, a plot of the relevant marginal posterior distribution will be employed, along with the posterior mean and variance of that parameter. Secondly, joint modal estimates of all the parameters will be calculated from an iterative technique.

Marginal Posterior Inferences

Consider making inferences for the error variance σ^2 of a mixed model, then one must plot the posterior marginal distribution of σ^2, either by numerical integration, via the joint posterior density (5.8) of τ and ρ, or via the approximate density (5.12).

Of course, in a similar way, inferences for the variance components are made by plotting the marginal posterior density of that parameter, via numerical integration from (4.8), or by plotting the approximate density (4.13).

Accompanying the plot of each parameter one may compute the posterior mean and variance. Again here one has a choice. Either compute these moments from the exact joint marginal posterior density (4.8), or from the approximate densities, (4.12) and (4.13). In the latter case, one may derive formulas for the mean and variance.

Posterior Means and Variances for Variance Components

The moments of σ^2 and σ_i^2 are difficult to obtain directly from the approximate densities (4.12) and (4.13), but may be obtained relatively easily by first computing the conditional moments of these parameters given b, which are exact, then averaging conditional moments over the distribution of b, using the normal approximation to b, (4.9).

For instance, the moments of σ^2 can be computed by noting that

(i) τ given b has a gamma distribution, from Theorem 4.2, and
(ii) b has an approximate normal distribution.

Using the formulas

(i) $E(\sigma^2) = E_b[E(\sigma^2|b)]$, and
(ii) $Var(\sigma^2) = Var_b[E(\sigma^2|b)] + E_b[Var(\sigma^2|b)]$,

where E_b denotes expectation with respect to b and Var_b means variance with respect to the distribution of b, one may show the marginal posterior mean of σ^2 is

$$E(\sigma^2|y) = (n - p + 2\alpha - 2)^{-1}[2\beta + y'Ry - \hat{b}'u'Ru\hat{b}$$
$$+ Tr(u'RuA*^{-1}) + (b* - \hat{b})'u'Ru(b* - \hat{b})], \qquad (4.14)$$

and the marginal posterior variance is

$$Var(\sigma^2|y) = (n - p + 2\alpha - 2)^{-2}(n - p + 2\alpha - 4)^{-1}$$
$$\times [2\beta + y'Ry - \hat{b}'u'Ru\hat{b} + Tr(u'RuA*^{-1})$$
$$+ (b* - \hat{b})'u'Ru(b* - \hat{b})]^2 + (n - p + 2\alpha - 2)^{-1}$$
$$\times (n - p + 2\alpha - 4)^{-1}$$
$$\times [2Tr(u'RuA*^{-1})^2 + 4(b* - \hat{b})'u'RuA*^{-1}$$
$$\times u'Ru(b* - \hat{b})], \qquad (4.15)$$

where $n - p > 4 - 2\alpha$.

In a similar way, the first two posterior moments of σ_i^2 are

$$E(\sigma_i^2|y) = (m_i + 2\alpha_i - 2)^{-1}[2\beta_i + Tr(A_i^{*-1}) + b_i^{*'}b_i^*] \qquad (4.16)$$

and

$$Var(\sigma_i^2|y) = 2(m_i + 2\alpha_i - 2)^{-2}(m_i + 2\alpha_i - 4)^{-1}[2\beta_i + Tr(A_i^{*-1})$$
$$+ b_i^{*'}b_i^*]^2 + (m_i + 2\alpha_i - 2)^{-1}(m_i + 2\alpha_i - 4)^{-1}$$
$$\times [2Tr(A_i^{*-1})^2 + 4b_i^{*'}A_i^{*-1}b_i^*] \qquad (4.17)$$

if $m_i > 4 - 2\alpha_i$, $i = 1, 2, \ldots, c$.

All the higher moments, provided they exist, of the variance components and error variance may be obtained in a similar fashion.

An Example of Marginal Posterior Analysis

A balanced one-way random model is considered to illustrate the preceding results. The linear model describing such a design is

$$y_{ij} = \mu + b_i + \epsilon_{ij}; \quad i = 1, 2, \ldots, m; \quad j = 1, 2, \ldots, t,$$

where b_i is normal with mean 0 and variance σ_1^2, and the ϵ_{ij} follow a normal distribution with zero mean and error variance σ^2 for all i and j. This is the mixed model (4.1) with $c = 1$, $m_1 = m$, $p = 1$, and $n = mt$. Also

$$y = (y_{11}, \ldots, y_{1t}; y_{21}, \ldots, y_{2t}; \ldots; y_{m1}, \ldots, y_{mt})',$$

$x = j_n$, a $n \times 1$ vector of ones,

$\theta = \mu$, a real scalar,

$U = \text{Diagonal}(j_t, j_t, \ldots, j_t)$, of order $n \times m$, and

$b = (b_1, b_2, \ldots, b_m)'$.

For this particular case, it can be shown that

$$R = I_n - (1/n)J_n, \quad J_n \text{ a } n \times n \text{ matrix of ones},$$

$$u'Ru = t[I_m - (1/m)J_m],$$

$$\overline{(u'Ru)} = (1/t)[I_m - (1/m)J_m],$$

$$\hat{b} = (\bar{y}_1 - \bar{y}, \bar{y}_2 - \bar{y}, \ldots, \bar{y}_m - \bar{y})',$$

$\hat{b}'u'Ru\hat{b} = s_1$, the between group sum of squares,

$$A_1 = (n - m + 2\alpha - 3)(2\beta + s_2)^{-1}[tI_m - (t/m)J_m],$$

where s_2 is the within group sum of squares,

$$A_2 = [(\alpha_1 - 1)/\beta_1]I_m,$$

$$A^* = A_1 + A_2 = (at + g)I_m - (at/m)J_m, \text{ where}$$

$$a = (n - m + 2\alpha - 3)(2\beta + s_2)^{-1}, \text{ and}$$

POSTERIOR INFERENCES

$$g = (\alpha_1 - 1)/\beta_1.$$

Using the above results, one may show that

$$(A^*)^{-1} = (at + g)^{-1}I_m + at(gm)^{-1}(at + g)^{-1}J_m,$$

$$b^* = at(at + g)^{-1}\hat{b},$$

$$u'Ru(A^*)^{-1} = t(at + g)^{-1}[I_m - (1/m)J_m],$$

$$b^{*'}b^* = a^2 t(at + g)^{-2}s_1,$$

$$(b^* - \hat{b})'u'Ru(b^* - \hat{b}) = g^2(at + g)^{-2}s_1.$$

The conditional Bayes estimators conditional on b^* are

$$\hat{\sigma}^2 = (n + 2\alpha - 3)^{-1}[2\beta + s_2 + g^2 s_1 (at + g)^{-2}],$$

$$\hat{\sigma}_1^2 = (m + 2\alpha_1 - 2)^{-1}[2\beta_1 + a^2 ts_1 (at + g)^{-2}]. \tag{4.18}$$

Approximate Bayes estimators, based on the approximations to the marginal posterior distribution of the parameters are given by (4.14) and (4.16) and reduce to the estimators

$$\hat{\sigma}^2 = (n + 2\alpha - 3)^{-1}[2\beta + s_2 + t(m - 1)(at + g)^{-1} + g^2 s_1 (at + g)^{-2}],$$

$$\hat{\sigma}_1^2 = (m + 2\alpha_1 - 2)^{-1}[2\beta_1 + (at + mg)g^{-1}(at + g)^{-1} + a^2 ts_1 (at + g)^{-2}]. \tag{4.19}$$

The approximate Bayes estimators (means of posterior densities) and the conditional Bayes estimators (means of the conditional posterior distribution) are very similar. For example, we see they each have two terms in common, only differing in the third term $t(m - 1) \cdot (at + g)^{-1}$ for the error variance and $(at + mg)g^{-1}(at + g)^{-1}$ for the between variance component.

The usual AOV estimators, see Box and Tiao (1973), are $\hat{\sigma}^2 = s_2 t^{-1}(m-1)^{-1}$ and $\hat{\sigma}_1^2 = [s_2 t^{-1}(m-1)^{-1} - s_1(m-1)^{-1}](t-1)$ and are quite different than the conditional Bayes and approximate Bayes estimators, which are not linear functions of the AOV sum of squares (as are the AOV estimators), but instead are ratios of integer powers of the AOV sum of squares. For this model, the AOV estimators are minimum variance quadratic unbiased estimators (but are negative with positive probability), see Graybill and Wortham (1956), thus it would be interesting to compare the sampling mean square errors of Bayes and AOV estimators.

We now continue with our example of the one-way random model by plotting the posterior density of the variance components and calculating the posterior mean and variance of these parameters for a set of data employed by Box and Tiao (1973). For the generated data the true value of σ^2 is 16 and is 4 for the between variance component σ_1^2.

Many different sets of values of the hyperparameters $(\alpha, \beta, \alpha_1, \beta_1)$ were taken so as to examine the effects of the prior parameter values on the two marginal posterior distributions and to see the effect on the closeness of the approximation.

For each set of hyperparameters, the posterior means and variances of the variance components were calculated in two ways. First from (4.14)–(4.17), the approximate moments were computed, then the true moments, mean and variance, were calculated from the exact posterior densities via numerical integration. The eigenvalues and other constants needed to evaluate the approximate densities (4.12) and (4.13) and their moments, (4.14)–(4.17), were computed with the MATRIX procedure in SAS (Statistical Analysis System). The true posterior distributions are given by (4.8), and the relevant marginal densities and their moments were evaluated using numerical integrations. This was done with a FORTRAN program (WATFIV). To evaluate the approximate densities (4.12) and (4.13) only 20 terms in the mixture were taken, because the contribution from the remaining terms was negligible. The constant c was chosen to be the smallest eigenvalue in order to make (4.12) and (4.13) mixtures. The calculations were done on an IBM 370/168 at Oklahoma State University by M. Rajagopalan (1980).

Table 4.1 provides one with the approximate and true values of the posterior means and variances of the variance components for various values of the hyperparameters.

The exact and approximate marginal posterior densities of the two parameters are plotted in Figures 4.1-4.6 for six sets of hyperparameters.

Results of the numerical study indicate the approximations are close, generally, to the true means and variance.

POSTERIOR INFERENCES

The closeness of the approximations increases with α, even for values as low as 8, and is independent of the β value. That this must be so follows from the fact that as the α parameter increases, the degrees of freedom of each factor of the poly-t density increase, thus each factor is well approximated by the normal density, and the posterior density of b is well approximated by a normal.

This insures the accuracy of the approximation, however a large α parameter also insures a small prior standard deviation for the parameter. For example, the prior standard deviation of σ^2 is $\beta(\alpha - 1)^{-1} \cdot (\alpha - 2)^{-1/2}$, thus the approximate posterior distribution of σ^2 is accurate if one has a strong informative prior for the variance component. When precise prior information is available the approximations are efficient. The accuracy of the approximations is confounded with one's prior beliefs.

Also, when the α-parameter is large and the prior means of the components are close to their true values, the posterior means are close to the true means. Thus in such cases, the two posterior distributions are centered very close to the true value of the variance components and the posterior variances are small.

Furthermore, when α is large, changes in the α parameter do not significantly affect the posterior distribution of the within component as compared to the between component. Informative priors influence the between component relatively more than the within. Hence, unless one is quite sure of what one is doing, then the value of the α parameter for the between variance component should not be selected as large even though it does not matter much for the within variance component. The reasons are explained by referring to the posterior density of b, (4.7), where in the first factor the degrees of freedom, which introduces prior information about σ^2 (or τ), is $(n - p + 2\alpha + m)$ which is likely to be large for most data, regardless of the value of α. Variations in the α parameter do not affect this factor, but on the other hand, the second part, consisting of c factors, of the density of b, and the degrees of freedom of the i-th factor, is only m_i, and variations in the value of α_i significantly affect the posterior distribution of the between component.

Lastly, when there is little prior information, it seems reasonable to let the value of the α parameter be close to 2 so the prior variance is large. Even when α is close to 2, the approximations are close, but whether the posterior distributions of the variance components are close to their true values depends on the β parameter. From Table 4.1, one can see that when $\alpha = 2$ and β varies from 2 to 20, the posterior mean of σ^2 goes from 13.06 to 14.82, a small change. Note the true value of $\sigma^2 = 16$. We conclude, therefore, that the posterior distribu-

Table 4.1
Mean and Variance of the True and the Approximate Marginal Posterior Distributions of the Within and Between Variance Components for Box's Data

Prior parameters				Within variance component				Between variance component			
α	β	α_1	β_1	mean		variance		mean		variance	
				true	app	true	app	true	app	true	app
2	2	2	2	12.96	13.06	11.72	12.24	1.22	1.62	0.84	1.07
2	5	2	5	13.14	13.56	11.15	13.67	2.40	3.01	1.97	2.66
2	8	2	8	13.49	13.81	12.84	14.22	3.29	3.69	2.63	4.02
2	20	2	20	14.59	14.82	16.16	17.02	7.44	8.97	16.87	24.40
2	8	5	5	13.35	13.39	12.30	12.68	1.11	1.17	0.30	0.30
3	3	3	3	12.21	12.28	9.92	10.13	1.17	1.33	0.56	0.60
3	5	3	5	12.32	12.50	9.63	10.73	1.79	2.05	1.07	1.23
3	10	3	10	12.75	12.97	10.96	11.67	3.07	3.36	2.12	2.61
3	20	3	20	13.56	13.86	12.92	13.58	5.77	6.50	9.37	11.90
3	50	3	50	15.80	16.18	17.86	19.08	12.07	12.23	24.22	26.73

3	3	5	5	12.21	12.26	9.91	10.07	1.12	1.16	0.30	0.30
4	5	4	5	11.59	11.69	8.24	8.74	1.38	1.50	0.56	0.56
4	10	4	10	11.96	12.08	9.23	9.40	2.54	2.77	1.72	2.00
4	20	4	20	12.68	12.85	10.52	10.77	4.64	5.03	5.15	6.08
4	50	4	50	14.76	14.85	14.53	14.81	10.25	10.70	18.29	20.76
5	5	5	5	10.95	11.00	7.23	7.24	1.11	1.17	0.30	0.30
5	10	5	10	11.26	11.34	7.67	7.79	2.07	2.20	0.95	1.03
5	20	5	20	11.91	12.04	8.70	8.89	3.85	4.08	2.98	3.35
5	50	5	50	13.85	13.98	11.98	12.15	8.76	9.13	13.86	14.37
8	8	8	8	9.48	9.53	4.98	4.65	1.08	1.09	0.16	0.16
10	10	20	10	8.79	8.81	3.55	3.57	0.52	0.52	0.01	0.01
10	40	20	50	10.10	10.14	4.77	4.82	2.51	2.52	0.32	0.32
10	100	20	100	12.86	12.94	7.55	7.87	4.90	4.89	1.10	1.25
10	200	20	200	17.56	17.31	14.62	14.90	9.70	9.74	4.65	4.71
20	100	50	100	8.82	8.84	2.49	2.50	2.01	2.02	0.08	0.06
20	400	50	200	18.14	18.02	10.58	11.26	4.00	4.00	0.43	0.33
32	500	20	80	15.49	15.51	5.52	5.55	3.98	3.96	0.80	0.79

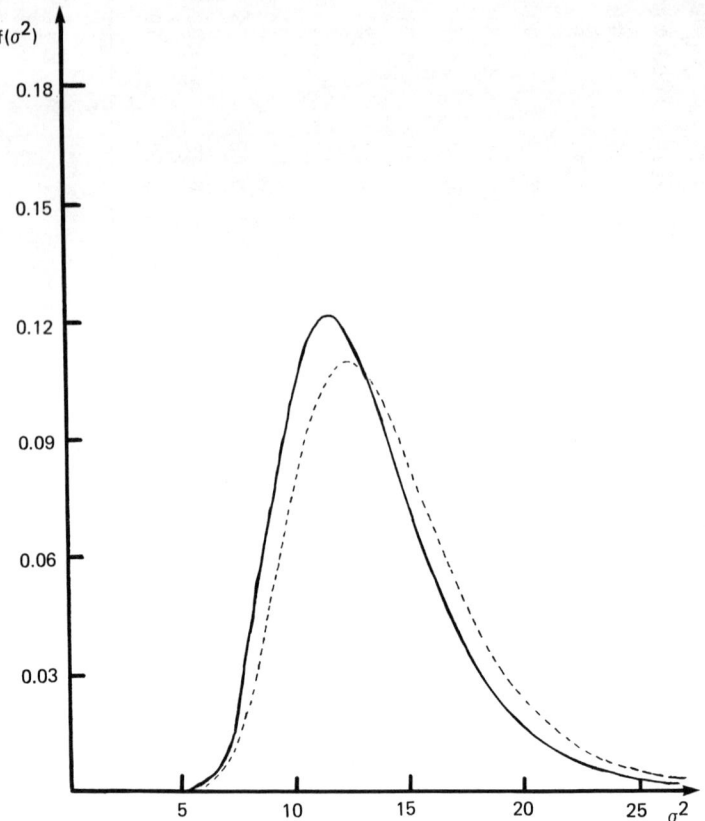

Figure 4.1. Marginal posterior density of the within-variance component; $\alpha = 2$, $\beta = 5$, $\alpha_1 = 2$, $\beta_1 = 5$; ---- = approximation, —— = true distribution.

tion of the within variance component is not very sensitive to changes in β, for values of α close to 2. This is not the case with the between component since when α is close to 2, the posterior mean and variance heavily depend on the β parameter. Thus a fairly good estimate of the true value is needed for a proper choice of the β parameter.

These conclusions are based on a limited study of one data set. More extensive numerical studies need to be done before one can formulate general rules on the closeness of the approximations and the choice of the hyperparameters.

Overall, one may say that the results of the numerical study indicate the approximations are good. Generally, the accuracy of the ap-

POSTERIOR INFERENCES

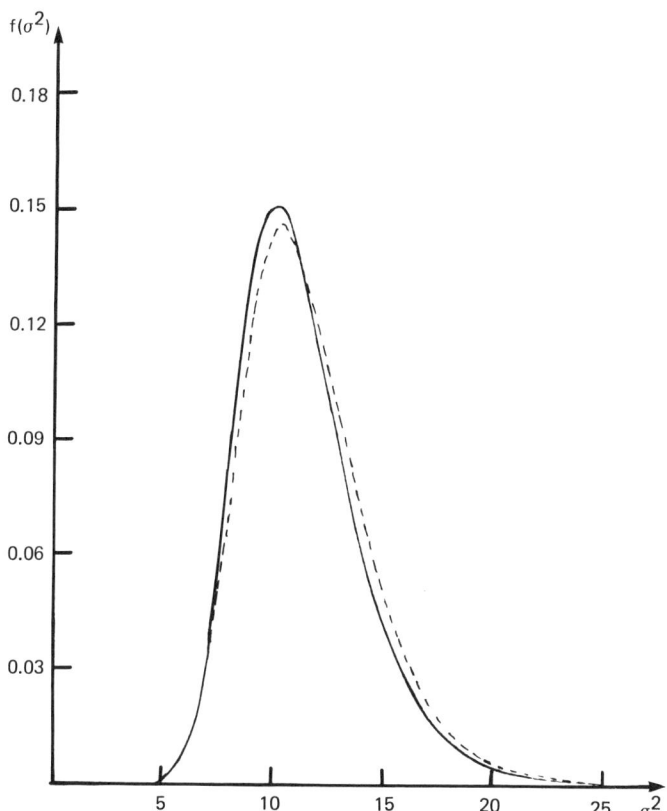

Figure 4.2. Marginal posterior density of the within-variance component; $\alpha = 5$, $\beta = 10$, $\alpha_1 = 5$, $\beta_1 = 10$; ---- = approximation, ——— = true distribution.

proximations increases with the α parameter (α or α_1) of the prior distribution of the variance components and the approximations are close for values of α as small as 8. The posterior distribution of σ^2 is less sensitive to variations in the β parameters of the gamma priors as compared to that of the between variance component. When precise prior information is available, it doesn't seem unreasonable to set the α parameter close to 2. Generally, in order to fix the β parameters, a good prior estimate of the true value of the variance components is needed. But, more detailed numerical studies are needed before one can put forward these conclusions as general rules.

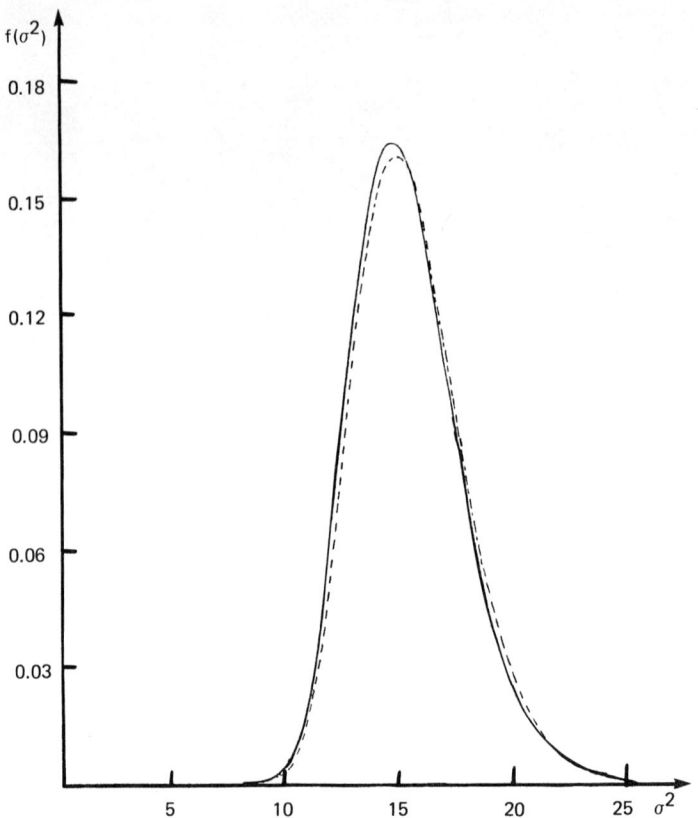

Figure 4.3. Marginal posterior density of the within-variance component; $\alpha = 32$, $\beta = 500$, $\alpha_1 = 20$, $\beta_1 = 80$; ---- = approximation, —— = true distribution.

Joint Modal Estimators of Parameters of Mixed Models

Another way to estimate the parameters of mixed linear models is with the mode of the joint posterior distribution of all the parameters. This approach is somewhat similar to the method of maximum likelihood estimation, which was first developed by Hartley and Rao (1968). The mode of the joint posterior distribution is found in the same way as done by Lindley and Smith (1972), who used iterative procedures to find modal estimates of the parameters. These modal estimators were viewed as large-sample approximations to posterior means.

The view taken here is to regard joint modal estimators as "natural" Bayesian estimators, since they correspond to points of highest pos-

POSTERIOR INFERENCES   167

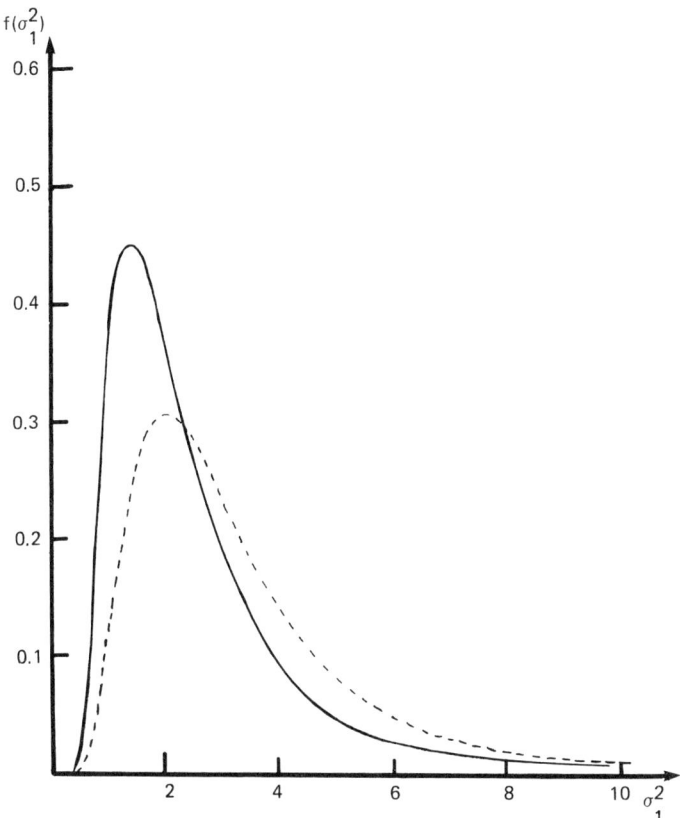

Figure 4.4. Marginal posterior density of the between-variance component; $\alpha = 2$, $\beta = 5$, $\alpha_1 = 2$, $\beta_1 = 5$; ---- = approximation, —— = true distribution.

terior density, and to use the posterior mean if a quadratic loss function is appropriate.

Of course, a complete Bayesian analysis dictates that one give the marginal posterior distributions of all the parameters, thus for a model with k parameters, there are $2^k - 1$ marginal distributions to determine. If one considers a one-way model, it has three parameters, excluding the random factor, one should specify seven marginal distributions, something which has not been done.

Instead, one is forced to use summary characteristics of the posterior distributions such as posterior means, modes, medians, variances and so on of the one- and two-dimensional marginal distribu-

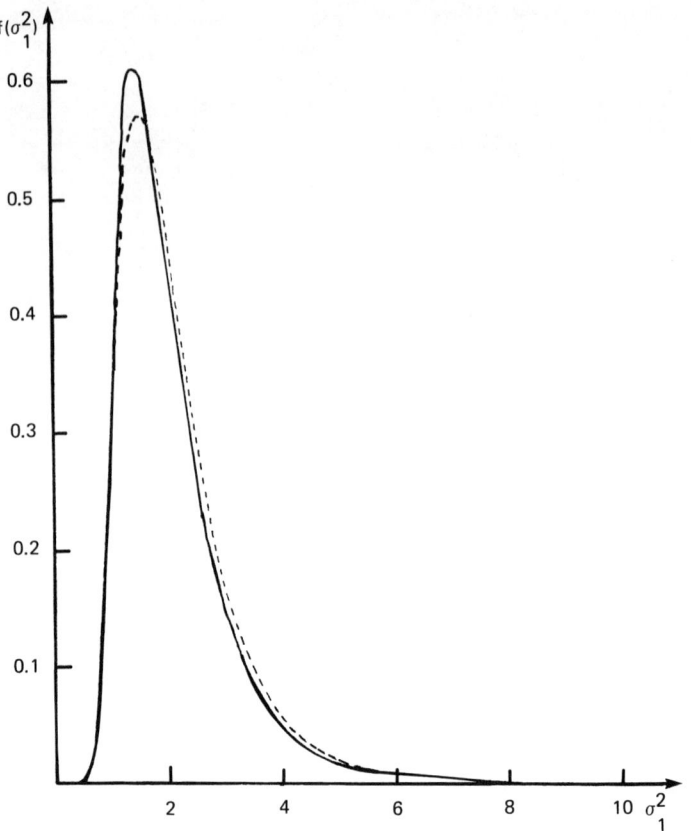

Figure 4.5. Marginal posterior density of the between-variance component; $\alpha = 5$, $\beta = 10$, $\alpha_1 = 5$, $\beta_1 = 10$; ---- = approximation, ——— = true distribution.

tions of the two parameters σ^2 and σ_1^2. This approach usually involves numerical integration of functions of several variables.

In a related study, Rajagopalan (1980) presents a way to compute all one-dimensional marginal densities of the parameters of mixed linear models and this method was presented in the previous sections of the chapter. One advantage of his approach is that it avoids numerical integration of functions of several variables.

Lindley and Smith (1972) and later Smith (1973) propose using the mode of the joint posterior density of the parameters in the model. They employed their method with a two-factor model with two variance components, the error variance and the row and column effects. They

POSTERIOR INFERENCES 169

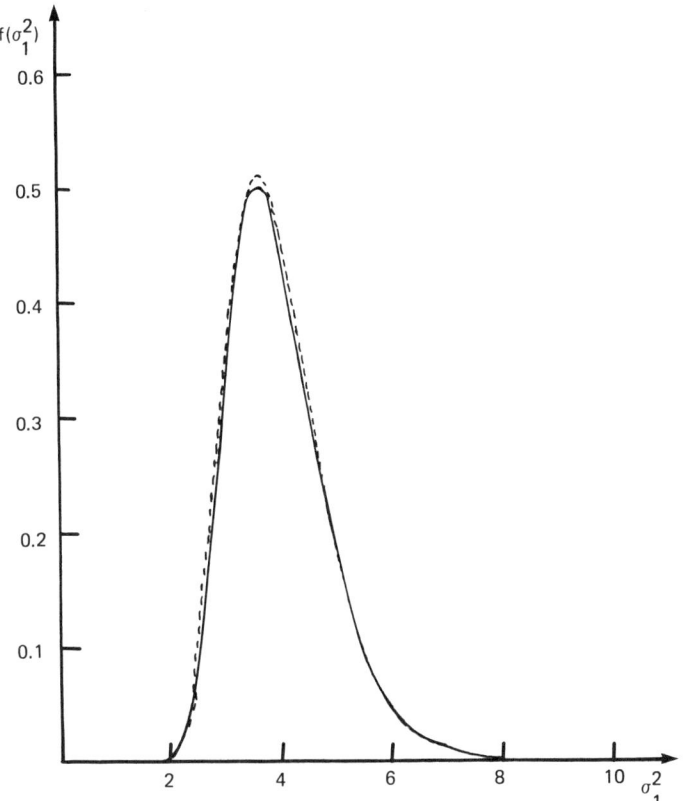

Figure 4.6. Marginal posterior density of the between-variance component; $\alpha = 32$, $\beta = 500$, $\alpha_1 = 20$, $\beta_1 = 80$; ---- = approximation, ──── = true distribution.

were primarily interested in estimating the main effect factors, but in principle, their method will work with any subset of parameters.

Consider the joint posterior density (4.5) of all the parameters of a mixed linear model. Theorem 4.1 gives the conditional posterior distributions of θ, b, τ, and ρ, given the other parameters of the model, thus all the conditional modes are known, namely

$$M(\theta|b,\tau,\rho) = (X'X)^{-1}X'(y - ub) \qquad (4.20)$$

is the conditional posterior mode of θ given b, τ, and ρ. Also,

$$M(b|\theta,\tau,\rho) = [\tau u'u + A(\rho)]^{-1}\tau u'(y - X\theta), \tag{4.21}$$

$$M(\tau^{-1}|\theta,b,\rho) = \frac{2\beta + (y - X\theta - ub)'(y - X\theta - ub)}{n + 2\alpha + 2} \tag{4.22}$$

and

$$M(\tau_i^{-1}|\theta,b,\tau) = \frac{2\beta_i + b_i'b_i}{m_i + 2\alpha_i + 2} \tag{4.23}$$

are the conditional modes of b, τ^{-1}, and τ_i^{-1}, $i = 1, 2, \ldots, c$, given the other parameters. The last two modes refer to the error variance τ^{-1} and the variance components τ_i^{-1}, and the four equations for the conditional modes will be used to find a mode of the joint posterior density, (4.5), of all the parameters θ, b, τ, and ρ.

Suppose a mode of the joint density is given by a solution to the equations

$$\frac{\partial}{\partial \theta} p(\theta,b,\tau,\rho|y) = 0, \tag{4.24}$$

$$\frac{\partial}{\partial b} p(\theta,b,\tau,\rho|y) = 0, \tag{4.25}$$

$$\frac{\partial}{\partial \tau} p(\theta,b,\tau,\rho|y) = 0, \tag{4.26}$$

and

$$\frac{\partial}{\partial \tau_i} p(\theta,b,\tau,\rho|y) = 0, \tag{4.27}$$

where $i = 1, 2, \ldots, c$.

Now consider (4.24), then it is known that the joint density factors as

$$p(\theta,b,\tau,\rho|y) = p_1(\theta|b,\tau,\rho,y)p_2(b,\tau,\rho|y),$$

where p_1 is the conditional density of θ given b, τ, and ρ and p_2 the marginal density of b, τ, and ρ. If $p_2 > 0$ for $b \in R^m$, $\tau > 0$, and all $\tau_i > 0$, the first equation (4.24) is equivalent to

POSTERIOR INFERENCES   171

$$\frac{\partial}{\partial \theta} p_1(\theta|b,\tau,\rho,y) = 0, \qquad (4.28)$$

and in a similar fashion

$$\frac{\partial}{\partial b} p_3(b|\theta,\tau,\rho,y) = 0, \qquad (4.29)$$

$$\frac{\partial}{\partial \tau} p_5(\tau|b,\theta,\rho,y) = 0, \qquad (4.30)$$

and

$$\frac{\partial}{\partial \tau_i} p_7(\tau_i|\rho,\theta,b,y) = 0, \quad i = 1,2,\ldots,c \qquad (4.31)$$

where p_3, p_5, and p_7 are the conditional densities of b, τ, and τ_i given the other parameters, and this system of four equations is equivalent to the system (4.24)-(4.27).

Since each conditional density is unimodal for all values of the conditioning variables, this system is equivalent to the four modal equations (4.20)-(4.23). In order to find the joint mode, this system must be solved simultaneously for θ, b, τ, and τ_i, $i = 1,2,\ldots,c$, over the space $b \in R^m$, $\theta \in R^p$, $\tau > 0$, and $\tau_i > 0$, $i = 1,2,\ldots,c$.

Starting with some initial value for b, one may find a θ value from the first modal equation (4.20), then a τ^{-1} value from (4.22), and finally c τ_i^{-1} values from (4.23), then the cycle is repeated, and if the process converges, the solution is a solution to the system (4.24)-(4.27), and a mode of the joint posterior density is obtained.

Let us see how this method works when applied to some simple mixed models. First, consider the one-way random balanced model of the previous section, and let the initial value of b be the least squares estimator of b, namely $_0b = (\bar{y}_1 - \bar{y}, \bar{y}_2 - \bar{y}, \ldots, \bar{y}_m - \bar{y})'$, then the first modal equation is $_0\mu = \bar{y}$ where $\theta = \mu$. Further substitution gives

$$_0\tau^{-1} = \frac{2\beta + \sum_{1}^{m}\sum_{1}^{t}(y_{ij} - \bar{y}_i)^2}{mt + 2\alpha + 2}$$

and

$$_0\tau_1^{-1} = \frac{2\beta_1 + \sum_1^m (\bar{y}_i - \bar{y})^2}{m + 2\alpha_1 + 2}$$

as the first stage estimates of the error variance and between variance component, respectively. Gharraf (1979) shows that $_0b$, $_0\mu$, $_0\tau$, and $_0\tau_1^{-1}$ are conditional Bayes (conditional posterior modes) estimators of b, μ, τ^{-1}, and τ_1^{-1}, respectively.

The cycle is repeated using $_1b = m(b\,|\,_0\mu, _0\tau^{-1}, _0\tau_1^{-1})$ and continued until the value of the estimates stabilizes. Note the first stage estimators depend on the familiar analysis of variance sum of squares.

Next consider a two-fold nested random balanced model

$$y_{ijk} = \theta + a_i + b_{ij} + \varepsilon_{ijk}, \tag{4.32}$$

where $i = 1, 2, \ldots, a$; $j = 1, 2, \ldots, g$; $k = 1, 2, \ldots, d$. In terms of the general mixed model, $n = agd$, $c = 2$ (two variance components), $m_1 = a$ and $m_2 = g$. The observations are $\{y_{ijk}\}$, θ is a real unknown scalar, the sets $\{a_i\}$, $\{b_{ij}\}$, and $\{\varepsilon_{ijk}\}$ are independent and the a_i are independent and $n(0, \sigma_1^2)$, the b_{ij} are independent and $n(0, \sigma_2^2)$, and the ε_{ijk} are independent and $n(0, \sigma^2)$. If the starting value of b is $_0b = (j'Ru)^- u'Ry$, the modal equations reduce to

$$_0\theta = \bar{y},$$

$$_0\sigma^2 = \frac{2\beta + \sum_1^a \sum_1^g \sum_1^d (y_{ijk} - \bar{y})^2}{agd + 2\alpha + 2}$$

$$_0\sigma_1^2 = \frac{2\beta_1 + g^2 \sum_1^a (\overline{y_{i..}} - \bar{y})^2/(g+1)^2}{a + 2\alpha_1 + 2}$$

and

$$_0\sigma_2^2 = \frac{2\beta_2 + \sum_1^a \sum_1^g [\bar{y}_{ij.} - g\bar{y}_i/(g+1) - \bar{y}/(g+1)]^2}{g + 2\alpha_2 + 2}$$

POSTERIOR INFERENCES

for the first stage estimates of the parameters. The cycle is repeated beginning with a new value of b, namely

$$_1b = [_0\tau u'y + A(_0\rho)]^{-1} {}_0\tau u'(y - X_0\theta],$$

where $_0\rho = (_0\tau_1, {}_0\tau_2)$ and $\tau_i^{-1} = \sigma_i^{-2}$, i = 1, 2. This model is given as an example by Box and Tiao (1973) to illustrate the Bayesian approach, by Hartley and Rao (1968), who use maximum likelihood estimation, and by Gharraf (1979), who showed the first stage estimators were conditional Bayes estimators.

The starting value of b was the least squares estimator of b, however other reasonable estimators of b will suffice.

There are some problems with this procedure. First, when does it converge and if it converges does it converge to the joint posterior mode? Also if it converges, how does the rate of convergence depend on the starting value? There are also some other computational problems with inverting large matrices. The generalized Moore-Penrose inverse of u'Ru and the inverse of $\tau u'u + A(\rho)$ are of order m × m, the total number of random effects of the model. This problem also occurs with maximum likelihood estimation, thus perhaps the W transform introduced by Hemmerle and Lorens (1976) is a way to calculate inverses of patterned matrices such as u'Ru.

This section is concluded with numerical examples for the one-way and twofold random models. Again, the data set generated by Box and Tiao (1973) is used as in the previous section. With regard to the first mode the parameters were set at $\tau^{-1} = 16$, $\tau_1^{-1} = 4$, and $\theta = 5$, and there are six groups, $m_1 = 6$, and five observations per group. The analysis here uses gamma prior densities for σ^2 and σ_1^2, which differs from Box and Tiao (1973), who place an improper vague prior on the analysis of variance expected mean squares. See pp. 246-276 of Box and Tiao (1973) for a detailed account. The analysis of variance estimators is $\hat{\sigma}^2 = 14.495$ and $\hat{\sigma}_1^2 = 1.3219$, and the joint posterior mode of (σ^2, σ_1^2) is (13.796, 0).

Table 4.2 gives a solution to the modal equations (4.20)-(4.23) corresponding to a one-way model using different values for the hyperparameters α, α_1, β, and β_1, and in all cases the starting value of b was the least squares estimator. The random effect estimates are not shown since b is a 6 × 1 vector and the other parameters are our primary concern. In all cases $\hat{\theta}$ is 5.666 because the conditional mode of θ is $\bar{y}$ and is not affected by prior information.

Table 4.2 demonstrates the effect of the prior distribution on the modal estimates. For example, the first row uses the "true" value of

Table 4.2
Modal Estimates of Parameters:
The One-Way Model

α	α_1	β	β_1	$\hat{\theta}$	$\hat{\sigma}^2$	$\hat{\sigma}^2_1$
2	2	4	4	5.6656	11.4861	.713785
2	2	2	2	5.6656	10.9041	.349768
2	2	10	10	5.6656	10.8208	1.8334
2	2	20	20	5.6656	12.9161	3.59343
5	5	4	4	5.6656	9.93803	.462915
5	5	2	2	5.6656	9.41301	.228348
5	5	50	20	5.6656	11.1399	2.35272
5	5	10	10	5.6656	9.3689	1.1868
10	10	16	4	5.6656	8.08161	.293158
10	10	2	2	5.6656	7.63936	.145248
10	10	15	20	5.6656	9.05468	1.49166
10	10	100	20	5.6656	11.0234	1.47858
10	10	10	10	5.6656	7.63557	.748002
1.1	1.1	1	1	5.6656	11.5485	.20321
1.01	1.01	.1	.1	5.6656	11.7507	.0200421
1.001	1.001	.01	.01	5.6656	11.7734	.0200043

the variance components as the prior mean of these parameters, and the estimates of the within and between components are 11.4861 and .713785, respectively. On the other hand, when the prior means are set at 10, the modal estimates are 11.7734 and .0200043.

The modal estimate of θ is not sensitive to prior information, but the modal estimates of the within and between components are sensitive; however, the within estimate is less sensitive than the between estimate. For instance, the ratio of the largest to smallest of the within estimates is only 1.5419 while the ratio is 179.6328 for the between estimates.

POSTERIOR INFERENCES   175

It is difficult to state any definitive conclusions about these estimates and a more thorough sensitivity analysis needs to be done. The estimates are closest to the true value of the parameters when the prior means are set at 15 and 20 and in all cases the estimates are smaller than the true values of the parameters.

The next example is the twofold nested model, (4.32), and is discussed by Box and Tiao (1973) beginning on page 282. This model has four parameters, the fixed effect θ, the error variance σ^2, and two variance components, σ_1^2 and σ_2^2. The authors generated the data and set the parameter values at $\theta = 0$, $\sigma_1^2 = 1.0$, $\sigma_2^2 = 4.0$, and $\sigma^2 = 2.25$. Also $a = 10$ and $g = d = 2$ and the analysis of variance estimators are $\hat{\sigma}_1^2 = 1.04$, $\hat{\sigma}_2^2 = 5.62$, and $\hat{\sigma}^2 = 3.60$. Assuming a vague prior distribution on the expected mean squares of the analysis of variance, Box and Tiao numerically determine the marginal distribution of the variance components.

Suppose one uses independent gamma densities for the precision components $\tau_1 (= \sigma_1^{-2})$, τ_2, and the error precision $\tau (= \sigma^{-2})$ with hyperparameters α_1, β_1, α_2, β_2, α, and β and a constant prior density for the fixed effect. For various values of the hyperparameters the model equations were solved and the solution listed in Table 4.3, where in all cases the starting value of b is the least squares estimator $(u'Ru)^{-}u'Ry$.

The conditional mode of θ is $\bar{y}$ and is not affected by prior information. The estimates $\hat{\sigma}_1^2$ and $\hat{\sigma}_2^2$ of the two variance components are more sensitive to prior information than the error variance estimate $\hat{\sigma}^2$. For example, the ratios of the largest to smallest estimates are 5.86, 5703.23, and 14.22 for $\hat{\sigma}^2$, $\hat{\sigma}_1^2$, and $\hat{\sigma}_2^2$, respectively. Definitive conclusions are difficult to state on the basis of such a limited study.

In this section, a mode of the posterior marginal distribution of all the parameters is found for a general mixed model. The estimates are solutions to the modal equations based on the conditional mode of each set of parameters, conditional on the remaining.

In all cases the iterative scheme rapidly converged and was illustrated with two random models. The estimates of the variance components were more sensitive to prior information than the error variance, and estimates of the fixed effects were not sensitive to prior information.

The program of the iterative procedure was written in SAS.

Table 4.3
Modal Estimates for Twofold Nested Model

α_1	α_2	β_1	β_2	α	β	$\hat{\theta}$	$\hat{\sigma}^2$	$\hat{\sigma}_1^2$	$\hat{\sigma}_2^2$
2	2	10	10	2	10	.488375	.94567	2.6596	3.09019
2	2	4	2.25	2	1	.488375	.510797	.681372	4.86223
2	2	.1	.1	2	.1	.488375	.466851	.0125578	.49854
1	1	2	2	1	2	.488375	.578268	.32538	6.26469
1.001	1.001	.01	.01	1.001	.01	.488375	.483015	.00142911	7.09156
2	2	50	50	2	50	.488375	2.73923	8.15055	5.65159
1	1	10	10	1	10	.488375	.984649	3.24959	3.24959

EXERCISES

SUMMARY AND CONCLUSIONS

This chapter has introduced a Bayesian analysis of the mixed model. Generally two methodologies were given. First, the one-dimensional marginal distributions of all the variance components and of the error variance are derived and are based on a normal approximation to the marginal posterior density of the random factors. This method applies to any mixed model, be it unbalanced or otherwise, and general formulas for the posterior means and variances are derived.

A numerical study of the one-way random model reveals the accuracy of the approximation and the sensitivity of the marginal posterior densities to the parameters of the prior density.

Second, modal estimates of all the parameters may be found by solving a set of modal equations based on the conditional posterior distributions given in Theorem 4.1. The sensitivity of the modal estimates to prior information was investigated for the one-way and twofold nested models.

The advantage of the first method is that one may plot each marginal posterior density, thus, giving us at a glance (see Figures 4.1-4.6) inferences about the parameters based on all the information in the sample. But, a major disadvantage is that this method is based on an approximation to the marginal posterior densities.

With regard to the second method, estimates of all the parameters, including the random and fixed effects, are easily found, however the posterior precision of the parameters is not available.

These two methodologies should both be done in a Bayesian analysis and are not competing methods, but should augment each other. The second is easier to program than the first.

As we have seen, making inferences for the parameters of a mixed model presents many challenges. With regard to the variance components, further research, perhaps, will give us an accurate approximation to the marginal distribution of b, in which case, the Bayesian solution will be complete.

EXERCISES

1. Formulas (4.14), (4.15), (4.16), and (4.17) give the approximate mean and variance for the posterior distributions of the error variance and variance components of any mixed model. Using the above formulas, find the approximate mean and variance of the posterior distribution of a balanced twofold nested random model (4.32).
2. Repeat the above procedure for a balanced two-way crossed random model without interaction and with one observation per cell.

3. For a balanced two-way crossed random model one observation per cell and with no interaction, find the conditional posterior mean of the error variance and the two variance components, given b the vector of random effects.
4. Consider the general mixed model (4.1) and the prior density (4.2) of the parameters. The conditional posterior density of θ given b is expressed by Theorem 4.6. Explain how you would make inferences about θ, from the marginal posterior distribution of θ. For example, what is the mean of the marginal distribution of θ?
5. Prove Theorems 4.1, 4.2, 4.3, and 4.4.
6. Consider Theorem 4.3, which gives the marginal distribution of b, the vector of random effects. Develop a normal approximation to the distribution of b.
7. Referring to Theorem 4.4 and equation (4.8), what is the joint posterior density of τ and τ_1, the error precision and precision component of the one-way random model given above? Can the marginal posterior density of τ be isolated?
8. Consider a twofold nested balanced random model (4.32) with error variance σ^2 and variance components σ_1^2 and σ_2^2. Using the conditional joint marginal posterior distribution of σ^2, σ_1^2, and σ_2^2 given b, find the conditional joint posterior distribution of

$$\gamma_1 = \frac{\sigma_1^2}{\sigma^2 + \sigma_1^2 + \sigma_2^2}$$

and

$$\gamma_2 = \frac{\sigma_2^2}{\sigma^2 + \sigma_1^2 + \sigma_2^2}$$

given b (the vector of random effects).

REFERENCES

Box, G. E. P. and G. C. Tiao (1973). *Bayesian Inference in Statistical Analysis*, Addison-Wesley, Reading, Massachusetts.

Daniels, H. E. (1939). "The estimation of components of variance," Journal of the Royal Statistical Society (Supp.), Vol. 6, pp. 186-197.

REFERENCES

Dreze, Jacque H. (1977). "Bayesian regression analysis using poly-t-distributions," from *New Developments in the Application of Bayesian Methods,* edited by Aykac and Brumat, North-Holland Publishing Co., Amsterdam.

Gharrah, M. K. (1979). A General Solution to Making Inferences about the Parameters of Mixed Linear Models. Ph.D. dissertation, Oklahoma State University.

Goel, Prem K. and Morris H. DeGroot (1981). "Information about hyperparameters in hierarchical models," Journal of the American Statistical Association, Vol. 76, pp. 140-147.

Graybill, F. A. (1954). "On quadratic estimates of variance components," Annals of Math. Statistics, Vol. 25, pp. 367-372.

Graybill, F. A., and R. A. Hultquist (1961). "Theorems on Eisenhart's model II," Annals of Math Statistics, Vol. 32, pp. 261-269.

Graybill, F. A. and A. W. Wortham (1956). "A note on uniformly best unbiased estimators for variance components," Journal of the American Statistical Association, Vol. 51, pp. 266-268.

Hartley, H. O. and J. N. K. Rao (1967). "Maximum likelihood estimation for the mixed analysis of variance model," Biometrika, Vol. 54, pp. 93-108.

Hemmerle, W. J. and J. A. Lorens (1976). "Improved algorithm for the W-transform in variance component estimation," Technometrics, Vol. 18, No. 2, pp. 207-212.

Henderson, C. R. (1953). "Estimation of variance and covariance components," Biometrics, Vol. 9, pp. 226-252.

Hill, B. M. (1965). "Inference about variance components in the one-way model," Journal of the American Statistical Association, Vol. 60, pp. 806-825.

Hill, B. M. (1967). "Correlated errors in the random model," Journal of the American Statistical Association, Vol. 62, pp. 1387-1400.

Lindley, D. V. and A. F. M. Smith (1972). "Bayes estimates for the linear model," Journal of the Royal Statistical Society, Series B, Vol. 34, pp. 1-18.

Rajagopalan, M. (1980). "Bayesian inference for the variance components in mixed linear models," Ph.D. dissertation, Oklahoma State University, Stillwater, Oklahoma.

Rajagopalan, M. and Lyle Broemeling (1983). "Bayesian inference for the variance components of the general mixed models," Communications in Statistics, Vol. 12, No. 6, pp. 701-724.

Ruben, H. (1960). "Probability content of regions under spherical normal distributions," Annals of Math. Statistics, Vol. 31, pp. 598-619.

Ruben, H. (1962). "Probability contents of regions under spherical normal distributions, IV: the distribution of homogeneous and non-homogeneous quadratic function of normal variables," Annals of Math. Statistics, Vol. 33, pp. 542-570.

Searle, S. R. (1971). *Linear Models*, John Wiley and Sons, Inc., New York.

Searle, S. R. (1978). "A summary of recently developed methods of estimating variance components," Technical Report, BU-338, Biometrics Unit, Cornell University, New York.

Smith, A. F. M. (1973). "Bayes estimates in one-way and two-way models," Biometrika, Vol. 60, pp. 319-329.

Stone, M. and B. G. F. Springer (1965). "A paradox involving quasi prior distributions," Biometrika, Vol. 52, p. 623.

Winsor, C. P. and G. L. Clarke (1940). "A statistical study of variation in the catch of plankton nets," Sears Foundation Journal of Marine Research, Vol. 3, pp. 1-34.

5
TIME SERIES MODELS

INTRODUCTION

Up to this point, our study of linear models has been confined to those which are usually studied in traditional courses on linear models. Regression models and models used to analyze designed experiments have been analyzed from a Bayesian viewpoint.

With the exception of the autoregressive model, very little has appeared from the Bayesian viewpoint and Zellner's (1971) book appears to be the only one devoted to the Bayesian analysis of selected parametric time series models. His book examines the autoregressive process and distributed lag models. This chapter introduces the prior, posterior, and predictive analysis of selected autoregressive, moving average, and autoregressive-moving average models as well as the distributed lag model, thus one may view this chapter as an extension of Zellner's work.

The literature about the theoretical and methodological aspects of ARMA models is quite large and the standard reference is Box and Jenkins (1970). They present the theory and methodology for the analysis of data which are generated by univariate and multivariate ARMA and transfer function models and models which incorporate seasonal and intervention effects. The methodology is iterative and consists of differencing the data to achieve stationarity, identifying the model from the sample autocorrelation and partial autocorrelation function, estimating the parameters via least squares or maximum likelihood, and forecasting future observations by linear combinations of past observations and forecasts. There is no doubt that the Box-Jenkins way is the prevailing methodology for analyzing ARMA processes. Their methods are not Bayesian and include many classical algorithms for identification, forecasting, and control.

On the other hand, the Bayesian literature devoted to ARMA models is sparse. When compared to Box-Jenkins and other classical techniques, such as those given by Granger and Newbold (1977), there are few Bayesian contributions. The Current Index to Statistics (1980) lists very few Bayesian articles, but among them are Smith (1979) and Akaike (1979). The first article is a generalization of the Harrison and Stevens (1976) steady-state forecasting model, and the second is a Bayesian interpretation of the author's identification method for autoregressive processes. As mentioned earlier, Zellner (1971) is the only book exclusively devoted to the Bayesian analysis of time series and econometric models but deals only with autoregressive and distributed lag models. Aoki's (1967) book is a Bayesian decision-theoretic study of the linear dynamic systems of engineering control.

Bayesian studies of time series have been mostly made by engineers and economists. For example, the econometric literature about structural change in linear models, including selected time series, is well developed and reviewed by Poirier (1976). More recently, Holbert and Broemeling (1977), Chin Choy and Broemeling (1980), Abraham and Wei (1979), Salazar, Broemeling, and Chi (1981), and Tsurumi (1980) have investigated structural change. Structural change in a model is similar to ARMA models which have intervention effects.

Bayesian analyses developed primarily for ARMA models are limited; however, Newbold (1973) derives the posterior distribution of the parameters of a univariate transfer function noise model. Since 1973 there has been very little work and Chatfield's (1977) review refers to one paper, the Harrison and Stevens (1976) paper. Control engineers are developing an interest in Bayesian inferential procedures and Perterka's (1981) article reviews Bayesian methodology and demonstrates its applicability to system identification. Most recently, Monahan (1981) has devised a Bayesian analysis of ARMA time series models. His approach is mostly numerical in that the posterior distribution of the parameters must be specified numerically by numerical integrations. In spite of these difficulties, Monahan's work is an important first step.

Why is there so little work on the Bayesian analysis of time series? It appears the likelihood function is analytically intractable for most ARMA models and this presents a problem, not only to Bayesians but also to those interested in maximum likelihood estimation. Many papers, such as those by Newbold (1974), Dent (1977), Ali (1977), Ansley (1979), Hillmer and Tiao (1979), Ljung and Box (1976), Nicholls and Hall (1979), and many others are investigations for simplifying the likelihood function so one may develop more efficient algorithms.

If the likelihood function can be represented in such a way as to produce analytically tractable posterior distributions, a complete Bayesian analysis is possible.

AUTOREGRESSIVE PROCESSES

Let Y_t: $t = 1, 2, \ldots$, be a time series and suppose

$$Y_t - \theta_1 Y_{t-1} = \varepsilon_t, \tag{5.1}$$

where θ_1 is a real unknown parameter, and suppose $\varepsilon_1, \varepsilon_2, \ldots$, are independent and identically distributed normal variables with mean zero and precision $\tau > 0$. Thus each observation is a linear function of the previous observation plus white noise. This model is called a first order autoregressive process and is stationary if $|\theta_1| < 1$.

A second order process, denoted by AR(2), is

$$Y_t - \theta_1 Y_{t-1} - \theta_2 Y_{t-2} = \varepsilon_t, \quad t = 1, 2, \ldots \tag{5.2}$$

where θ_1 and θ_2 are real unknown parameters and as with the AR(1) model, the ε_t are i.i.d. $n(0, \tau^{-1})$ random variables. The process is stationary if $(\theta_1, \theta_2) \in T$, where T is the triangle with vertices $(-2, 0)$, $(2, 0)$, and $(0, 1)$.

In general an AR(p) process is given by

$$\theta(B) Y_t = \varepsilon_t, \quad t = 1, 2, \ldots \tag{5.3}$$

where the ε_t are white noise with precision τ,

$$\theta(B) = 1 - \theta_1 B - \theta_2 B^2 - \cdots - \theta_p B^p \tag{5.4}$$

and $B^i Y_t = Y_{t-i}$, $i = 1, 2, \ldots$. The process is stationary if the roots of $\theta(B) = 0$ lie outside the unit circle.

Suppose there are n observations $Y_1, Y_2, \ldots, Y_n$ from an AR(p) process, then

$$\theta(B) Y_t = \varepsilon_t, \quad t = 1, 2, \ldots, n \tag{5.5}$$

where $y_0, y_{-1}, \ldots, y_{1-p}$ are initial observations assumed to be known constants. Since $\varepsilon_1, \varepsilon_2, \ldots, \varepsilon_n$ are i.i.d. $n(0, \tau^{-1})$, the joint density of the n observations is

$$f(y_1, y_2, \ldots, y_n | \theta, \tau) \propto \tau^{n/2} \exp -\frac{\tau}{2}$$

$$\times \sum_{t=1}^{n} (y_t - \theta_1 y_{t-1} - \cdots - \theta_p y_{t-p})^2 \quad (5.6)$$

where $\theta_i \in R$, $i = 1, 2, \ldots, p$, $\tau > 0$, and $y_t \in R$, $t = 1, 2, \ldots, n$. Suppose stationarity is not assumed and one wants to make inferences about the parameters $\theta = (\theta_1, \theta_2, \ldots, \theta_p) \in R^p$ and $\tau > 0$. The y_i are observed values of Y_i, and (5.6) is the likelihood function of θ and τ based on the observations.

What does one do about prior information for the parameters? It is well known that the conjugate class of prior distributions is the normal-gamma family, i.e., θ given τ is normal and τ is gamma, thus let

$$p(\theta|\tau) \propto \tau^{p/2} \exp -\frac{\tau}{2}(\theta - \mu)'P(\theta - \mu), \quad \theta \in R^p \quad (5.7)$$

be the conditional prior density of θ given τ and

$$p(\tau) \propto \tau^{\alpha-1} e^{-\tau\beta}, \quad \tau > 0 \quad (5.8)$$

be the marginal prior density of τ. The known hyperparameters are μ, P, α and β, where $\mu \in R^p$, $\alpha > 0$, $\beta > 0$, and P is a p-th order positive definite matrix. Since we are working with the conjugate prior class for the parameters, the joint prior distribution of the parameters is also normal-gamma, and one must find the posterior parameters.

Using Bayes' theorem, the posterior density of the parameters is

$$p(\theta, \tau | s_n) \propto \tau^{(n+2\alpha+p)/2-1} \exp -\frac{\tau}{2}[(\theta - A^{-1}C)'A(\theta - A^{-1}C)$$

$$+ D], \quad (5.9)$$

where $\theta \in R^p$ and $\tau > 0$. The other terms in this expression are A, C, and D, where A is a known $p \times p$ matrix

$$A = P + (a_{ij}),$$

$$C = P\mu + (c_j),$$

AUTOREGRESSIVE PROCESSES

and

$$D = 2\beta + \mu'P\mu + \sum_{1}^{n} y_t^2 - C'A^{-1}C.$$

Also, (a_{ij}) is $p \times p$ with ij-th element $a_{ij} = \Sigma_1^n y_{t-i} y_{t-j}$ and (c_j) is a $p \times 1$ vector with j-th element $\Sigma_{t=1}^n y_t y_{t-j}$.

Integrating the joint density with respect to τ,

$$p(\theta|s_n) \propto \frac{1}{[(\theta - A^{-1}C)'A(\theta - A^{-1}C) + D]^{(n+2\alpha+p)/2}}, \quad \theta \in R^p$$

(5.10)

is the marginal posterior density of θ, thus θ has a t density with $n + 2\alpha$ degrees of freedom, location $A^{-1}C$, and precision matrix $(n + 2\alpha)A^{-1}D$. The marginal posterior density of τ is

$$p(\tau|s_n) \propto \tau^{(n+2\alpha)/2-1} \exp \frac{\tau}{2} [2\beta + \mu'P\mu + \sum_{1}^{n} y_t^2 - C'A^{-1}C],$$

$$\tau > 0 \quad (5.11)$$

where $s_n = (y_1, y_2, \ldots, y_n)$ is the sample observations, thus τ has a gamma distribution with parameters $n + 2\alpha$ and $[2\beta + \mu'P\mu + \Sigma_1^n y_t^2 - C'A^{-1}C]/2$.

Thus, with regard to the autoregressive model, the posterior analysis is similar to the general linear model.

Usually with time series models, the main objective is to predict future observations $Y_{n+1}, Y_{n+2}, \ldots$, of the process, and as has been shown the Bayesian predictive distribution provides a convenient forecasting procedure.

Consider one future observation Y_{n+1}, then the density of Y_{n+1} given s_n, τ, and θ is

$$f(y_{n+1}|s_n, \tau, \theta) \propto \tau^{1/2} \exp -\frac{\tau}{2} [y_{n+1} - \theta_1 y_n - \theta_2 y_{n-1} - \cdots - \theta_p y_{n-(p-1)}]^2,$$

where $y_{n+1} \in R$. Now this density may be written as

$$f(y_{n+1}|s_n, \tau, \theta) \propto \tau^{1/2} \exp - \frac{\tau}{2}[y_{n+1}^2 + \theta'E\theta - 2\theta'F],$$

$$y_{n+1} \in R \quad (5.12)$$

where E is p × p with ij-th element $y_{n-(i-1)}y_{n-(j-1)}$ and F is p × 1 with j-th element $y_{n+1}y_{n-(j+1)}$. Now multiplying (5.9) with (5.12) gives for the exponent, apart from the $\tau/2$ factor, a perfect square in θ, then integrating with respect to θ and τ,

$$g(y_{n+1}|s_n) \propto \frac{1}{[y_{n+1}^2 + D + C'A^{-1}C - (C+F)'(A+E)^{-1}(C+F)]^{(n+2\alpha+1)/2}},$$

$$(5.13)$$

for $y_{n+1} \in R$, is the predictive density of the one-step-ahead forecast. This, in turn, can be simplified to give a t density on Y_{n+1}, with $n + 2\alpha$ degrees of freedom, location

$$E[Y_{n+1}|s_n] = [1 - F^{*\prime}(A+E)^{-1}F^*]^{-1}F^{*\prime}(A+E)^{-1}C$$

and precision

$$P[Y_{n+1}|s_n] = [1 - F^{*\prime}(A+E)^{-1}F^*]\{D + C'A^{-1}C - C'(A+E)^{-1}$$

$$\times F^*[1 - F^{*\prime}(A+E)^{-1}F^*]^{-1}F^{*\prime}(A+E)^{-1}C\},$$

where $F = y_{n+1}F^*$, and F^* is p × 1 with j-th component $y_{n-(j-1)}$, j = 1, 2, ..., p.

Of course, this can be generalized to more than one future observation, but the joint predictive density of two future observations is not the usual bivariate t distribution.

We have seen that the posterior and predictive analysis are straightforward for an AR(p), autoregressive model if one assumes a conjugate prior density for the parameters and if one does not assume stationarity. What if one assumes stationarity?

Consider an AR(1) model

$$Y_t - \theta_1 Y_{t-1} = \varepsilon_t,$$

MOVING AVERAGE MODELS

where $|\theta_1| < 1$, then how does one analyze this stationary model?

The normal-gamma family is no longer the conjugate class since the parameter space is interval $(-1,1) \times (0,\infty)$ and not as before $(-\infty,\infty) \times (0,\infty)$.

One approach is to ignore the stationarity assumption and proceed to use the normal-gamma family as a prior distribution for the parameters, but with an adjustment in the prior parameters to allow for approximate stationarity. To see how this works with an AR(1) model, use a normal-gamma prior with parameters μ, ξ, α, and β, then the marginal prior density of θ_1 and τ is a t distribution with 2α degrees of freedom, location μ, and precision $\alpha\xi/\beta$, and one may choose μ, α, β, and ξ in such a way so that $100\gamma\%$ of the marginal prior probability for θ_1 is contained in the interval $(-1,1)$, for any $0 < \gamma < 1$. By letting γ be "close" to 1, one is assuming "almost" stationarity. It should be noted that the α and β parameters are the parameters of the marginal prior distribution of τ, thus choosing an (α,β) pair in the marginal prior distribution of θ_1 will affect one's prior information about τ.

Another approach is to put a prior distribution on the parameter space $(-1,1) \times (0,\infty)$, but this way will produce analytically intractable posterior distributions, and a purely numerical procedure is required for a complete analysis. In spite of these disadvantages, Lahiff (1980) has developed an interesting algorithm to deal with this difficult problem, and Monahan (1981) has taken a similar approach with ARMA(p,q) models, where p,q = 0, 1, 2.

The identification problem of determining the order of an AR(p) process has not been explored yet, but Diaz and Farah (1981) and Monahan (1981) have begun some initial studies. Except for these two papers, Bayesian results have been limited.

With regard to the forecasting problem, very little has been done in finding the joint predictive density of future observations. Zellner (1971) has developed the distribution for first and second order AR processes and Land (1981) derived the forecasting distribution of AR(p) models with k future observations, but both authors deal with the non-stationary case. In her dissertation, Land shows that the forecast procedure may be done by a series of conditional t-forecasts.

For more information about autoregressive processes, the reader is referred to Zellner (1971), where interesting examples are explained in detail.

MOVING AVERAGE MODELS

A first order moving average process, MA(1), expresses each observation Y_t as

$$Y_t = \varepsilon_t - \phi_1 \varepsilon_{t-1}, \quad t = 1, 2, \ldots, \tag{5.14}$$

which is a linear combination of white noise, that is $\varepsilon_0, \varepsilon_1, \ldots,$ are n.i.d. $(0, \tau^{-1})$, where $\tau > 0$ and ϕ_1 is an unknown real moving average parameter.

A MA(2) model is given by

$$Y_t = \varepsilon_t - \phi_1 \varepsilon_{t-1} - \phi_2 \varepsilon_{t-2}, \quad t = 1, 2, \ldots, \tag{5.15}$$

where ϕ_1 and ϕ_2 are moving average parameters and the ε_t are white noise with precision τ.

In general, a q-th order moving average process is given by

$$Y_t = \phi(B)\varepsilon_t, \quad t = 1, 2, \ldots, \tag{5.16}$$

where

$$\phi(B) = 1 - \phi_1 B - \phi_2 B^2 - \cdots - \phi_q B^q,$$

and B is the backshift operator such that $B^k \varepsilon_t = \varepsilon_{t-k}$ and the moving average parameters are $\phi_1, \phi_2, \ldots, \phi_q$, assumed to be real unknown parameters. Moving average processes are always stationary.

Suppose there are n observations $(y_1, y_2, \ldots, y_n)' = y$ from a MA(1) process and let

$$y = A(\phi_1)\varepsilon, \tag{5.17}$$

where $\varepsilon = (\varepsilon_0, \varepsilon_1, \ldots, \varepsilon_n)$, $A(\phi_1)$ is a n × (n + 1) matrix given by

$$A(\phi_1) = \begin{pmatrix} -\phi_1, & 1, & 0, & 0, & \ldots, & 0 \\ 0, & -\phi_1, & 1, & 0, & \ldots, & 0 \\ 0, & 0, & \ldots, & -\phi_1, & 1 \end{pmatrix}$$

Since $\varepsilon \sim N[0, \tau^{-1} I_{n+1}]$, $y \sim N\{0, \tau[A(\phi_1)A'(\phi_1)]^{-1}\}$, and

MOVING AVERAGE MODELS

$$A(\phi_1)A'(\phi_1) = \begin{pmatrix} 1 + \phi_1^2, & -\phi_1, & 0, & \ldots, & 0 \\ -\phi_1, & 1 + \phi_1^2, & -\phi_1, & \ldots, & 0 \\ & & & & -\phi_1 \\ 0, & 0, & \ldots, & -\phi_1, & 1 + \phi_1^2 \end{pmatrix}$$

which is a tridiagonal Toeplitz matrix with $-\phi_1$ for the super- and subdiagonal components and $1 + \phi_1^2$ for the principal diagonal elements. The joint density of the observations is given by

$$f[y|\phi_1, \tau] \propto \tau^{n/2} |A(\phi_1)A'(\phi_1)|^{-1/2} \exp -\frac{\tau}{2} y'[A(\phi_1)A'(\phi_1)]^{-1} y,$$

$$y \in R^n, \quad (5.18)$$

Since $A(\phi_1)A'(\phi_1)$ is symmetric and positive definite, there exists an orthogonal matrix Q such that $D(\phi_1) = Q'A(\phi_1)A'(\phi_1)Q$ is diagonal with diagonal elements $1 + \phi_1^2 - 2\phi_1 \cos i\pi(n + 1)^{-1} = d_i(\phi_1)$, where $i = 1, 2, \ldots, n$. It can be shown, see Gregory and Karney (1969), that Q is independent of ϕ_1 and has ij-th element $\sqrt{2/(n + 1)} \cdot \sin ij (n + 1)^{-1}$. We see that $A(\phi_1)A'(\phi_1) = QD(\phi_1)Q'$, thus $A(\phi_1)A'(\phi_1) = QD^{-1}(\phi_1)Q'$ and the density (5.18) may be written as

$$f[y|\phi_1, \tau] \propto \tau^{n/2} \prod_{i=1}^{n} d_i(\phi_1)^{-1/2} \exp -\frac{\tau}{2} y' QD^{-1}(\phi_1)Q'y,$$

$$y \in R^n, \quad (5.19)$$

which is a simplification over the other form (5.18), because $A(\phi_1)A'(\phi_1)$ does not have to be inverted since its inverse is in terms of its eigenvalues and eigenvectors.

How does one analyze this MA(1) process? First a prior density for ϕ_1 and τ must be chosen, but this may be difficult because the form of the density (5.19) does not suggest a well-known family of distributions.

One choice is to let ϕ_1 and τ be independent, a priori, and choose the marginal density of ϕ_1 to be an arbitrary proper density $p(\phi_1)$ and let

$$p(\tau) \propto \tau^{\alpha-1} e^{-\tau\beta}, \quad \tau > 0$$

be the marginal prior density of τ, namely gamma with parameters α and β, which combines easily with the likelihood function for ϕ_1 and τ, (5.19).

By applying Bayes' theorem, the joint density of ϕ_1 and τ is

$$p(\phi_1, \tau | y) \propto \tau^{(n+2\alpha)/2 - 1} p(\phi_1) \prod_{i=1}^{n} d_i^{-1/2}(\phi_1) \exp - \frac{\tau}{2}$$

$$\times [2\beta + y'QD^{-1}(\phi_1)Qy], \quad (5.20)$$

where $\phi_1 \in R$ and $\tau > 0$, and the marginal density of ϕ_1 is

$$p(\phi_1 | y) \propto \frac{p(\phi_1) \prod_{i=1}^{n} d_i^{-1/2}(\phi_1)}{[2\beta + y'QD^{-1}(\phi_1)Qy]^{(n+2\alpha)/2}}, \quad \theta \in R \quad (5.21)$$

which, since ϕ_1 is scalar, is easy to plot, and it is an easy task numerically to determine the posterior moments of ϕ_1.

The marginal posterior density of τ must be determined numerically from the joint density (5.20), because the integration with respect to ϕ_1 of the joint density is not known analytically.

The conditional density of τ given ϕ_1 is gamma with parameters $(n + 2\alpha)/2$ and $[2\beta + y'QD^{-1}(\phi_1)Qy]/2$, hence the conditional mean and variance of $\sigma^2 = \tau^{-1}$ are

$$E(\sigma^2 | \phi_1, y) = \frac{2\beta + y'QD^{-1}(\phi_1)Q'y}{n + 2\alpha - 2},$$

$$\text{Var}(\sigma^2 | \phi_1, y) = \frac{[2\beta + y'QD^{-1}(\phi_1)Q'y]^2}{(n + 2\alpha - 2)^2 (n + 2\alpha - 4)}.$$

(5.22)

Thus the marginal mean and variance of σ^2 may be found numerically by averaging their moments with respect to the marginal posterior density of ϕ_1, (5.21). Since

MOVING AVERAGE MODELS

$$E(\sigma^2|y) = E_{\phi_1} E(\sigma^2|\phi_1,y)$$

and

$$\text{Var}(\sigma^2|y) = E_{\phi_1} \text{Var}(\sigma^2|\phi_1,y) + \text{Var}_{\phi_1} E(\sigma^2|\phi_1,y),$$

where E_{ϕ_1} and Var_{ϕ_1} denote mean and variance with respect to the marginal density of ϕ_1, the marginal moments of σ^2 may be found from the conditional moments of ϕ_1.

It is now seen that the posterior analysis of a MA(1) model is somewhat more messy than the posterior analysis of an AR(p) model in that the joint and marginal posterior distributions of the parameters are not standard distributions, nevertheless, it is possible to determine the marginal distribution of ϕ_1, the conditional of σ^2 (or τ) given ϕ_1, and the marginal moments of σ^2.

How does one forecast with a moving average model? With the autoregressive model, the Bayesian predictive density was employed, thus the same method will be adopted for moving average models. Suppose we first consider a first order model (5.14) with the gamma prior for τ and the posterior density (5.20). Let Y_{n+1} be the future observation, then the joint density of $Y = (Y_1, \ldots, Y_n)$, the past observations, and the future observation Y_{n+1}, given ϕ_1 and τ is from (5.19)

$$f(y, y_{n+1}|\phi_1, \tau) \propto \tau^{(n+1)/2} \prod_{1}^{n+1} d_i(\phi_1)^{-1/2} \exp -\frac{\tau}{2}(y', y_{n+1})$$

$$\times Q*D^{-1}(\phi_1) Q*'(y', y_{n+1})', \qquad (5.23)$$

where $y \in R^n$, $y_{n+1} \in R$, $\phi_1 \in R$, and $\tau > 0$. $Q*$ and $D*(\phi_1)$ are analogous to the quantities Q and $D(\phi_1)$ of (5.19), that is

$$Q*'A*(\phi_1)A*'(\phi_1)Q* = D*(\phi_1),$$

where $A*(\phi_1)$ is the $(n + 1) \times (n + 2)$ analog of (5.17), and $D*(\phi_1)$ is diagonal with the characteristic roots $1 + \phi_1^2 - 2\phi_1 \cos i\pi(n + 2)^{-1}$, where $i = 1, 2, \ldots, n + 1$, of $A*(\phi_1)A*'(\phi_1)$ as diagonal elements. The ij-th element of Q^* is $\sqrt{2/(n+2)}\sin ij(n+2)^{-1}$, where $i,j = 1, 2, \ldots, n+1$. Consider the exponent of (5.23) and let $Q^*D^{*-1}(\phi_1)Q^{*\prime} = E(\phi_1)$, where $E(\phi_1)$ is partitioned as

$$E(\phi_1) = \begin{pmatrix} E_{11}(\phi_1) & E_{12}(\phi_1) \\ E_{21}(\phi_1) & E_{22}(\phi_1) \end{pmatrix}$$

and $E_{11}(\phi_1)$ is $n \times n$, $E_{22}(\phi_1)$ is a scalar. The exponent is now written as $y'E_{11}(\phi_1)y - 2y'E_{12}(\phi_1)y_{n+1} + y_{n+1}^2 E_{22}(\phi_1)$.

Multiply the density (5.23) by the prior density for ϕ_1 and τ and integrate the product of the two densities with respect to τ, then

$$f(y_{n+1}|y,\phi_1) \propto \frac{1}{[y_{n+1}^2 E_{22}(\phi_1) - 2y_{n+1}y'E_{12}(\phi_1) + y'E_{11}(\phi_1)y + 2\beta]^{(n+1+2\alpha)/2}}, \quad (5.24)$$

$y_{n+1} \in R$ is the conditional density of the future observation given the past observations y and ϕ_1. Complete the square in y_{n+1} by letting $A(\phi_1) = E_{22}(\phi_1)$, $B(\phi_1) = y'E_{12}(\phi_1)$, and $C(\phi_1) = y'E_{11}(\phi_1)y + 2\beta$, then the conditional density (5.24) is

$$f(y_{n+1}|y,\phi_1) \propto \frac{1}{\{[y_{n+1} - A^{-1}(\phi_1)B(\phi_1)]^2 A(\phi_1) + C(\phi_1) - B'(\phi_1)A^{-1}(\phi_1)B(\phi_1)\}^{(n+2\alpha+1)/2}}.$$

$$(5.25)$$

where $y_{n+1} \in R$.

This is the conditional predictive density of Y_{n+1} given ϕ_1, not the Bayesian predictive density, but we see that Y_{n+1} given ϕ_1 is a t distribution with $n + 2\alpha$ degrees of freedom, location

$$E(Y_{n+1}|\phi_1, y) = A^{-1}(\phi_1)B(\phi_1), \quad (5.26)$$

MOVING AVERAGE MODELS   193

and precision

$$P(Y_{n+1}|\phi_1,y) = A(\phi_1)(n + 2\alpha)[C(\phi_1) - B'(\phi_1)A^{-1}(\phi_1)B(\phi_1)]^{-1}.$$

(5.27)

As with the posterior analysis, the marginal moments can be computed from the conditional moments. For example, suppose a point forecast of Y_{n+1} is needed, then the mean of the marginal distribution of Y_{n+1} is given by

$$E(Y_{n+1}|y) = E_{\phi_1} E(Y_{n+1}|\phi_1,y),$$

(5.28)

where E_{ϕ_1} is the expectation with respect to the marginal posterior density of ϕ_1, given by (5.21). In a similar way the marginal variance $Var(Y_{n+1}|y)$ of the predictive distribution of Y_{n+1} is computed by first finding the conditional variance $Var(Y_{n+1}|\phi_1,y)$ of Y_{n+1} given ϕ_1 from (5.27), and using

$$Var(Y_{n+1}|y) = E_{\phi_1} Var(Y_{n+1}|\phi_1,y) + Var_{\phi_1} E(Y_{n+1}|\phi_1,y),$$

(5.29)

where Var_{ϕ_1} means the variance is computed over the marginal posterior distribution of ϕ_1.

Since ϕ_1 cannot be eliminated analytically from the joint density of Y_{n+1} and ϕ_1, there is no closed form expression for the predictive density of Y_{n+1}; nevertheless, one may plot the predictive density via one-dimensional numerical integrations of (5.25) and compute the forecast mean and variance from the formulas (5.28) and (5.29).

Can the above be extended to a second order model (5.15) with n observations $Y = (y_1, y_2, \ldots, y_n)$? Again, let

$$Y = A(\phi_1, \phi_2)\varepsilon,$$

where $\varepsilon = (\varepsilon_{-1}, \varepsilon_0, \varepsilon_1, \ldots, \varepsilon_n)$ and $A(\phi_1, \phi_2)$ is $n \times (n + 2)$ and

$$A(\phi_1,\phi_2) = \begin{pmatrix} -\phi_2, & -\phi_1, & 1, & 0, & \ldots, & 0 \\ 0, & -\phi_2, & -\phi_1, & 1, & \ldots, & 0 \\ \vdots & & & & & \vdots \\ 0, & & \ldots, & -\phi_2, & -\phi_1, & 1 \end{pmatrix}.$$

Since ε is normally distributed, y is also normal with mean zero and dispersion matrix $A(\phi_1,\phi_2)A'(\phi_1,\phi_2)\sigma^2$ which is $n \times n$ and has five diagonals ($n \geq 5$). The principal diagonal has common element (apart from τ) $1 + \phi_1^2 + \phi_2^2$, and the first off-diagonals have common element $\phi_1(\phi_2 - 1)$, while the second off-diagonals have common element $-\phi_2$. The conditional density of Y given ϕ_1, ϕ_2, and τ is

$$f(y|\phi_1,\phi_2,\tau) \propto \tau^{n/2} |B(\phi_1,\phi_2)|^{-1/2} \exp{-\frac{\tau}{2} y'B^{-1}(\phi_1,\phi_2)y},$$

$$y \in R^n, \quad (5.30)$$

where

$$B(\phi_1,\phi_2) = A(\phi_1,\phi_2)A'(\phi_1,\phi_2).$$

Suppose $(\phi_1,\phi_2) \in R^2$ and let $p(\phi_1,\phi_2)$ be an arbitrary proper density on R^2 and suppose τ is independent of (ϕ_1,ϕ_2) and has a gamma prior distribution with parameters α and β. Of course, the posterior density of the parameters is the product of the likelihood function (5.30) and the prior density $p(\phi_1,\phi_2)\tau^{\alpha-1} e^{-\tau\beta}$, $\tau > 0$, $(\phi_1,\phi_2) \in R^2$, but can $B^{-1}(\phi_1,\phi_2)$ be expressed in closed form, as was the case for the first order model? If so, the posterior analysis for the second order model is a straightforward generalization of that of the first order model, but in any case, the marginal posterior density of (ϕ_1,ϕ_2) is

$$p(\phi_1,\phi_2|y) \propto \frac{|B(\phi_1,\phi_2)|^{-1/2} p(\phi_1,\phi_2)}{[2\beta + y'B^{-1}(\phi_1,\phi_2)y]^{(n+2\alpha)/2}}, \quad (\phi_1,\phi_2) \in R^2.$$

$$(5.31)$$

MOVING AVERAGE MODELS

If $B^{-1}(\phi_1, \phi_2)$ cannot be expressed in closed form, this density is difficult to work with, because the inverse and determinant of $B(\phi_1, \phi_2)$ must be evaluated for each pair (ϕ_1, ϕ_2) of R^2. On the other hand, if $B(\phi_1, \phi_2)$ has known eigenvalues and its eigenvectors are not functions of ϕ_1 and ϕ_2 (as was the case with the first order model), then one may plot the density (5.31) and compute its moments relatively easily.

The marginal density of τ must be plotted numerically by a series of bivariate numerical integrations of the joint density

$$p(\phi_1, \phi_2, \tau | y) \propto \tau^{(n+2\alpha)/2 - 1} p(\phi_1, \phi_2) |B(\phi_1, \phi_2)|^{-1/2}$$

$$\times \exp \frac{\tau}{2} [2\beta + y'B^{-1}(\phi_1, \phi_2)y] \qquad (5.32)$$

with respect to (ϕ_1, ϕ_2) for each value of $\tau > 0$. The marginal moments of τ may be computed directly from the marginal density of τ or indirectly by noting the conditional density of τ, given (ϕ_1, ϕ_2) is gamma with parameters $(n + 2\alpha)/2$ and $[2\beta + y'B^{-1}(\phi_1, \phi_2)y]/2$, thus the conditional mean of τ^{-1} is

$$E(\tau^{-1} | \phi_1, \phi_2, y) = \frac{2\beta + y'B^{-1}(\phi_1, \phi_2)y}{n + 2\alpha - 2} \qquad (5.33)$$

The marginal mean of τ^{-1} is

$$E(\tau^{-1} | y) = E_{\phi_1, \phi_2} E(\tau^{-1} | \phi_1, \phi_2, y), \qquad (5.34)$$

where E_{ϕ_1, ϕ_2} denotes expectation with respect to (ϕ_1, ϕ_2), with density (5.31), and the other marginal moments of τ^{-1} may be computed in a similar way.

We have seen that the moving average model presents computational and theoretical problems that do not arise with the autoregressive model. The marginal posterior distributions of the parameters of an autoregressive model are standard distributions, namely the multivariate t for the autoregressive parameters and a gamma for the precision of the white noise. Also, the predictive distribution for a future observation is a univariate t distribution.

With regard to the moving average model, the marginal posterior distributions are not standard; however, the conditional posterior distributions of the noise precision given the moving average parameters are a gamma. Although the marginal distribution of the moving average parameters is not standard, it may be plotted numerically and the moments completed. The marginal moments of the noise precision are computed from the conditional moments by averaging the latter over the marginal posterior distribution of the moving average parameters.

AUTOREGRESSIVE MOVING AVERAGE MODELS

The AR(p) and MA(q) processes may be coupled to form an ARMA(p,q) model,

$$\theta(B)Y_t = \Phi(B)\varepsilon_t, \quad t = 1, 2, \ldots, \qquad (5.35)$$

where

$$\theta(B) = 1 - \theta_1 B - \cdots - \theta_p B^p,$$

and

$$\Phi(B) = 1 - \phi_1 B - \cdots - \phi_q B^q$$

The observations are Y_t, the ε_t are white noise with precision $\tau > 0$, the autoregressive coefficients are $\theta_1, \theta_2, \ldots, \theta_p$, and $\phi_1, \phi_2, \ldots, \phi_q$ are the moving average parameters. The parameter space is $\Omega = R^p \times R^q \times (0, \infty)$, and the process is stationary if the roots of $\theta(B) = 0$ are outside the unit circle, while if all the roots of $\Phi(B) = 0$ are outside the unit circle, the process is invertible. In particular an ARMA(1,1) model is given by

$$Y_t - \theta_1 Y_{t-1} = \varepsilon_t - \phi_1 \varepsilon_{t-1}, \quad t = 1, 2, \ldots, \qquad (5.36)$$

and is stationary and invertible if $|\phi_1| < 1$ and $|\theta_1| < 1$. If $\phi_1 = 0$, the process is AR(1) and is MA(1) if $\theta_1 = 0$.

Before proceeding with the posterior analysis of an ARMA(1,1) let us derive the likelihood function of the three parameters ϕ_1, θ_1, and τ, but not assuming stationarity or invertibility, and letting Y_0 be a known constant.

AUTOREGRESSIVE MOVING AVERAGE MODELS

First let

$$z_t = \varepsilon_t - \phi_1 \varepsilon_{t-1}, \quad t = 1, 2, \ldots, n$$

where $z_t = Y_t - \theta_1 Y_{t-1}$, then first the joint density of $z_1, z_2, \ldots, z_n$ will be determined and then the joint density of $Y_1, \ldots, Y_n$ will be derived.

The density of $z = (z_1, z_2, \ldots, z_n)'$ is that of the observations of a MA(1) process, and has been derived in the previous section. To recapitulate, the density of z is (5.19), namely

$$f(z|\phi_1, \tau) \propto \tau^{n/2} \left[\prod_1^n d_i(\phi_1) \right]^{-1/2} \exp -\frac{\tau}{2} z'QD^{-1}(\phi_1)Q'z,$$

$$z \in R^n \quad (5.37)$$

where $D(\phi_1)$ is a $n \times n$ diagonal matrix, the elements of which are the eigenvalues of $A(\phi_1)A'(\phi_1)$ of (5.17). The eigenvalues of this matrix are

$$d_i(\phi_1) = 1 + \phi_1^2 - 2\phi_1 \cos i\pi (n+1)^{-1}, \quad i = 1, 2, \ldots, n,$$

and the eigenvectors of the matrix are the columns of the $n \times n$ matrix Q, which has ij-th element $\sqrt{2/(n+1)} \sin ij (n+1)^{-1}$ and is independent of ϕ_1. Let $Y = (Y_1, \ldots, Y_n)'$ be the vector of observations, then since $Z = Y - \theta_1 Y^*(Y)$, where $Y^*(Y) = (Y_0, \ldots, Y_{n-1})$, the density of Y is

$$g(y|\theta_1, \phi_1, \tau) \propto \tau^{n/2} \left[\prod_1^n d_i(\phi_1) \right]^{-1/2} \exp -\frac{\tau}{2} [y - \theta_1 y^*(y)]'QD^{-1}$$

$$\times (\phi_1)Q'[y - \theta_1 y^*(y)] \quad (5.38)$$

where $y^*(y) = (y_0, \ldots, y_{n-1}) \in R^n$ and $y \in R^n$, $\theta_1 \in R$, $\phi_1 \in R$, and $\tau > 0$. This can be simplified by letting $\hat{y} = Q'y$ and $\tilde{y} = Q'y^*(y)$, thus the likelihood function for θ_1, ϕ_1, and τ is

$$L(\theta_1,\phi_1,\tau|y) \propto \tau^{n/2}\left[\prod_1^n d_i(\phi_1)\right]^{-1/2} \exp -\frac{\tau}{2}(\hat{y} - \theta_1\tilde{y})'D^{-1}$$

$$\times (\phi_1)(\hat{y} - \theta_1\tilde{y}), \qquad (5.39)$$

for $\theta_1 \in R$, $\phi_1 \in R$, and $\tau > 0$. The form of the likelihood function suggests the following will be a convenient form for the prior distribution of θ_1, ϕ_1, and τ. Suppose the prior density is

$$p(\theta_1,\phi_1,\tau) \propto p(\theta_1|\tau)p(\tau)p(\phi_1), \qquad (\theta_1,\phi_1) \in R^2, \quad \tau > 0,$$

(5.40)

where

$$p(\theta_1|\tau) \propto \tau^{1/2}\exp -\frac{\tau\xi}{2}(\theta_1 - \mu)^2, \qquad \theta \in R,$$

$$p(\tau) \propto \tau^{\alpha-1} e^{-\tau\beta}, \qquad \tau > 0,$$

and $p(\phi_1)$ is a proper prior density over R and $\mu \in R$, $\alpha > 0$, $\beta > 0$, $\xi > 0$ are prior parameters.

By Bayes' theorem, the marginal posterior density of ϕ_1 is

$$p(\phi_1|y) \propto \frac{p(\phi_1)\prod_1^n d_i^{-1/2}(\phi_1)}{A^{1/2}(\phi_1)\{C(\phi_1) - B^2(\phi_1)A^{-1}(\phi_1)\}^{(n+2\alpha+1)/2}}, \quad \phi_1 \in R$$

(5.41)

where

$$A(\phi_1) = y^{*'}(y)D^{-1}(\phi_1)y^*(y) + \xi,$$

(5.42)

$$B(\phi_1) = \hat{y}'D^{-1}(\phi_1)y^*(y) + \xi\mu,$$

and

$$C(\phi_1) = \hat{y}'D^{-1}(\phi_1)\hat{y} + 2\beta + \xi\mu^2.$$

Furthermore, the conditional density of θ_1 given ϕ_1 is

$$p(\theta_1|\phi_1,y) \propto \frac{1}{\{[\theta_1 - A^{-1}(\phi_1)B(\phi_1)]^2 A(\phi_1) + C(\phi_1) - B^2(\phi_1)A^{-1}(\phi_1)\}^{(n+2\alpha+1)/2}},$$

$$\theta_1 \in R \quad (5.43)$$

where $A(\phi_1)$, $B(\phi_1)$, and $C(\phi_1)$ are defined as above, and it is seen the conditional distribution of θ_1 given ϕ_1 is a t with $n + 2\alpha$ degrees of freedom, location $A^{-1}(\phi_1)B(\phi_1)$, and precision $A(\phi_1)(n + 2\alpha)[C(\phi_1) - B^2(\phi_1)A^{-1}(\phi_1)]$.

In addition, the conditional distribution of τ, given ϕ_1 is gamma with parameters $(n + 2\alpha)/2$ and $2^{-1}[C(\phi_1) - B^2(\phi_1)A^{-1}(\phi_1)]$.

Thus, the marginal density of ϕ_1 is not a standard density, nevertheless it should be easy to plot and to compute the marginal moments of ϕ_1. Of course, the marginal distributions of τ and θ_1 are also not standard, however the conditional of θ_1 given ϕ_1 and the conditional of τ given ϕ_1 are well-known distributions.

The marginal moments of τ and ϕ_1 may be computed from their conditional moments. For example, the marginal posterior mean of θ_1 is

$$E(\theta_1|y) = E_{\phi_1}\{E(\theta_1|\phi_1,y)\}, \quad (5.44)$$

where

$$E(\theta_1|\phi_1,y) = A^{-1}(\phi_1)B(\phi_1).$$

The marginal mean of τ^{-1} is

$$E(\tau^{-1}|y) = E_{\phi_1} E(\tau^{-1}|\phi_1,y), \quad (5.45)$$

where

$$E(\tau^{-1}|\phi_1, y) = \frac{C(\phi_1) - B^2(\phi_1)A^{-1}(\phi_1)}{n + 2\alpha - 2}.$$

The marginal posterior variances (provided they exist) of τ and θ_1 can be computed in a similar fashion.

Suppose one wants point estimates of θ_1, ϕ_1, and τ, then one could compute the marginal means of θ_1 and τ, via (5.44) and (5.45) respectively, but first one would need to compute the marginal posterior density of ϕ_1, from (5.41), then average the conditional means with respect to the marginal posterior density of ϕ_1. The marginal posterior variances of θ_1 and τ will allow one to express their confidence in the point estimates of these parameters.

The marginal posterior densities of θ_1 and τ cannot be expressed in closed form, thus one-dimensional numerical integration is necessary, in order to find them, but should not be too much of a problem. If the marginal density of τ, say, is found in this manner, additional numerical integrations will provide the marginal posterior mean and variance and this would be an alternative to finding these moments by averaging their conditional mean and variance.

The preceding derivation did not assume stationarity, $|\theta_1| < 1$, and invertibility, $|\phi_1| < 1$. How should the Bayesian approach this problem? Recall the prior density of θ_1, ϕ_1, and τ, then it may be shown that the marginal prior density of θ_1 is

$$p(\theta_1) \propto \frac{1}{[2\beta + \xi(\theta - \mu)^2]^{(2\alpha+1)/2}}, \quad \theta_1 \in R$$

where α, β, ξ, and μ are hyperparameters, and they may be chosen so that $100\gamma\%$, $0 < \gamma < 1$, of the marginal prior density of θ_1 is concentrated on $(-1, 1)$, the interval of stationarity. One disadvantage of this way is that the α and β parameters are also the parameters of the marginal prior distribution of τ, hence one must compromise one's prior information between these two distributions. The prior density of ϕ_1 should be concentrated over $(-1, +1)$, the region of invertibility, for example with a uniform distribution over this interval. Monahan (1981) develops an alternative solution for stationary ARMA models.

The ARMA(1,1) model has been explored but will the above derivation generalize to higher order models? Consider the general ARMA(p,q) model, (5.35).

AUTOREGRESSIVE MOVING AVERAGE MODELS

It seems reasonable that the above results can be extended to the ARMA(p,q) model if one does not assume stationarity and invertibility and if one uses the prior density $p(\theta|\tau)p(\tau)p(\phi)$, where $p(\theta|\tau) \propto \tau^{p/2} \exp - \tau/2(\theta-\mu)'P(\theta-\mu)$, $\theta \in R^p$, $p(\tau) \propto \tau^{\alpha-1} e^{-\tau\beta}$, $\tau > 0$, and $p(\phi)$ is a proper density over R^q, where $\theta = (\theta_1, \theta_2, \ldots, \theta_p)'$ is the vector of autoregressive parameters and ϕ the vector of the q moving average parameters.

If these assumptions are satisfied, it is conjectured the conditional distribution of τ given ϕ is a gamma, the conditional posterior distribution of θ given ϕ is a p-variate multiple t distribution, and the marginal posterior distribution of ϕ, although not a standard density, may be explicitly determined in closed form. If this is true, the marginal posterior moments of τ and θ can be computed from the conditional posterior moments of these parameters.

The section on ARMA processes is concluded with the Bayesian procedure for forecasting future observations of an ARMA(1,1) model, thus consider the one-step-ahead forecast of Y_{n+1}. The Bayesian predictive density of Y_{n+1} will be found by deriving the joint marginal density of $Y = (Y_1, Y_2, \ldots, Y_n)$ and Y_{n+1}, then identifying the conditional density of Y_{n+1} given $Y = y$. First consider the joint conditional density of Y and Y_{n+1} given θ_1, ϕ_1, and τ, which is (5.38) with the following changes. Let

$$y^{(1)} = \begin{pmatrix} y \\ y_{n+1} \end{pmatrix}$$

and

$$y^{(2)} = \begin{pmatrix} y^*(y) \\ y_n \end{pmatrix}$$

and replace y by $y^{(1)}$ and $y^*(y)$ by $y^{(2)}$ in (5.38), then the joint density of Y and Y_{n+1} given θ_1, ϕ_1, and τ is

$$g(y, y_{n+1}|\theta_1, \phi_1, \tau) \propto \tau^{(n+1)/2} \left[\prod_1^{n+1} d_i(\phi_1)\right]^{-1/2}$$

$$\times \exp -\frac{\tau}{2}[y^{(1)} - \theta_1 y^{(2)}]'R(\phi_1)$$

$$\times [y^{(1)} - \theta_1 y^{(2)}], \qquad (5.46)$$

where $y \in R^n$, $y_{n+1} \in R$, and

$$R(\phi_1) = QD^{-1}(\phi_1)Q'$$

but now $R(\phi_1)$ is $(n+1) \times (n+1)$ and Q and $D(\phi_1)$ are defined as in (5.38) but with $i = 1, 2, \ldots, n+1$. Now multiplying the above density by the prior density and completing the square in θ_1 in the exponent, and integrating the result with respect to τ and θ_1 gives the joint density of Y and Y_{n+1}, given ϕ_1, namely

$$g(y, y_{n+1} | \phi_1) \propto \frac{d_{n+1}^{-1/2}(\phi_1) p(\phi_1)}{\{[y_{n+1} - A^{-1}(\phi_1) B(\phi_1)]^2 A(\phi_1) + C(\phi_1) - B^2(\phi_1) A^{-1}(\phi_1)\}^{(n+2\alpha+1)/2}}$$

$$(5.47)$$

for $y \in R^n$ and $y_{n+1} \in R$, where

$$A(\phi_1) = r_{22}(\phi_1) - [R_{21}(\phi_1)y^* + r_{22}(\phi_1)y_n]^2[\xi + y'^{(2)}R(\phi_1)y^{(2)}],$$

$$B(\phi_1) = [y'R_{11}(\phi_1)y^*(y) + y'R_{12}(\phi_1)y_n + \xi\mu]$$

$$\times [R_{21}(\phi_1)y^*(y) + r_{22}(\phi_1)y_n][\xi + y'^{(2)}R(\phi_1)y^{(2)}]^{-1}$$

$$- y'R_{12}(\phi_1),$$

and

$$C(\phi_1) = \mu'\xi\mu + 2\beta + y'R_{11}(\phi_1)y - [y'R_{11}(\phi_1)y^* + y'R_{12}(\phi_1)y_n$$

$$+ \xi\mu]^2[\xi + y'^{(2)}R(\phi_1)y^{(2)}]^{-1},$$

where the $(n + 1) \times (n + 1)$ matrix $R(\phi_1)$ is partitioned as

$$R(\phi_1) = \begin{pmatrix} R_{11}(\phi_1) & R_{12}(\phi_1) \\ R_{21}(\phi_1) & r_{22}(\phi_1) \end{pmatrix},$$

and $R_{11}(\phi_1)$ is of order n.

From (5.47), one may conclude the conditional density of y_{n+1} given y and ϕ_1 is a t distribution with $n + 2\alpha$ degrees of freedom, location $A^{-1}(\phi_1)B(\phi_1)$ and precision $A(\phi_1)(n + 2\alpha)[C(\phi_1) - B^2(\phi_1) \times A^{-1}(\phi_1)]^{-1}$.

Suppose the mean of the predictive distribution of Y_{n+1} is used as a point forecast of the future observation, then since

$$E(Y_{n+1}|y,\phi_1) = A^{-1}(\phi_1)B(\phi_1) \qquad (5.48)$$

is the conditional mean of Y_{n+1} given ϕ_1 and y, this conditional moment can be averaged with respect to the marginal posterior density of ϕ_1, given by (5.41).

It would be interesting to see if the forecasting procedure of the ARMA(1,1) model can be extended to the general ARMA(p,q) model.

THE IDENTIFICATION PROBLEM

In the preceding Bayesian analysis of ARMA models the orders p and q of the autoregressive and moving average components of the model were assumed known; however, in practice one does not usually know these values and they must be identified or estimated. Box and Jenkins (1970) identify the order of an ARMA process from the sample autocorrelation and partial autocorrelation functions, while the method of Akaike (1979) is popular in estimating the order of an AR(p) model.

A purely Bayesian procedure to identify the order of an ARMA(p,q) process has yet to be examined; however Diaz and Farah (1981) have derived the exact posterior mass function of the order p of an autoregressive process and their derivation will be given.

Recall that for an AR(p) process with n observations $y = (y_1, y_2, \ldots, y_n)$ and assuming the initial observations $y_0, y_{-1}, \ldots, y_{1-p}$ are known, the likelihood function is

$$L[\theta^{(p)}, p, \tau | y] \propto \tau^{n/2} \exp - \frac{\tau}{2} \sum_{t=1}^{n} \left(y_t - \sum_{i=1}^{p} \theta_{pi} y_{t-i} \right)^2,$$

where $\theta^{(p)} = (\theta_{p1}, \ldots, \theta_{pp})' \in R^p$, $\tau > 0$, and $p = 1, 2, \ldots, k$, where k is the largest possible order of the process. The likelihood may be expressed as

$$L(\theta^{(p)}, p, \tau | y) \propto \tau^{n/2} \exp - \frac{\tau}{2} [\theta^{(p)'} A(p,y) \theta^{(p)}$$

$$- 2\theta^{(p)'} B(p,y) + C(y)], \qquad (5.49)$$

where $A(p,y)$ is $p \times p$ with i-th diagonal element $\Sigma_{t=1}^{n} y_{t-i}^2$ and ij-th off-diagonal element $\Sigma_{t=1}^{n} y_{t-i} y_{t-j}$, $B(p,y)$ is a $p \times 1$ vector with i-th component $\Sigma_{t=1}^{n} y_{t-i}$, and $C(y) = \Sigma_{i=1}^{n} y_i^2$.

The form of the likelihood suggests letting $\theta^{(p)}$ have a conditional normal distribution, a priori, given p and τ with mean vector $\mu^{(p)}$ and precision matrix Q_p. Suppose τ and p are independent, where τ is gamma with parameters α and β, and p is uniform over the integers 1, 2, ..., k. Thus, the conditional prior density of $\theta^{(p)}$ given p and τ is

$$g[\theta^{(p)} | p, \tau] \propto \frac{\tau^{p/2}}{(2\pi)^{p/2}} \exp - \frac{\tau}{2} [\theta^{(p)} - \mu^{(p)}]' Q_p [\theta^{(p)} - \mu^{(p)}],$$

$$\theta^{(p)} \in R^p, \qquad (5.50)$$

the marginal prior density of τ is

$$g(\tau) \propto \tau^{\alpha-1} e^{-\tau\beta}, \quad \tau > 0, \qquad (5.51)$$

and the marginal prior mass function of p is $g(p) = k^{-1}$, $p = 1, 2, \ldots, k$.

Now, combining the prior of $\theta^{(p)}$, p, and τ with the likelihood function,

$$h[\theta^{(p)}, p, \tau | y] \propto \frac{\tau^{(n+2\alpha+p)/2-1}}{(2\pi)^{p/2}} \exp - \frac{\tau}{2} \{[\theta^{(p)} - A*^{-1}(p,y) B*$$

$$\times (p,y)]'A^*(p,y)[\theta^{(p)} - A^{*-1}(p,y)B^*(p,y)]$$

$$+ C^*(p,y) - B^{*\prime}(p,y)A^{-1}*(p,y)B^*(p,y)\}$$

(5.52)

is the joint posterior density of $\theta^{(p)}$, p, and τ, where

$$A^*(p,y) = Q_p + A(p,y),$$

$$B^*(p,y) = B(p,y) + Q_p\mu^{(p)},$$

and

$$C^*(p,y) = C(y) + 2\beta + \mu'^{(p)}Q_p\mu^{(p)}.$$

Eliminating first $\theta^{(p)}$, then τ from (5.52) gives

$$h(p|y) \propto \frac{|A^*(p,y)|^{-1/2}}{|Q_p|^{-1/2}} [C^*(p,y) - B'^*(p,y)A^{*-1}(p,y)$$

$$\times B^*(p,y)]^{-(n+2\alpha)/2} \qquad (5.53)$$

p = 1,2, ..., k, as the marginal posterior mass function of p.

This identification procedure assigns a posterior probability to each of the k autoregressive models

$$y_t - \sum_{i=1}^{p} \theta_{pi} y_{t-i} = \varepsilon_t, \quad p = 1,2, \ldots, k.$$

where θ_{pi} is the coefficient of the i-th lagged value of the dependent variable in the p-th model. Formula (5.53) allows one to compute a plot of the posterior mass function of p over all possible orders of the model.

To identify the model, one should inspect the plot of the posterior mass function of p, and select an order of the model. Notice for each value of p, the inverse and determinant of $A(p,y) + Q_p$ must be computed.

How does one make inferences about the vector $\theta^{(p)}$ of autoregressive parameters? One could identify p from its posterior mass function, then use the conditional posterior desnity of $\theta^{(p)}$ given $p = \hat{p}$, where $\hat{p}$ is the identified value of p.

The conditional posterior density of $\theta^{(p)}$, $h(\theta^{(p)}|p,y)$ given p is a multivariate t with $(n + 2\alpha)$ degrees of freedom, θ^* is the location vector and $(n + 2\alpha)(A(p,y) + Q_p)/s_p$ is the precision parameter, where $\theta^* = A^{-1}*(p,y)B*(p,y)$ and $s_p = C*(p,y) - B'*(p,y)A^{-1}* \times (p,y)B*(p,y)$.

Note, to use the above analyses of an AR(p) model, one must specify a sequence of hyperparameters, namely the means $\mu^{(p)}$ and precisions Q_p of θ for $p = 1, 2, \ldots, k$. The conditional prior density of $\theta^{(p)}$ given p is

$$g[\theta^{(p)}|p] \propto \frac{1}{\{2\beta + [\theta^{(p)} - \mu^{(p)}]'Q_p[\theta^{(p)} - \mu^{(p)}]\}^{(p+2\alpha)/2}},$$

$$\theta^{(p)} \in R^p \quad (5.54)$$

which is a t density with 2α degrees of freedom, location μ_p and dispersion matrix $\beta Q_p^{-1}(\alpha - 1)^{-1}$. Now, how should one choose α, β, $\mu^{(p)}$, and Q_p^{-1}, $p = 1, 2, \ldots, k$? The parameters α and β control the prior gamma distribution of the noise variance and the prior precision of all the conditional distributions of θ given $p = 1, 2, \ldots, k$.

One way advocated by Litterman (1980) is to let the first component of μ_p always be one and the others zero and let the prior variances of θ_{11}, θ_{12}, θ_{1p} decrease, e.g., let the i-th component of the dispersion matrix of $\theta^{(p)}$ given p be $\beta(\alpha - 1)^{-1}i^{-1}$, $i = 1, 2, \ldots, p$. This is equivalent to setting the i-th diagonal element of Q_p^{-1} at i^{-1}. The off-diagonal elements of Q_p^{-1} control the prior covariance between the p components of $\theta^{(p)}$.

The above analysis selects the order with the largest posterior probability then inferences about the autoregressive parameters are based on the conditional posterior distributions of $\theta^{(p)}$ given p.

Is it possible to extend Bayesian identification to moving average and autoregressive moving average models? It appears that it is possible at least numerically to develop an identification analysis for moving average models, but one should expect analytical difficulties in the posterior analysis.

OTHER TIME SERIES MODELS

The previous sections of this chapter have been devoted to moving average, autoregressive, and autoregressive moving average models. The regression model with autocorrelated errors and the distributed lag models will now be considered.

Regression models with autocorrelated errors are often used with time series data and they are simple generalizations of the regression model with independent errors, thus suppose the observations Y_t are given by

$$Y_t = \beta x_t + a_t \qquad (5.55)$$

and

$$a_t = \rho a_{t-1} + \varepsilon_t, \qquad t = 1, 2, \ldots, n,$$

where β is a real scalar unknown parameter, ρ, the autocorrelation parameter is a real unknown scalar parameter, and the ε_t are independent and identically distributed $n(0, \tau^{-1})$ random variables with precision parameter $\tau > 0$. The process is explosive if $|\rho| > 1$, otherwise it is called a nonexplosive process. If $|\rho| < 1$, the correlation between adjacent observations is ρ, i.e., the correlation between Y_t and Y_{t-1} is ρ, and in general the correlation between Y_t and Y_{t-k} is ρ^k, thus in the non-explosive case the autocorrelation function becomes zero in a geometric fashion as the lag increases.

The model may be expressed as

$$Y_t = \rho Y_{t-1} + \beta(X_t - \rho X_{t-1}) + \varepsilon_t,$$

where Y_0 is a known initial value. Since the ε_t are n.i.d. $(0, \tau^{-1})$, the joint density of $Y_1, \ldots, Y_n$, conditional on the parameters is

$$f(y|\tau, \beta, \rho) \propto \tau^{n/2} \exp - \frac{\tau}{2} \sum_{t=1}^{n} [y_t - \rho y_{t-1} - \beta(x_t - \rho x_{t-1})]^2,$$

$$y \in R^n, \quad (5.56)$$

where $y = (y_1, y_2, \ldots, y_n)$, $\tau > 0$, $\beta \in R$, and $\rho \in R$.

Zellner (1971) analyzes this model with a vague improper prior distribution, however, our analysis will use a proper prior density for β, ρ, and τ.

Suppose (β,τ) is independent of ρ, and that (β,τ) has a normal-gamma distribution with density

$$p(\beta,\tau) \propto \tau^{\alpha-1} e^{-\tau\gamma} \tau^{1/2} \exp{-\frac{\tau}{2}(\beta-\mu)'\xi(\beta-\mu)},$$

$$\beta \in R, \quad \tau > 0, \quad (5.57)$$

where $\alpha > 0$, $\gamma > 0$, $\mu \in R$, and $\xi > 0$.

Let ρ have a constant density over R, then combining the likelihood function with the prior density gives

$$p(\beta,\rho,\tau|y) \propto \tau^{(1+n+2\alpha)/2} \exp{-\frac{\tau}{2}\left\{2\gamma + \sum_{t=1}^{t=n}[y_t - \rho y_{t-1} - \beta(x_t - \rho x_{t-1})]^2\right\}}, \quad (5.58)$$

where $\beta \in R$, $\rho \in R$, and $\tau > 0$, as the posterior density of the parameters. The addition of ρ as an extra parameter to the usual regression model adds an extra complication to the posterior analysis.

The marginal densities of ρ and β are easily derived by using the properties of the gamma and univariate t densities. For example the marginal density of ρ is

$$p(\rho|y) \propto \frac{1}{A_1^{1/2}(\rho)[A_3(\rho) - A_2^2(\rho)A_1^{-1}(\rho)]^{(n+2\alpha)/2}}, \quad \rho \in R,$$

$$(5.59)$$

where

$$A_1(\rho) = \sum_1^n (x_t - \rho x_{t-1})^2,$$

$$A_2(\rho) = \sum_1^n (y_t - \rho y_{t-1})(x_t - \rho x_{t-1}),$$

and

$$A_3(\rho) = 2\gamma + \sum_1^n (y_t - \rho y_{t-1})^2,$$

OTHER TIME SERIES MODELS 209

thus the marginal density of ρ is not a standard density, nevertheless the density may be easily plotted and the posterior moments calculated.

In a similar fashion, the marginal density of β is

$$p(\beta|y) \propto \frac{1}{A_1^{1/2}(\beta)[A_3(\beta) - A_2^2(\beta)A_1^{-1}(\beta)]^{(n+2\alpha)/2}}, \quad \beta \in R,$$

(5.60)

where

$$A_1(\beta) = \sum_1^n (y_{t-1} - \beta x_{t-1})^2,$$

$$A_2(\beta) = \sum_1^n (y_{t-1} - \beta x_{t-1})(y_t - \beta x_t)$$

and

$$A_3(\beta) = 2\gamma + \sum_1^n (y_t - \beta x_t)^2.$$

The marginal density of β is nonstandard and must be determined numerically, however since it is a univariate function its moments are easily calculated by numerical integration.

The marginal density of the precision τ cannot be expressed in closed form but must be found numerically by integrating the joint density of (τ, β) or (τ, ρ) with respect to β and ρ, respectively, however one may compute the marginal posterior moments of τ from either the conditional moments of τ given β or the conditional distribution of τ given ρ.

Consider the conditional density of τ given ρ by integrating the joint density (5.58) with respect to β, then as a result the conditional distribution of τ given ρ is a gamma distribution with parameters $(n + 2\alpha)/2$ and $[A_3(\rho) - A_2^2(\rho)A_1^{-1}(\rho)]/2$, where A_1, A_2, and A_3 are functions of ρ given by (5.59).

The conditional mean of τ^{-1} given ρ is

$$E(\tau^{-1}|\rho,y) = \frac{A_3(\rho) - A_2(\rho)A_1^{-1}(\rho)}{n + 2\alpha - 2},$$

and the marginal mean of τ^{-1} is

$$E(\tau^{-1}|y) = E_\rho[E(\tau^{-1}|\rho,y)], \qquad (5.61)$$

where the expectation E_ρ is with respect to the marginal density of ρ, (5.59).

The conditional variance of τ^{-1} given ρ is

$$\mathrm{Var}(\tau^{-1}|\rho,y) = E^2(\tau^{-1}|\rho,y)/(n + 2\alpha - 4), \qquad (5.62)$$

hence the marginal variance of τ^{-1} may be computed.

The marginal mean and variance of β and ρ can be computed from the marginal density functions (5.60) and (5.59), respectively, but suppose the posterior mode of all the parameters is needed. How is the overall mode found?

Recall the method of Lindley and Smith (1972) for finding the mode of the parameters of the linear mixed model, where the conditional distribution of each parameter given the others is found in order to set up the modal equations (4.24) through (4.27) of Chapter 4. From the posterior density (5.58), the conditional distributions of β, ρ, and τ given the other parameters are easily identified.

The conditional distribution of τ given β and ρ is gamma with parameters $\alpha^* = (n + 2\alpha + 1)/2$ and $\beta^*(\rho,\beta) = [2\gamma + \Sigma(y_t - \rho y_{t-1})^2 + \beta^2\Sigma(x_t - \rho x_{t-1})^2 - 2\beta\Sigma(y_t - \beta y_{t-1})(x_t - \rho x_{t-1})]/2$, the conditional distribution of β given τ and ρ is normal with mean $A_1^{-1}(\rho)A_2(\rho)$, where A_1 and A_2 are given by (5.59), and precision $\tau A_1(\rho)[A_3(\rho) - A_1^{-1}(\rho)A_2(\rho)]^{-1}$, while a normal distribution with mean $A_1^{-1}(\beta)$. $A_2(\beta)$ and precision $\tau A_1(\beta)[A_3(\beta) - A_1^{-1}(\beta)A_2(\beta)]^{-1}$ is the conditional distribution of ρ given β and τ.

It follows that the three conditional modes are

$$M(\tau^{-1}|y,\beta,\rho) = \beta^*(\rho,\beta)/(\alpha^* - 1),$$

$$M(\beta|y,\tau,\rho) = A_1^{-1}(\rho)A_2(\rho), \qquad (5.63)$$

$$M(\rho|y,\beta,\tau) = A_1^{-1}(\beta)A_2(\beta).$$

OTHER TIME SERIES MODELS 211

Beginning with some initial starting value of β, say the usual least squares estimator

$$_0\beta = (X'X)^{-1}X'Y,$$

where X is the n × 1 vector with x_i as the i-th component, the third equation of the system (5.63) may be solved for ρ, say $_0\rho$, then if $_0\beta$ and $_0\rho$ are substituted into the first equation, one may solve for the conditional mode of τ^{-1}, namely $\beta^*(_0\rho,_0\beta)/[(n+2\alpha+1)/2)-1]$. If $_0\rho$ is substituted into the second equation, one has a new value for β, say $_1\beta$. Continuing in this fashion, until the estimates stabilize, gives an estimate of the posterior mode of (τ^{-1},β,ρ). Of course the mode of the joint posterior density does not have to be the modes of the one-dimensional marginal densities. Thus, in addition to computing the marginal means and variances of each of the three parameters one may compute the overall mode.

Now consider forecasting a future observation given a sample y from a regression model with autocorrelated errors and employing the same prior distribution of the preceding posterior analysis. Suppose initial values x_0 and y_0 are known and that $w_1 = y_{n+1}$ is the observation to be predicted and that the future value x_{n+1} of the regressor variable is known, then what is the predictive density of w_1 given y?

First consider the joint density of y and w_1 given τ, β, and ρ, namely

$$f(y,w_1|\tau,\beta,\rho) \propto \tau^{(n+1)/2} \exp -\frac{\tau}{2}\left\{\sum_1^n [(y_t - \rho y_{t-1}) - \beta(x_t - \rho x_{t-1})]^2 + [(w_1 - \rho y_n) - \beta(x_{n+1} - \rho x_n)]^2\right\}, \quad (5.64)$$

where $y \in R^n$, $w_1 \in R$, $\tau > 0$, $\beta \in R$, and $\rho \in R$. Now if this is multiplied by the prior density for (β,ρ,τ), the product is the joint density of y, w_1, and (β,ρ,τ), then if the product is integrated with respect to β, ρ, and τ, the result is the joint marginal density of y and w_1, from which the predictive density of w_1 can be identified.

If the product of (5.64) and the prior density of (β,ρ,τ) is integrated with respect to τ and β, the result is the joint marginal density of y, w_1, and ρ, which is

$$f(y,w_1,\rho) \propto \frac{1}{\{C_3(\rho) - C_2^2(\rho)C_1^{-1}(\rho)\}^{(n+2\alpha+1)/2} C_1^{1/2}(\rho)}, \quad (5.65)$$

where $y \in R^n$, $w_1 \in R$, and $\rho \in R$, where

$$C_1(\rho) = \sum_1^{n+1} (x_t - x_{t-1})^2,$$

$$C_2(\rho) = \sum_1^n (y_t - \rho y_{t-1})(x_t - \rho x_{t-1}) + (w_1 - \rho y_n)(x_{n+1} - \rho x_n),$$

and

$$C_3(\rho) = 2\gamma + \sum_1^n (y_t - \rho y_{t-1})^2 + (w_1 - \rho y_n)^2.$$

Unfortunately, this density cannot be integrated with respect to ρ, however, the conditional distribution of w_1 given ρ and y can be identified. By completing the square in $C_3(\rho) - C_2^2(\rho)C_1^{-1}(\rho)$ with respect to w_1, one may identify the conditional distribution of w_1 given y and ρ as a t-distribution with $n + 2\alpha$ degrees of freedom, location $D_1^{-1}(\rho)D_2(\rho)$ and precision $D_1(\rho)(n + 2\alpha)[D_3(\rho) - D_2^2(\rho)D_1^{-1}(\rho)]^{-1}$ where

$$D_1(\rho) = 1 - C_1^{-1}(\rho)(x_{n+1} - \rho x_n),$$

$$D_2(\rho) = \rho y_n + C_1^{-1}(\rho)\rho y_n(x_{n+1} - \rho x_n)^2$$

$$+ (x_{n+1} - \rho x_n)C_1^{-1}(\rho) \sum_1^n (y_t - \rho y_{t-1})(x_t - \rho x_{t-1}),$$

and

$$D_3(\rho) = 2\gamma + \sum_1^n (y_t - \rho y_{t-1})^2 + \rho^2 y_n^2 - C_1^{-1}(\rho)$$

$$\times \rho^2 y_n^2 (x_{n+1} - \rho x_n)^2 - C_1^{-1}(\rho) \sum_1^n (y_t - \rho y_{t-1})$$

$$\times (x_t - \rho x_{t-1}) + 2C_1^{-1}(\rho) \rho y_n (x_{n+1} - \rho x_n)$$

$$\times \sum_1^n (y_t - \rho y_{t-1})(x_t - \rho x_{t-1}).$$

One may plot the predictive density of w_1 by integrating (5.65) numerically with respect to ρ, for each value of w_1, then once having tabulated the conditional density of w_1 given y, compute the predictive moments of w_1.

The conditional mean of w_1 given ρ is $E[w_1|\rho,y] = D_1^{-1}(\rho)D_2(\rho)$, which if averaged with respect to the marginal density of ρ, (5.59), will give the conditional mean of w_1, given y, which may be used to forecast a future value of Y_{n+1}. If one computes the conditional variance of Y_{n+1}, given y, one has some idea of the precision of one's forecast of Y_{n+1}.

Zellner (1971) gives a more extensive examination of the regression model with autocorrelated errors and presents a numerical analysis based on some generated data where the regressor variable is rescaled investment expenditure taken from a paper by Haavelmo (1947). He examines explosive and nonexplosive cases by plotting the marginal densities of β and ρ, and the conditional distribution of β given ρ, for various values of ρ to see the effect of the autocorrelation parameter on inferences for β. His results can be obtained if one lets $\alpha \to -1/2$, $\beta \to 0$, and $\xi \to 0$ in the posterior analysis. Also, Zellner generalizes the model to include several regressor variables but does not develop a prediction analysis, as was done here.

The last time series model to be considered is the distributed lag model

$$y_t = \sum_{i=0}^{\infty} \lambda^i \chi_{t-i} + u_t, \qquad t = 1, 2, \ldots, n \qquad (5.66)$$

where y_t is the value of the dependent variable at time t, χ_{t-i} the value of the stimulus at time $t - i$, λ is an unknown real parameter, $0 < \lambda \leq 1$, and the u_t is white noise with precision $\tau > 0$. An equivalent representation of this model is obtained by subtracting λy_{t-1} from both sides to obtain

$$y_t = \lambda y_{t-1} + \alpha \chi_t + u_t - \lambda u_{t-1}, \qquad (5.67)$$

which is an ARMA(1,1) model containing a concomitant variable χ.

It can be shown, see Zellner, Chapter 7, that the dispersion matrix of $Y_1, Y_2, \ldots, Y_n$ is $G(\lambda)$, where $G(\lambda)$ is tridiagonal with $(1 + \lambda^2)\tau^{-1}$ as the principal diagonal elements and $-\lambda \tau^{-1}$ as the super- and subdiagonal elements, and all other elements are zero. This is the same matrix which was encountered with the MA(1) model of this chapter. So as before, let Q be an orthogonal matrix which diagonalizes $G(\lambda)$, i.e., $Q'G(\lambda)Q = D(\lambda)$, where $D(\lambda)$ is diagonal with diagonal elements $1 + \lambda^2 - 2\lambda \cos i\pi (n + 1)^{-1} = d_i(\lambda)$, $i, 2, \ldots, n$ and Q is independent of λ with ij-th element $\sqrt{2/(n+1)} \sin ij\pi (n+1)^{-1}$. Our analysis of the distributed lag model will be quite like the analysis of the ARMA(1,1) process. The likelihood function for τ, λ, and α is

$$\ell(\tau, \lambda, \alpha | y) \propto \tau^{n/2} |G(\lambda)|^{-1/2} \exp -\frac{\tau}{2} (y - \lambda y_{-1} - \alpha \chi)' G^{-1}$$
$$\times (\lambda)(y - \lambda y_{-1} - \alpha \chi), \qquad (5.68)$$

where $\tau > 0$, $0 \leq \lambda \leq 1$, $\alpha \in R$, $y_{-1} = (y_{-1}, y_0, \ldots, y_{n-1})'$, $y = (y_1, y_2, \ldots, y_n)'$, and $\chi = (\chi_1, \chi_2, \ldots, \chi_n)'$. Since $G^{-1}(\lambda) = QD^{-1}(\lambda)Q'$, the likelihood function reduces to

$$\ell(\tau, \lambda, \alpha | y) \propto \tau^{n/2} \prod_{i=1}^{n} d_i^{-1/2}(\lambda) \exp -\frac{\tau}{2} (y - \lambda y_{-1} - \alpha \chi)' QD^{-1}$$
$$\times (\lambda) Q' (y - \lambda y_{-1} - \alpha \chi). \qquad (5.69)$$

Let $y^* = Q'y$, $y^*_{-1} = Q'y_{-1}$ and $\chi^* = Q'\chi$, then

OTHER TIME SERIES MODELS

$$\ell(\tau,\lambda,\alpha|y) \propto \tau^{n/2} \prod_{i=1}^{n} d_i^{-1/2}(\lambda) \exp -\frac{\tau}{2}(y^* - \lambda y^*_{-1} - \alpha\chi^*)'D^{-1}$$

$$\times (\lambda)(y^* - \lambda y^*_{-1} - \alpha\chi^*)$$

gives a convenient form of the likelihood function to combine with the prior density of the parameters.

Suppose, a priori, that (α,τ) is independent of λ and that (α,τ) is normal-gamma with parameters μ, ξ, α, and β and λ has some proper density $g(\lambda)$ over $(0,1]$. The posterior density of the parameters is by Bayes theorem

$$p(\lambda,\tau,\alpha|y) \propto \tau^{(1+n+2\alpha)/2-1} \prod_{1}^{n} d_i^{-1/2}(\lambda) \exp -\frac{\tau}{2}\{2\beta + (\alpha-\mu)^2\xi$$

$$+ (y^* - \lambda y^*_{-1} - \alpha\chi^*)'D^{-1}(\lambda)(y^* - \lambda y^*_{-1} - \alpha\chi^*)],$$

(5.70)

where $0 \leq \lambda \leq 1$, $\tau > 0$, and $\alpha \in R$.

It will be shown that the marginal posterior density of λ can be isolated and determined analytically, and that the conditional posterior distributions of α given λ and τ given λ are t and gamma respectively.

Integrating with respect to α and τ gives

$$p(\lambda|y) \propto \frac{1}{[A(\lambda,y)]^{1/2}} \frac{p(\lambda)}{[C(\lambda,y) - B'(\lambda,y)A^{-1}(\lambda,y)B(\lambda,y)]^{(n+2\alpha)/2}},$$

$$0 \leq \lambda \leq 1 \quad (5.71)$$

for the marginal posterior density of λ, and furthermore it can be shown the conditional distribution of α given λ is a t with $n + 2\alpha$ degrees of freedom, location

$$E(\alpha|\lambda,y) = A^{-1}(\lambda,y)B(\lambda,y) \quad (5.72)$$

and precision $A(\lambda,y)(n + 2\alpha)[C(\lambda,y) - B'(\lambda,y)A^{-1}(\lambda,y)B(\lambda,y)]^{-1}$, where A, B, and C are the appropriate functions of λ and y.

The conditional density of τ given λ is a gamma with parameters $\alpha^* = (n + 2\alpha)/2$ and β^*, where $2\beta^* = C(\lambda,y) - B'(\lambda,y)A^{-1}(\lambda,y)B(\lambda,y)$.

The marginal posterior moments of α and τ may be computed by averaging their conditional moments with respect to the marginal density of λ, (5.71).

It should be relatively easy to develop the Bayesian predictive density of future observations for the distributed lag model, since the analysis is so similar to the ARMA(1,1) model.

There are many versions of the distributed lag model, and the reader is referred to Zellner (1971) for his study of the model. Also Zellner provides an interesting numerical example.

A NUMERICAL STUDY OF AUTOREGRESSIVE PROCESSES WHICH ARE ALMOST STATIONARY

The autoregressive process was examined on pp. 183-187, where the posterior and predictive analyses were derived, and then on pp. 203-206 the identification of the order of an autoregressive model was explained.

Suppose we consider a first-order model

$$Y_t = \theta Y_{t-1} + \varepsilon_t \tag{5.1}$$

where the ε_t are i.i.d. $n(0, \tau^{-1})$, $\theta \in R$, Y_t is the observation at time t, and θ and $\tau(>0)$ are unknown.

By using a normal-gamma prior density for θ and τ, the posterior density of θ and τ is normal-gamma and given by formula (5.9), and the marginal posterior density of θ by (5.10). Also, (5.11) gives the marginal posterior density of the precision τ, and the one-step-ahead predictive density of a future observation Y_{n+1} is given by (5.13).

All these derivations are for a general AR(p) model for p = 1,2, ... and no assumption of stationarity was made.

With the first-order model (p = 1), the process is stationary if $|\theta| < 1$, thus if one knows the stationarity assumption is true, the posterior and predictive analyses of pp. 183-187 do not apply.

If stationarity is indeed the case, the parameter space is $\Omega^* = \{(\theta, \tau), |\theta| < |, \tau > 0\}$ and one should develop a Bayesian analysis with a prior distribution defined on Ω^*. This has been done by Lahiff (1980) and Monahan (1981).

Suppose that one is not certain that the first-order model is stationarity but is, say, 95% confident the stationarity assumption holds? If this is the case, can one use a normal-gamma prior density for (θ, τ) in the analysis of a first-order model? Suppose one chooses a normal-gamma prior density with parameters μ, ξ, α and β, i.e., let

$$p(\theta, \tau) \propto \tau^{1/2} e^{-\tau \xi / 2 (\theta - \mu)^2} \tau^{\alpha - 1} e^{-\tau b}, \quad \theta \in R, \tau > 0$$

be the prior density of θ and τ, then

$$p(\theta) \propto \frac{1}{\left[1 + \frac{\xi}{2\beta}(\theta - \mu)^2\right]^{(2\alpha+1)/2}}, \quad \theta \in R \quad (5.73)$$

is the marginal prior density of θ, which is a t density with 2α degrees of freedom, location μ and precision $\alpha\xi/\beta$. The marginal posterior density of τ is gamma with parameters α and β, which are also involved in the degrees of freedom and precision of the marginal distribution of θ.

Suppose one must choose the parameters of the prior distribution in such a way that, say, $(1 - \gamma)$ 100% of the marginal prior probability of θ is contained in $(-1, +1)$, then since

$$Pr\left[\left|\frac{\theta - \mu}{\sqrt{\beta/(\alpha - 1)\xi}}\right| < t_{\gamma/2, 2\alpha}\right] = (1 - \gamma), \quad 0 < \gamma < 1,$$

μ, α, β and ξ may be chosen so that

$$1 = \mu + t_{\gamma/2, 2\alpha} \sqrt{\beta/(\alpha - 1)\xi}.$$

If μ, α and β are given, ξ must be

$$\xi = \frac{\beta t_{\gamma/2, 2\alpha}^2}{(1 - \mu)^2(\alpha - 1)}, \quad (5.74)$$

where $-1 < \mu < 1$, $\beta > 0$, and $\alpha > 1$, and in this way $(1 - \gamma)$ 100% of the marginal prior density of θ is contained in $(-1, +1)$, and one would be $(1 - \gamma)$ 100% confident the AR(1) process is stationary. For example, if $\gamma = .05$, $\beta = \alpha - 1$, and $|\mu| < 1$, then by letting

$$\xi = \frac{t_{.025, 2\alpha}^2}{(1 - \mu)^2}, \quad (5.75)$$

one would be 95% confident the model is stationary and the prior mean of θ would be μ, and the prior information about τ would be a gamma distribution with parameters α and β where $\beta = \alpha - 1$, thus $E(\tau^{-1}) = \beta/(\alpha - 1) = 1$, $\alpha > 1$. By choosing γ, α and β first, then μ, and finally ξ, according to (5.74), approximate stationarity, a priori, may be induced. However, there are some problems, as $\mu \to 1(-1)$, $\xi \to \infty$, thus if one puts μ close to one of the end points, one is forced to put a large prior precision on the marginal prior density of θ.

A numerical study is now introduced to investigate the effect of the sample size n and the prior mean μ on the marginal posterior

Table 5.1

AR(1), $Y_t = \varepsilon_t$, $\varepsilon_t \sim$ i.i.d. $N(0,1)$, $y_0 = 0$, N-G Prior $\alpha = 10$, $\beta = 9$

			Prior information			Posterior information				Predictive information	
μ	n	$E(\phi)$	$V(\phi)$	$E(\tau)$	$V(\tau)$	$E(\phi\|S)$	$V(\phi\|S)$	$E(\tau\|S)$	$V(\tau\|S)$	$E(Y_{n+1})$	$V(Y_{n+1})$
0.00	25	0	.2298	1.1111	.1235	.1020	.0326	.8548	.0325	.1373	1.2834
	50	0	.2298	1.1111	.1235	.1805	.0180	.9211	.0242	-.2759	1.1596
	100	0	.2298	1.1111	.1235	.1425	.0094	.9948	.0165	.7584	1.0250
	750	0	.2298	1.1111	.1235	.0418	.0013	1.0216	.0027	.8453	.9868
0.25	25	.25	.1293	1.1111	.1235	.1408	.0299	.8538	.0324	.1895	1.2800
	50	.25	.1293	1.1111	.1235	.2008	.0171	.9227	.0243	-.3068	1.1554
	100	.25	.1293	1.1111	.1235	.1554	.0091	.9949	.0165	.8273	1.0247
	750	.25	.1293	1.1111	.1235	.0442	.0013	1.0212	.0027	.0894	.9872

ALMOST-STATIONARY PROCESSES

0.50	25	.5	.0575	1.1111	.1235	.2476	.0249	.8290	.0305	.3334	1.3076
	50	.5	.0575	1.1111	.1235	.2651	.0151	.9080	.0236	-.4051	1.1690
	100	.5	.0575	1.1111	.1235	.1986	.0085	.9805	.0160	.1057	1.0396
	750	.5	.0575	1.1111	.1235	.0527	.0013	1.0168	.0027	.1065	.9914
0.75	25	.75	.0144	1.1111	.1235	.5450	.0139	.7290	.0237	.7339	1.4591
	50	.75	.0144	1.1111	.1235	.4982	.0099	.8179	.0191	-.7613	1.2817
	100	.75	.0144	1.1111	.1235	.3889	.0066	.8845	.0130	.2070	1.1516
	750	.75	.0144	1.1111	.1235	.1034	.0013	.9802	.0025	.2093	1.0281
0.90	25	.9	.0023	1.1111	.1235	.8443	.0036	.6285	.0176	1.1370	1.6715
	50	.9	.0023	1.1111	.1235	.8174	.0030	.6916	.0137	-1.2492	1.4955
	100	.9	.0023	1.1111	.1235	.7543	.0026	.7136	.0085	.4015	1.4258
	750	.9	.0023	1.1111	.1235	.3617	.0011	.8065	.0017	.7318	1.2474

Table 5.2

AR(1), $Y_t = .25 Y_{t-1} + \varepsilon_t$; $\varepsilon_t \sim$ i.i.d. $N(0,1)$, $y_0 = 0$, N-G Prior $\alpha = 10$, $\beta = 9$

			Prior information			Posterior information				Predictive information	
μ	n	$E(\phi)$	$V(\phi)$	$E(\tau)$	$V(\tau)$	$E(\phi\|S)$	$V(\phi\|S)$	$E(\tau\|S)$	$V(\tau\|S)$	$E(Y_{n+1})$	$V(Y_{n+1})$
0.00	25	0	.2298	1.1111	.1235	.3004	.0299	.8458	.0318	.4189	1.2954
	50	0	.2298	1.1111	.1235	.3892	.0156	.9104	.0237	-.5178	1.1583
	100	0	.2298	1.1111	.1235	.3684	.0083	.9887	.0163	.2017	1.0311
	750	0	.2298	1.1111	.1235	.2803	.0012	1.0205	.0027	.4964	.9863
0.25	25	.25	.1293	1.1111	.1235	.3209	.0275	.8521	.0323	.4475	1.2815
	50	.25	.1293	1.1111	.1235	.3973	.0148	.9166	.0240	-.5286	1.1493
	100	.25	.1293	1.1111	.1235	.3738	.0080	.9927	.0164	.2047	1.0269
	750	.25	.1293	1.1111	.1235	.2815	.0012	1.0210	.0027	.4986	.9859

ALMOST-STATIONARY PROCESSES 221

0.50	25	.5	.0575	1.1111	.1235	.3883	.0227	.8477	.0319	.5414	1.2786
	50	.5	.0575	1.1111	.1235	.4315	.0131	.9175	.0241	−.5741	1.1451
	100	.5	.0575	1.1111	.1235	.3968	.0075	.9920	.0164	.2172	1.0274
	750	.5	.0575	1.1111	.1235	.2865	.0012	1.0199	.0027	.5074	.9869
0.75	25	.75	.0144	1.1111	.1235	.6063	.0124	.7908	.0278	.8453	1.3476
	50	.75	.0144	1.1111	.1235	.5839	.0086	.8743	.0218	−.7768	1.1926
	100	.75	.0144	1.1111	.1235	.5170	.0057	.9469	.0149	.2831	1.1057
	750	.75	.0144	1.1111	.1235	.3194	.0012	1.0023	.0026	.5658	1.0039
0.90	25	.9	.0023	1.1111	.1235	.8558	.0031	.7074	.0222	1.1932	1.4855
	50	.9	.0023	1.1111	.1235	.8343	.0026	.7770	.0173	−1.1099	1.3295
	100	.9	.0023	1.1111	.1235	.7880	.0023	.8219	.0113	.4314	1.2380
	750	.9	.0023	1.1111	.1235	.5002	.0009	.8936	.0021	.8851	1.1249

Table 5.3

AR(1), $Y_t = .5 Y_{t-1} + \varepsilon_t$, $\varepsilon_t \sim$ i.i.d. $N(0,1)$, $y_0 = 0$, N-G Prior $\alpha = 10$, $\beta = 9$

μ	n	Prior information				Posterior information				Predictive information	
		$E(\phi)$	$V(\phi)$	$E(\tau)$	$V(\tau)$	$E(\phi\|S)$	$V(\phi\|S)$	$E(\tau\|S)$	$V(\tau\|S)$	$E(Y_{n+1})$	$V(Y_{n+1})$
0.00	25	0	.2298	1.1111	.1235	.4854	.0254	.8316	.0307	.7905	1.3259
	50	0	.2298	1.1111	.1235	.5846	.0121	.8926	.0228	−.8737	1.1783
	100	0	.2298	1.1111	.1235	.5896	.0062	.9788	.0160	.3234	1.0408
	750	0	.2298	1.1111	.1235	.5208	.0010	1.0190	.0027	.8017	.9862
0.25	25	.25	.1293	1.1111	.1235	.4909	.0235	.8409	.0314	.7995	1.3070
	50	.25	.1293	1.1111	.1235	.5842	.0116	.8998	.0231	−.8399	1.1681
	100	.25	.1293	1.1111	.1235	.5893	.0061	.9838	.0161	.3232	1.0355
	750	.25	.1293	1.1111	.1235	.5209	.0010	1.0198	.0027	.8019	.9854

ALMOST-STATIONARY PROCESSES 223

0.50	25	.5	.0575	1.1111	.1235	.5232	.0197	.8490	.0320	.8522	1.2849
	50	.5	.0575	1.1111	.1235	.5945	.0105	.9085	.0236	-.8547	1.1547
	100	.5	.0575	1.1111	.1235	.5952	.0057	.9900	.0163	.3265	1.0290
	750	.5	.0575	1.1111	.1235	.5226	.0010	1.0206	.0027	.8045	.9846
0.75	25	.75	.0144	1.1111	.1235	.6644	.0110	.8289	.0305	1.0821	1.2917
	50	.75	.0144	1.1111	.1235	.6722	.0071	.9020	.0232	-.9664	1.1560
	100	.75	.0144	1.1111	.1235	.6489	.0045	.9831	.0161	.3559	1.0357
	750	.75	.0144	1.1111	.1235	.5376	.0009	1.0161	.0027	.8277	.9889
0.90	25	.9	.0023	1.1111	.1235	.8654	.0028	.7690	.0263	1.4096	1.3684
	50	.9	.0023	1.1111	.1235	.8503	.0023	.8426	.0203	-1.2225	1.2265
	100	.9	.0023	1.1111	.1235	.8200	.0019	.9140	.0139	.4498	1.1132
	750	.9	.0023	1.1111	.1235	.6365	.0007	.9653	.0024	.9798	1.0403

Table 5.4
AR(1), $Y_t = .75\, Y_{t-1} + \varepsilon_t$, ε_t i.i.d. $N(0,1)$, $y_0 = 0$, N-G Prior $\alpha = 10$, $\beta = 9$

			Prior information			Posterior information					Predictive information	
μ	n	$E(\phi)$	$V(\phi)$	$E(\tau)$	$V(\tau)$	$E(\phi\|S)$	$V(\phi\|S)$	$E(\tau\|S)$	$V(\tau\|S)$		$E(Y_{n+1})$	$V(Y_{n+1})$
0.00	25	0	.2298	1.1111	.1235	.7231	.0168	.8119	.0293		1.6301	1.3742
	50	0	.2298	1.1111	.1235	.7987	.0069	.8743	.0218		-1.7879	1.2121
	100	0	.2298	1.1111	.1235	.8100	.0033	.9667	.0156		.2774	1.0524
	750	0	.2298	1.1111	.1235	.7591	.0006	1.0168	.0027		1.0167	.9871
0.25	25	.25	.1293	1.1111	.1235	.7167	.0159	.8200	.0299		1.6157	1.3570
	50	.25	.1293	1.1111	.1235	.7943	.0068	.8793	.0221		-1.7779	1.2046
	100	.25	.1293	1.1111	.1235	.8075	.0032	.9700	.0157		.2765	1.0487
	750	.25	.1293	1.1111	.1235	.7587	.0006	1.0175	.0027		1.0162	.9864

ALMOST-STATIONARY PROCESSES 225

0.50	25	.5	.0575	1.1111	.1235	.7149	.0140	.8331	.0309	1.6116	1.3270
	50	.5	.0575	1.1111	.1235	.7893	.0063	.8881	.0225	−1.7668	1.1909
	100	.5	.0575	1.1111	.1235	.8044	.0031	.9760	.0159	.2755	1.0423
	750	.5	.0575	1.1111	.1235	.7584	.0006	1.0186	.0027	1.0158	.9853
0.75	25	.75	.0144	1.1111	.1235	.7584	.0087	.8487	.0320	1.7097	1.2772
	50	.75	.0144	1.1111	.1235	.7991	.0048	.9038	.0233	−1.7887	1.1632
	100	.75	.0144	1.1111	.1235	.8083	.0027	.9874	.0162	.2768	1.0302
	750	.75	.0144	1.1111	.1235	.7605	.0005	1.0201	.0027	1.0187	.9838
0.90	25	.9	.0023	1.1111	.1235	.8809	.0025	.8314	.0307	1.9858	1.2713
	50	.9	.0023	1.1111	.1235	.8779	.0019	.8976	.0230	−1.9651	1.1564
	100	.9	.0023	1.1111	.1235	.8668	.0014	.9805	.0160	.2969	1.0373
	750	.9	.0023	1.1111	.1235	.7887	.0005	1.011	.0027	1.0504	.9924

Table 5.5

AR(1), $Y_t = .9 Y_{t-1} + \varepsilon_t$, $\varepsilon_t \sim$ i.i.d. $N(0,1)$, $y_0 = 0$, N-G Prior $\alpha = 10$, $\beta = 9$

			Prior information				Posterior information				Predictive information	
μ	n	$E(\phi)$	$V(\phi)$	$E(\tau)$	$V(\tau)$	$E(\phi\|S)$	$V(\phi\|S)$	$E(\tau\|S)$	$V(\tau\|S)$	$E(Y_{n+1})$	$V(Y_{n+1})$
0.00	25	0	.2298	1.1111	.1235	.9144	.0082	.7961	.0282	3.7187	1.4498
	50	0	.2298	1.1111	.1235	.9332	.0025	.8641	.0213	-2.0811	1.2035
	100	0	.2298	1.1111	.1235	.9292	.0013	.9549	.0152	.2090	1.0651
	750	0	.2298	1.1111	.1235	.9036	.0002	1.0152	.0027	.9144	.9878
0.25	25	.25	.1293	1.1111	.1235	.9073	.0080	.7994	.0284	3.6900	1.4411
	50	.25	.1293	1.1111	.1235	.9307	.0024	.8661	.0214	-2.0755	1.2007
	100	.25	.1293	1.1111	.1235	.9277	.0013	.9563	.0152	.2087	1.0635
	750	.25	.1293	1.1111	.1235	.9033	.0002	1.0155	.0027	.9142	.9875

ALMOST-STATIONARY PROCESSES

0.50	25	.5	.0575	1.1111	.1235	.8958	.0075	.8062	.0289	3.0431	1.4216
	50	.5	.0575	1.1111	.1235	.9262	.0024	.8700	.0216	−2.0655	1.1951
	100	.5	.0575	1.1111	.1235	.9251	.0013	.9592	.0153	.2081	1.0603
	750	.5	.0575	1.1111	.1235	.9028	.0002	1.0162	.0027	.9137	.9869
0.75	25	.75	.0144	1.1111	.1235	.8814	.0056	.8253	.0303	3.5844	1.3508
	50	.75	.0144	1.1111	.1235	.9174	.0021	.8814	.0222	−2.0458	1.1785
	100	.75	.0144	1.1111	.1235	.9197	.0012	.9672	.0156	.2069	1.0515
	750	.75	.0144	1.1111	.1235	.9019	.0002	1.0178	.0027	.9128	.9853
0.90	25	.9	.0023	1.1111	.1235	.9105	.0021	.8495	.0321	3.7030	1.2663
	50	.9	.0023	1.1111	.1235	.9218	.0012	.9023	.0233	−2.0558	1.1471
	100	.9	.0023	1.1111	.1235	.9223	.0008	.9818	.0161	.2075	1.0358
	750	.9	.0023	1.1111	.1235	.9041	.0002	1.0199	.0027	.9150	.9832

density of θ (5.10), the marginal posterior density of τ (5.11), and the Bayesian predictive density of Y_{n+1}, given by (5.13).

Series of length n = 25, 50, 100, and 750 were generated from AR(1) models with θ = 0, .25, .5, .75 and .9, initial value of Y_0 = 0 and τ = 1. The parameters of the normal-gamma prior density were chosen as α = 10, β = 9, μ = 0, .25, .75, .9, and ξ was calculated from (5.74), and in this way 95% of the prior probability for θ was contained in (-1, 1).

Tables 5.1-5.5 give the prior mean and variance of θ and τ, the posterior mean and variance of θ and τ, and the predictive mean and variance of Y_{n+1}. For example, Table 5.1 is based on data generated from an AR(1) model with θ = 0, Y_0 = 0, and τ =1, for samples of size n = 25, 50, 100, 750, and the prior mean of θ, namely μ = 0, .25, .5, .75 and .9, α = 10, β = 9, and ξ was calculated from (5.74). The prior variance of θ is $\beta/\xi(\alpha - 1) = 1/\xi$ since $\beta = \alpha - 1$. From Table 5.1 if μ = 0, the prior variance of θ is V(θ) = .2298.

Consider n = 50, μ = .25, then from Table 5.1, the prior mean of θ is .25, the prior variance .1293, the prior mean of τ is 1.1111 and its variance .1235, the posterior mean and variance of θ are .2008 and .0171 respectively, the posterior mean and variance of τ are .9227 and .0243, respectively, and the mean and variance of the predictive distribution of Y_{n+1} are -.3068 and 1.1554, respectively. Tables 5.2-5.5 are now self-explanatory.

Several graphs are also given and Figure 5.1 is a graph of the prior and posterior marginal t densities of θ corresponding to the first row of Table 5.3, that is when the series was generated from an AR(1) model with θ = .5 and a length of n = 25. The prior parameters were μ = 0 and a prior variance for θ of .2298. Figure 5.2 is for the same case as above but is a plot of the prior and posterior gamma densities of the precision τ, while Figure 5.3 gives a plot of the predictive t density of the future observation Y_{26}.

What may be concluded from this study? The tables show the anticipated reduction of the posterior variances of θ and τ and predictive variance Y_{n+1} as the sample size increases. Also, the effect of the prior mean of θ on the posterior mean of θ is clearly demonstrated. As μ becomes closer to the "true" θ value so does the posterior mean of θ and as μ departs further from the "true" value of θ, the posterior mean of θ does the same. Also, as the sample size increases, the effect of the prior mean μ of θ on the posterior mean of θ becomes less and less.

A very interesting result of this study shows that values of θ close to one are easier to estimate (with the posterior mean of θ), than θ values close to zero. Table 5.5 reveals remarkably small changes

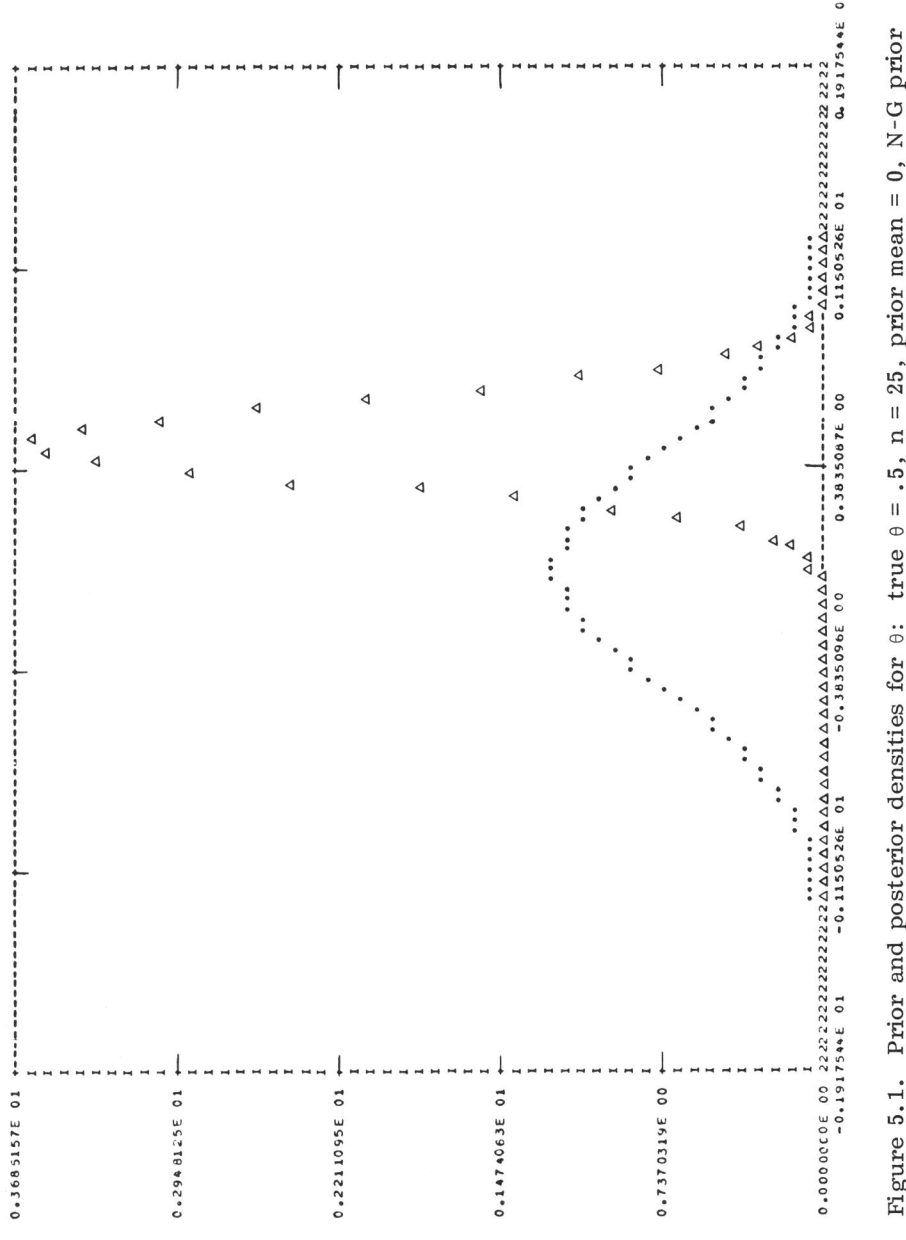

Figure 5.1. Prior and posterior densities for θ: true θ = .5, n = 25, prior mean = 0, N-G prior for (θ, τ); · = prior, △ = posterior.

230 TIME SERIES MODELS

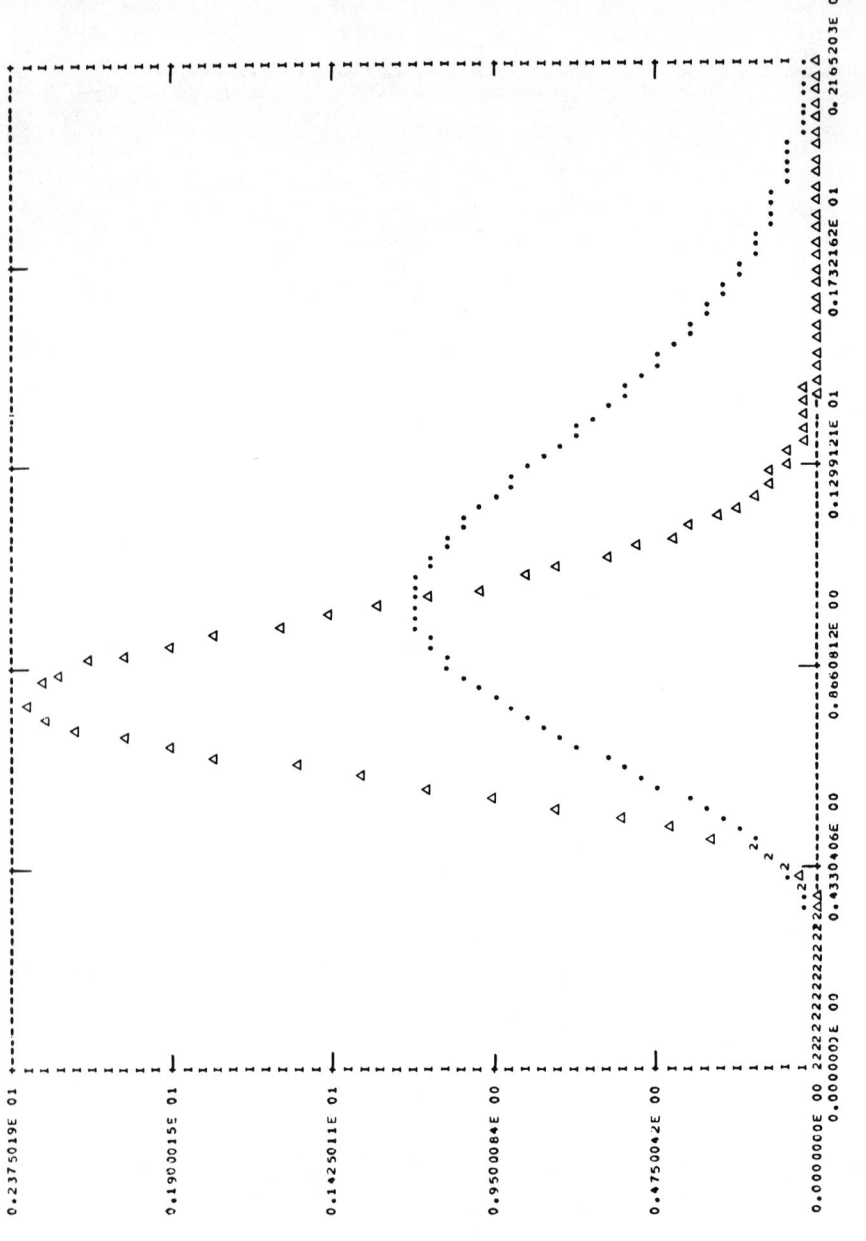

Figure 5.2. Prior and posterior densities for τ: true $\tau = 1$, true $\theta = .5$, $n = 25$, $E(\tau) = 1.1111$, $E(\theta) = 0$, N-G prior for (θ, τ); . = prior, $\triangle$ = posterior.

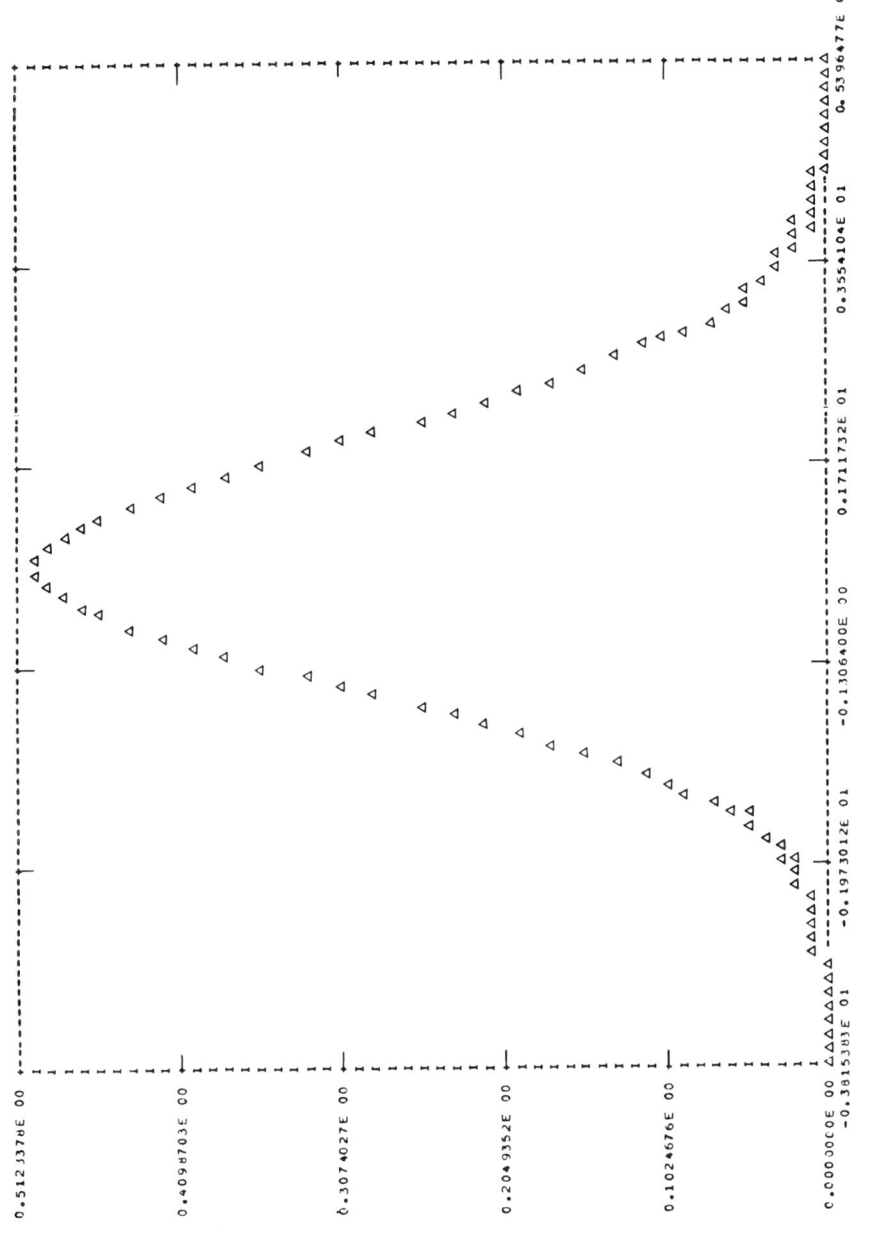

Figure 5.3. Predictive probability density of Y_{n+1}; $n = 25$, $\theta = .5$, $\mu = 0$.

in the posterior means $E(\theta|s)$ for changes in n (= 25, 50, 100, 750) and changes in μ (= 0, .25, .75, .9). There is little effect on the posterior mean of θ from the prior distribution of θ and the sample size when the "true" value of θ is close to one. On the other hand, Table 5.1 shows θ is more difficult to estimate when it is close to zero in the sense that the posterior mean of θ is more sensitive to the prior mean of θ and the sample size (than when θ is close to one).

If θ is close to one, one could use a small sample size and not have an accurate prior estimate u of θ and still have a reasonably good estimate of θ (at least for this data set), but if θ is close to zero, one should have precise prior information or a large sample size in order to obtain good estimates of θ. This phenomenon is confirmed by the sampling theory approach to time series analysis. The sampling variance of the usual estimator of θ is approximately $(1 - \theta^2)/n$, thus as $\theta \to 1$, one obtains better estimates. See Box and Jenkins (1970) for the sampling theory approach to time series analysis.

As a last observation, we see as the prior mean of θ increases the posterior variance of θ decreases, and this should occur because as μ increases, the prior precision of θ increases according to (5.74); consequently the posterior variances of θ decrease along with the prior variance of θ.

Is the normal-gamma distribution adequate to express prior knowledge about AR(1) processes, which are almost stationary? We have seen one limitation with this approach, namely, the prior variance of θ is a function of the prior mean of θ. In order to keep 95% of the prior probability of θ in $(-1, +1)$, for values of μ close to one, the prior precision is forced to be high. This is not a problem when the "true" value of θ is close to one (see Table 5.5), but can be a problem when θ is close to zero (see Table 5.1). Note in these tables that $\phi = \theta$.

EXERCISES

1. Consider an autoregressive process of order one and let Y_{n+1}, Y_{n+2} be two future observations, then if the "initial" observation is given and one adopts a normal-gamma prior distribution, the marginal posterior density of θ_1 is given by (5.10) and the predictive density of Y_{n+1} by (5.13). Derive the conditional predictive density of Y_{n+2}, given Y_{n+1}. Describe how you should predict Y_{n+2} with this density.

EXERCISES 233

2. Equation (5.2) is that of an AR(2) process with parameters θ_1, θ_2, and τ. Using a normal-gamma prior density and assuming Y_0 and Y_{-1} are known, find:

   (a) The marginal posterior density of θ_2 and,

   (b) The predictive density of a future observation Y_{n+1}.

   Do not assume the process is stationary.

3. Why is it so difficult to develop the posterior and predictive analysis of a moving average process? Explain carefully.

4. The section above entitled "The Identification Problem" (pp. 203-207) develops a way to identify the order of an autoregressive process when one uses the marginal posterior mass function of the order parameter p given by equation (5.53). Propose an alternative method by which a Bayesian may identify the order of an autoregressive process.

5. Verify that equation (5.59) is the marginal posterior density of ρ, the autocorrelation parameter of a regression model (5.55) with autocorrelated errors.

6. Describe the similarities between the density (5.60) and the univariate t density.

7. From equation (5.65), show the conditional predictive density of w_1 given ρ is a univariate t distribution and present a way to forecast a future observation.

8. Given an AR(1) process with a normal-gamma prior distribution and assuming Y_0 is given, find the joint predictive density of $Y_{n+1}, Y_{n+2}, \ldots, Y_{n+k}$, where $k = 1, 2, \ldots$.

9. Consider a MA(1) process (5.14)

   $$Y_t = \varepsilon_t - \theta_1 \varepsilon_{t-1}$$

   where $t = 1, 2, \ldots, n$, $\theta_1 \in R$ and $\varepsilon_0, \varepsilon_1, \ldots, \varepsilon_n$ are n.i.d. $(0, \tau^{-1})$, where $\tau > 0$. Letting $\varepsilon_0 = 0$, develop an approximation to the likelihood function of θ_1 and τ based on $Y_1, Y_2, \ldots, Y_n$. Discuss the merits of the approximation and why it should provide an easier way to study moving average processes.

10. (Continuation of Problem 9.) If $\varepsilon_0 = 0$, one may calculate the residuals recursively as

    $$\varepsilon_t = Y_t + \theta_1 \varepsilon_{t-1}, \quad t = 1, 2, \ldots, n$$

    corresponding to the n observations. Thus, one may estimate the moving average parameter by finding the value of θ_1, say $\hat{\theta}_1$ which minimizes the residual sum of squares

$$g(\theta_1) = \sum_{t=1}^{n} \varepsilon_t^2$$

with respect to θ_1 over R.

Using the approximate likelihood function of Problem 9 and a normal-gamma prior density for θ_1 and τ, develop a t-approximation to the marginal posterior distribution of θ_1. Also show the marginal posterior distribution of τ is approximately a gamma.

11. Referring to Problem 10 above, find a t-approximation to the predictive distribution of a future observation y_{n+1}.

REFERENCES

Abraham, Bovas and William W. S. Wei (1979). "Inferences in a switching time series," American Statistical Association Proceedings of the Business and Economic Statistics Section, pp. 354-358.

Akaike, H. (1979). "A Bayesian extension of the minimum AIC procedure of autoregressive model building," Biometrika, Vol. 66, No. 2, pp. 237-242.

Ali, M. M. (1977). "Analysis of autoregressive moving average models: estimation and prediction," Biometrika, Vol. 64, pp. 535-545.

Ansley, C. F. (1979). "An algorithm for the exact likelihood of mixed autoregressive moving average process," Biometrika, Vol. 66, pp. 59-65.

Aoki, Masanao (1967). *Optimization of Stochastic Systems, Topics in Discrete Time Systems*, Academic Press, New York.

Box, George E. P. and Gwilym M. Jenkins (1970). *Time Series Analysis, Forecasting and Control*, Holden-Day, San Francisco.

Broemeling, Lyle D. (1977). "Forecasting future values of changing sequences," Communications in Statistics, Theory and Methods, Vol. A6(1), pp. 87-102.

Chatfield, C. (1977). "Some recent developments in time series analysis," Journal of the Royal Statistical Society, Series A, Vol. 140, pp. 492-510.

Chin Choy, J. H. and L. D. Broemeling (1980). "Some Bayesian inferences for a changing linear model," Technometrics, Vol. 22, No. 1, pp. 71-78.

Current Index to Statistics, Appplications, Methods, and Theory, Volume 5, 1979 (1980).

Dent, W. T. (1977). "Computation of the exact likelihood function for an ARIMA process," Journal of Statistical Computations and Simulation, Vol. 5, pp. 193-206.

REFERENCES

Diaz, Jaoquin and Jose Luis Farah (1981). "Bayesian identification of autoregressive processes," 22nd NBER-NSF Seminar on Bayesian Inference in Econometrics.

Granger, C. W. J. and Paul Newbold (1977). *Forecasting Economic Time Series*, Academic Press, New York.

Haavelmo, T. (1953). "Methods of measuring the marginal propensity to consume," Journal of the American Statistical Assoc., Vol. 42, pp. 105-122.

Harrison, P. J. and Stevens, C. F. (1976). "Bayesian forecasting" (with discussion), Journal of the Royal Statistical Society, Series B, Vol. 38, pp. 205-247.

Hillmer, Steven C., and George C. Tiao (1979). "Likelihood function of stationary multiple autoregressive moving average models," Journal of the American Statistical Association, Vol. 74, No. 367, pp. 652-660.

Lahiff, Maureen (1980). "Time series forecasting with informative prior distributions," Technical Report No. 111, Department of Statistics, The University of Chicago.

Land, Margaret Foster (1981). Bayesian Forecasting for Switching Regression and Autoregressive Processes, Ph.D. dissertation, Oklahoma State University, Stillwater, Oklahoma.

Lindley, D. V. and A. F. M. Smith (1972). "Bayes estimates for the linear model," Journal of the Royal Statistical Society, Series B, Vol. 34, pp. 1-18.

Litterman, Robert B. (1980). "A Bayesian procedure for forecasting with vector autoregressions," An unpublished photocopy paper, Massachusetts Institute of Technology, Cambridge, Massachusetts.

Ljung, G. M. and G. E. P. Box (1976). "Studies in the modeling of discrete time series—maximum likelihood estimation in the autoregressive moving average model," Technical Report No. 476, Department of Statistics, University of Wisconsin, Madison.

Monahan, John F. (1981). "Computations for Bayesian time series analysis," a paper presented at the 141st Annual Meeting of the American Statistical Association, August 12, 1981.

Newbold, Paul (1973). "Bayesian estimation of Box Jenkins transfer function noise models," Journal of the Royal Statistical Society, Series B, Vol. 35, No. 2, pp. 323-336.

Newbold, Paul (1974). "The exact likelihood function for a mixed autoregressive moving average process," Biometrika, Vol. 61, pp. 423-426.

Nicholls, D. F. and A. D. Hall (1979). "The exact likelihood function of multivariate autoregressive moving average model," Biometrika, Vol. 66, No. 2, pp. 259-264.

Perterka, V. (1981). "Bayesian system identification," Automatica, Vol. 17, No. 1, pp. 41-53.

Poirier, Dale J. (1976). *The Econometrics of Structural Change*, North-Holland Publishing Company, Amsterdam.

Salazar, D., Broemeling, L., and A. Chi (1981). "Parameter changes in a regression model with autocorrelated errors," Communications in Statistics, Part A—Theory and Methods, Vol. A10, Number 17, pp. 1751-1758.

Smith, J. O. (1979). "A generalization of the Bayesian steady forecasting model," Journal of the Royal Statistical Society, Series B, Vol. 41, No. 3, pp. 375-387.

Tsurumi, H. (1980). "A Bayesian estimation of structural shifts by gradual switching regression with an application to the US gasoline market," in *Bayesian Analysis in Econometrics and Statistics: Essays in Honor of Harold Jeffreys,* edited by A. Zellner, North-Holland, Amsterdam.

Zellner, A. (1971). *An Introduction to Bayesian Inference in Econometrics,* John Wiley and Sons, Inc., New York.

6
LINEAR DYNAMIC SYSTEMS

INTRODUCTION

Linear dynamic systems are used by communications and control engineers to monitor and control the state of a system as it evolves through time. To determine whether a system is operating satisfactorily and later to control the system one must know the behavior of the system at any time.

For example, in the navigation of a submarine, spacecraft, or aircraft, the state of the system is the position and velocity of the craft. The state of the system may be random because physical systems are often subject to random disturbances and one uses measuring instruments to take observations on the system, but the measurements are often corrupted with noise because of the electrical and mechanical components of the measuring device.

Finding the state of the system from noisy measurements is called estimation and estimation is essential to monitoring the system as it evolves through time.

This chapter will give the reader a brief introduction to linear dynamic systems. The engineering literature on this subject is vast and one should consult some of the many books on the subject. For example Aoki (1967), Maybeck (1979), Jazwinski (1970), Sawaragi et al. (1967), and Harrison and Stevens (1976) provide one with the essentials for understanding linear dynamic systems. The above references are for the most part written by and for engineers except the Harrison and Stevens paper which is statistical in nature and emphasizes business and economic forecasting. Maybeck is a good general introduction to estimation (filtering, prediction, smoothing), control, adaptive estimation, and nonlinear estimation. For an unusually excellent account of filtering, the book by Jazwinski is recommended and for a thorough treatment of the Bayesian approach to control strategies one should read Aoki. Sawaragi et al. should appeal to statisticians because it shows the relationship between statistical decision theory and adaptive control systems.

One will find that Bayesian ideas have played an important role in the development of estimation, identification, and control of linear dynamic systems. Indeed, the Kalman (1960) filter is based on a Bayesian estimation technique and many of the resulting methodologies have a Bayesian foundation, thus it is interesting that the theory and methodology of linear dynamic models is not very familiar to statisticians. Recently, however, statisticians are beginning to study the linear dynamic model and use it with some of the standard statistical problems. For example, Sallas and Harville (1981) estimate the parameters of the mixed models with best recursive estimators, while Downing, Pike, and Morrison (1980) apply the Kalman filter to inventory control. Shumway, Olsen, and Levy (1981) employ mixed model estimation to a linear dynamic model, which in a way is the reverse of that done by Sallas and Harville.

This chapter begins with an explanation of the linear dynamic model and reviews the Kalman filter, then develops the Bayesian analysis for control, adaptive estimation, and nonlinear filtering.

DISCRETE TIME LINEAR DYNAMIC SYSTEMS

Consider an observation equation

$$Y(t) = F(t)\theta(t) + U(t), \quad t = 1, 2, \ldots \quad (6.1)$$

where $Y(t)$ is an $n \times 1$ observation vector, $F(t)$ a $n \times p$ nonrandom $n \times p$ matrix, $\theta(t)$ a $p \times 1$ real unknown parameter vector, and $U(t)$ a $n \times 1$ observation error vector. The observation equation (6.1) looks like a sequence of usual type linear models, where $F(t)$ is a $n \times p$ design matrix. The parameter $\theta(t)$ is called the state of the system at time t and how the states evolve is given by the system equation.

$$\theta(t) = G\theta(t-1) + V(t), \quad t = 1, 2, \ldots \quad (6.2)$$

where G is a $p \times p$ nonrandom systems transition matrix, $\theta(0)$ is the initial state of the system, and $V(t)$ is a $p \times 1$ systems error vector.

Some measuring device records at each time point $Y(t)$, which indirectly measures the state of the system at time t and is linearly related to the actual state $\theta(t)$, but since the signal $Y(t)$ is corrupted by noise, the additive error term $U(t)$ is put into the observation equation. The system equation is a first-order autoregressive process, but of course other stochastic processes could represent the states of the system.

ESTIMATION

Estimation is the activity of estimating the states θ(t) of the system as observations Y(1), Y(2), ..., become available, and each observation adds one parameter to be estimated. Now suppose one has k observations Y(1), Y(2), ..., Y(k), then there are k parameters θ(1), θ(2), ..., θ(k) to be estimated.

Filtering is the estimation of the current state θ(k) and in the next section the Kalman filter will be explained. Smoothing is the estimation of the previous states θ(1), θ(2), ..., θ(k − 1) and prediction is the forecasting of the future states θ(k + 1), θ(k + 2), ... As we will see, prediction is important when one wants to control a future state, say θ(k + 1) at some known target value T(k + 1).

The next three sections explain filtering, smoothing, and prediction.

Filtering

The Kalman filter is a recursive algorithm to estimate the current states of the system at each time point. Suppose F(t) and G are known, and that θ(0) is random with known mean m(0) and known precision matrix p(0), that U(t) is normal with mean zero and known precision matrix $P_u(t)$, that V(t) is normal with mean zero and precision matrix $P_v(t)$, and assume θ(0), U(1), U(2), ..., V(1), V(2), ..., are stochastically independent, then given Y(1), Y(2), ..., Y(k), the posterior distribution of θ(k) is normal with mean m(k) and precision matrix P(k), where

$$m(k) = P^{-1}(k)\{F'(k)P_u(k)Y(k) + P_v(k)G[G'P_v(k)G + P(k-1)]^{-1}$$
$$\times P(k-1)m(k-1)\} \qquad (6.3)$$

and

$$P(k) = F'(k)P_u(k)F(k) + [GP^{-1}(k-1)G' + P_v^{-1}(k)]^{-1},$$
$$\text{for } k = 1, 2, \ldots \qquad (6.4)$$

Thus the current state θ(k) is recursively estimated by m(k) and P(k) is expressed as a function of the current Y(k), the known current parameters $P_u(k)$ and $P_v(k)$, and the parameters m(k − 1) and P(k − 1) of the posterior distribution of the previous state θ(k − 1).

This is summarized by the following theorem.

Theorem 6.1 (The Kalman Filter). Consider the linear dynamic system

$$Y(t) = F(t)\theta(t) + U(t),$$
$$\theta(t) = G\theta(t-1) + V(t),$$
$$t = 1, 2, \ldots, k$$

and suppose for each k, k = 1, 2, ..., that F(t) and G are known and suppose $\theta(0), U(1), \ldots, U(k), V(1), V(2), \ldots, V(k)$ are stochastically independent. Let $U(t) \sim N[0, P_u^{-1}(t)]$, $V(t) \sim N[0, P_v^{-1}(t)]$ for t = 1, 2, ..., k, and $\theta(0) \sim N[m(0), P^{-1}(0)]$, the marginal posterior distribution of $\theta(k)$ is normal with mean m(k) and precision matrix P(k), where m(k) and P(k) are given by (6.3) and (6.4), respectively.

To prove this theorem, note the prior distribution of $\theta(k)$ is induced by the posterior distribution of $\theta(k-1)$ via the system equation (6.2) and at each stage only p × p matrices need to be inverted.

Plackett (1950) gave an earlier version of these formulas, but Kalman's main contribution is the recursive nature of the algorithm, (6.3) and (6.4), which provides one with an efficient computational procedure.

Smoothing

Suppose the assumptions of Theorem 6.1 are satisfied, then how should the previous states $\theta(0), \theta(1), \ldots, \theta(k-1)$, given k observations $Y(1), Y(2), \ldots, Y(k)$, be estimated? To a Bayesian, one should estimate these parameters from the posterior distribution of $\theta^{(k-1)} = [\theta'(0), \theta'(1), \ldots, \theta'(k-1)]'$, which is easily derived from the posterior distribution of $\theta^{(k)}$, namely the conditional distribution of $\theta(0), \ldots, \theta(k)$, given $Y^{(k)} = [Y'(1), \ldots, Y'(k)]'$. In this way, all of the previous observations are available to estimate the previous states.

First, consider the posterior distribution of $\theta^{(k)}$, so suppose the assumptions of Theorem 6.1 hold, then the joint density of $Y^{(k)}$ and $\theta^{(k)}$ is

$$\pi[Y^{(k)}, \theta^{(k)}] = \pi_1[Y^{(k)} | \theta^{(k)}] \pi_2[\theta^{(k)}], \qquad (6.5)$$

ESTIMATION

where $Y^{(k)} \in R^{nk}, \theta^{(k)} \in R^{p(k+1)}$, and

$$\pi_1[Y^{(k)}|\theta^{(k)}] \propto \exp -\frac{1}{2}\sum_{t=1}^{k} [Y(t) - F(t)\theta(t)]'P_u(t)$$

$$\times [Y(t) - F(t)\theta(t)] \qquad (6.6)$$

and

$$\pi_2[\theta^{(k)}] \propto \exp -\frac{1}{2}\sum_{t=1}^{k} [\theta(t) - G\theta(t-1)]'P_v(t)$$

$$\times [\theta(t) - G\theta(t-1)]$$

$$\times \exp -\frac{1}{2}[\theta(0) - m(0)]'P(0)[\theta(0) - m(0)]. \qquad (6.7)$$

Combining the exponents of π_1 and π_2, the resulting exponent is quadratic in the $k+1$ states $\theta(0), \ldots, \theta(k)$, thus π is the density of a multivariate normal distribution and the mean vector and precision matrix need to be identified.

It can be shown that $\theta^{(k)}$ has density

$$\pi[\theta^{(k)}|Y^{(k)}] \propto \exp -\frac{1}{2}[\theta^{(k)} - A^{-1}B]'A[\theta^{(k)} - A^{-1}B] \qquad (6.8)$$

where A is $p(k+1) \times p(k+1)$ symmetric and patterned as follows:

$$A = \begin{pmatrix} A_{11} & A_{12} & A_{13} \\ A_{21} & A_{22} & A_{23} \\ A_{31} & A_{32} & A_{33} \end{pmatrix},$$

$A_{13} = 0(p \times p) = A_{31}$, A_{22} is quasi-tridiagonal of order $(k-1)p$ and the i-th diagonal matrix is $F'(i)P_u(i)F(i) + P_v(i) + G'P_v(i+1)G$ for $i = 1, 2, \ldots, k-1$ and the super- and subdiagonal matrices are $-G'P_v(i)$, $i = 2, \ldots, k-1$. A_{23} is $(k-1)p \times p$ and

$$A_{23} = \begin{pmatrix} 0(p \times p) \\ 0(p \times p) \\ \vdots \\ 0(p \times p) \\ -G'P_v(k) \end{pmatrix},$$

and

$$B = \begin{pmatrix} P(0)m(0) \\ F'(1)P_u(1)Y(1) \\ F'(2)P_u(2)Y(2) \\ \vdots \\ F'(k)P_u(k)Y(k) \end{pmatrix}$$

is a $(k+1)p \times 1$ vector. The scalar C is given by

$$C = \sum_{1}^{k} Y'(t)P_u(t)Y(t) + m'(0)P(0)m(0),$$

thus $\theta^{(k)}$ has a normal distribution with mean vector $A^{-1}B = \mu$ and precision matrix $A = T$.

How does one estimate the previous states $\theta(0), \theta(1), \ldots, \theta(k-1)$?

Obviously all inferences should be based on the marginal distribution of these states, thus let

$$\phi = [I, 0]\theta^{(k)}, \qquad (6.9)$$

where 0 is a $pk \times p$ zero matrix and I is the identity matrix of order pk, then ϕ is the $pk \times 1$ vector $\theta^{(k-1)}$ of the previous states, i.e. those previous to $\theta(k)$. From (6.8) and (6.9), $\theta^{(k-1)}$ has a normal distribution with mean $[I, 0]\mu$ and precision $\{[I, 0]T^{-1}[I, 0]'\}^{-1}$, and in a similar way any subvector of $\theta^{(k-1)}$ has a normal distribution.

The previous results are summarized as follows by

ESTIMATION

Theorem 6.2. If the assumptions of Theorem 6.1 are true and one has k observations $Y^{(k)} = [Y'(1), Y'(2), \ldots, Y'(k)]'$, then the conditional distribution of the previous states $\theta^{(k-1)}$ is normal with mean $[I,0]\mu$ and precision matrix $\{[I,0]T^{-1}[I,0]'\}^{-1}$, where μ and T are the mean vector and precision matrix of $\theta^{(k)}$, which has density given by (6.8).

It should be noted that the posterior distribution of $\theta(i)$, $1 \leq i \leq k-1$ given by Theorem 6.2 is not the same as that given by the Kalman filter, because Theorem 6.1 gives the posterior distribution of $\theta(i)$ based on i observations $Y(1), Y(2), \ldots, Y(i)$, but Theorem 6.2 provides the conditional distribution of $\theta(i)$, $0 \leq i \leq k-1$ based on k observations.

Smoothing is a way to improve one's estimate of the previous states beyond that given by the Kalman filter and is a post-mortem operation.

Prediction

Suppose one has available k observations $Y(1), Y(2), \ldots, Y(k)$ and we want to forecast $\theta(k+1)$, before observing $Y(k+1)$, then the appropriate distribution with which to forecast $\theta(k+1)$ is the prior (prior to $Y(k+1)$) distribution of $\theta(k+1)$. The prior distribution of $\theta(k+1)$ is induced by the posterior distribution of $\theta(k)$, where

$$\theta(k+1) = G\theta(k) + V(k+1)$$

but the posterior distribution of $\theta(k)$ is given by the Theorem 6.1 (the Kalman filter), where $m(k)$ is the mean of $\theta(k)$ and $P(k)$ the precision matrix of $\theta(k)$, thus by the Bayes theorem, we have

Theorem 6.3. Under the assumptions of Theorem 6.1, the prior distribution of $\theta(k+1)$ is normal with mean

$$m^*(k+1) = P^{*-1}(k+1)\{P_V(k+1)G[G'P_V(k+1)G + P(k)]^{-1}$$

$$\times P(k)m(k)\} \qquad (6.10)$$

and precision matrix

$$P^*(k+1) = [GP^{-1}(k)G' + P_V^{-1}(k+1)]^{-1}. \qquad (6.11)$$

Note that $m^*(k+1)$ is a function of the first k observations because $m(k)$ is a function of these observations.

By repeated application of this theorem one may predict $\theta(k + 2)$, $\theta(k + 3)$, and so on, since the prior distribution of $\theta(k + 2)$ is induced by the prior distribution of $\theta(k + 1)$, via the systems equation $\theta(k + 2) = G\theta(k + 1) + V(k + 2)$. Thus, the prior distribution of $\theta(k + 2)$ is normal with mean $m^*(k + 2)$ and precision matrix $P^*(k + 2)$ where

$$m^*(k + 2) = P^{*-1}(k + 2)P_v(k + 2)G[G'P_v(k + 2)G + P^*(k + 1)]^{-1}$$
$$\times P^*(k + 1)m^*(k + 1) \qquad (6.12)$$

and

$$P^*(k + 2) = [GP^{*-1}(k + 1)G' + P_v^{-1}(k + 2)]^{-1}. \qquad (6.13)$$

Again $m^*(k + 2)$ is a function of the first k observations, because $m^*(k + 2)$ is a function of $m^*(k + 1)$, which is a function of $m(k)$ (see Theorem 6.3), which is the mean of the posterior distribution of $\theta(k)$.

Prediction in the context of linear dynamic systems means something other than the way prediction is used with time series analysis. In time series, to predict means to forecast future observations (see Chapter 5), while with linear dynamic systems, to predict means to forecast the future states of the system.

In the next section on the control problem, the predictive distribution of future observations will be employed in the strategy of choosing control variables.

CONTROL STRATEGIES

This section gives only a brief introduction to the control of linear dynamic systems (6.1) and (6.2). For a more thorough treatment of control problems, Aoki (1967) provides one with a Bayesian analysis of various strategies to control the state of a physical system.

Consider the linear dynamic system with observation equation

$$Y(t) = F(t)\theta(t) + U(t) \qquad (6.1)$$

and with system equation

$$\theta(t) = G\theta(t - 1) + H(t)X(t - 1) + V(t), \qquad (6.14)$$

CONTROL STRATEGIES

where the term $H(t)X(t-1)$ has been added to the system equation (6.2), and $H(t)$ is a known $p \times q$ matrix and $X(t-1)$ is a $q \times 1$ vector of control variables. That is one can set these vectors at various values to control the process.

The control problem is: how does one choose $X(0), X(1), \ldots$, in such a way that the corresponding states $\theta(1), \theta(2), \ldots$, are close to preassigned target values $T(1), T(2), \ldots$? Initially, one has yet to observe $Y(1)$, and $X(0)$ must be chosen in order for $\theta(1)$ to be "close" to $T(1)$, then having observed $Y(1) = y(1)$ but not yet $Y(2)$, $X(1)$ is chosen so that $\theta(2)$ is "close" to $T(2)$, and so on. Choosing a set of control values $\hat{X}(0), \hat{X}(1), \ldots$, for the control variables $X(0), X(1), \ldots$, is called a strategy, and one's strategy depends on how one measures the closeness of $\theta(i)$ to $T(i)$, $i = 1, 2, \ldots$.

The strategy presented here is based on the predictive distribution of the posterior means and is just one of many criteria that could have been used. Aoki (1969) gives us many control strategies, and the reader should refer to him and Maybeck (1979) for more detailed accounts of the control problem.

The strategy here is to choose the control variables $X(0), X(1), \ldots$, in such a way that at the first stage

$$T(1) = \underset{1}{E}E[\theta(1)|Y(1), X(0)], \tag{6.15}$$

at the second stage

$$T(2) = \underset{2}{E}E[\theta(2)|y(1), Y(2), \hat{X}(0), X(1)] \tag{6.16}$$

and in general

$$T(i) = \underset{i}{E}E[\theta(i)|y(1), \ldots, y(i-1), Y(i), \hat{X}(0), \hat{X}(1), \ldots,$$

$$\hat{X}(i-2), X(i-1)] \tag{6.17}$$

for $i = 1, 2, \ldots$, and $\underset{i}{E}$ denotes expectation with respect to the conditional predictive distribution of $Y(i)$ given $Y(j) = y(j)$, $j = 1, 2, \ldots, i-1$, where $y(j)$ is the observed value of $Y(j)$, and $\hat{X}(i)$ is the value chosen for $X(i)$ so that $T(i)$ is close to the posterior mean of $\theta(i)$, in the sense of (6.17). Thus, $X(0)$ is chosen so that the average value of the posterior mean of $\theta(1)$ is the target value $T(1)$, and the average is taken with respect to the predictive distribution of $Y(1)$ (which has yet to be observed). At the second stage $X(1)$ is chosen before $Y(2)$ is observed but after $Y(1)$ is observed to be $y(1)$, in such a way the target value T(2) is equal to the average value of the posterior mean of $\theta(2)$. Because $Y(1) = y(1)$ and $Y(2)$ is yet to be observed the average E is with respect to the conditional predictive distribution of $Y(2)^2$ given $Y(1) = y(1)$, and so on. Solving the system (6.17) with respect to $X(0), X(1), \ldots$, gives us as the solution $\hat{X}(0), \hat{X}(1), \ldots$, for the control strategy.

Let us return to the initial stage of the control problem, where $X(0)$ must be chosen, then one must be able to first find the posterior mean of $\theta(1)$, then the predictive distribution of $Y(1)$, for the linear dynamic system (6.1) and (6.14). To do this, one must stipulate assumptions comparable to those given in Theorem 6.1, thus suppose

$$\theta(0) \sim N[m(0), P^{-1}(0)], \quad U(t) \sim N[0, P_u^{-1}(t)], \quad V(t) \sim N[0, P_v^{-1}(t)]$$

and that $\theta(0)$; $U(1), U(2), \ldots$; $V(1), V(2), \ldots$, are independent and that $F(t)$, $H(t)$, $P_u(t)$, and $P_v(t)$ are known for $t = 1, 2, \ldots$, and that also G is known. Now under these assumptions how should $X(0)$ be chosen so that (6.15) is satisfied?

First, it can be shown the posterior mean of $\theta(1)$ is

$$E[\theta(1)|Y(1), X(0)] = P^{-1}(1)\{[GP^{-1}(0)G' + P_v^{-1}(1)]^{-1}$$

$$\times H(1)X(0) + F'(1)P_u(1)Y(1)$$

$$+ P_v(1)G[P(0) + G'P_v(1)G]^{-1}\}, \quad (6.18)$$

where

$$P(1) = F'(1)P_u(1)F(1) + [G'P^{-1}(0)G + P_v^{-1}(1)]^{-1} \quad (6.19)$$

and is the posterior precision matrix of $\theta(1)$.

It can also be shown that the predictive distribution of $Y(1)$ is normal with mean

$$E[Y(1)] = Q^{-1}(1)P_u(1)F'(1)P^{-1}(1)\{[GP^{-1}(0)G' + P_v^{-1}(1)]^{-1}$$

$$\times H(1)X(0) + P_v(1)G[P(0) + G'P_v(1)G]^{-1}P(0)m(0)\}$$

$$(6.20)$$

CONTROL STRATEGIES 247

and precision matrix

$$Q(1) = P_u(1) - P_u(1)F(1)P^{-1}(1)F'(1)P_u(1). \quad (6.21)$$

Both the predictive mean $E[Y(1)]$ and the posterior mean $E[\theta(1)|Y(1),X(0)]$ may be derived from the joint distribution of $Y(1)$, $\theta(1)$, and $\theta(0)$, which, from the assumptions given above, have density

$$\pi[Y(1),\theta(1),\theta(0)] = \pi_1[Y(1)|\theta(1)]\pi_2[\theta(1)|\theta(0)]\pi_3[\theta_0], \quad (6.22)$$

where

$$\pi_1[Y(1)|\theta(1)] \propto \exp -\frac{1}{2}[Y(1) - F(1)\theta(1)]'P_u(1)$$

$$\times [Y(1) - F(1)\theta(1)], \quad Y(1) \in R^n,$$

$$\pi_2[\theta(1)|\theta(0)] \propto \exp -\frac{1}{2}[\theta(1) - G\theta(0) - H(1)X(0)]'P_v(1)$$

$$\times [\theta(1) - G\theta(0) - H(1)X(0)], \quad \theta(1) \in R^p,$$

and

$$\pi_3[\theta(0)] \propto \exp -\frac{1}{2}[\theta(0) - m(0)]'P(0)[\theta(0) - m(0)],$$

$$\theta(0) \in R^p.$$

If one eliminates $\theta(1)$ and $\theta(0)$ from (6.22), one has a normal predictive distribution for $Y(1)$, then if one integrates (6.22) with respect to $\theta(0)$, the result is the posterior density of $\theta(1)$.

The posterior mean of $\theta(1)$, (6.18), is a linear function of $Y(1)$ and $X(0)$, thus let

$$E[\theta(1)|Y(1),X(0)] = AX(0) + BY(1) + C \quad (6.23)$$

where

$$A = P^{-1}(1)[GP^{-1}(0)G' + P_v^{-1}(1)]^{-1}H(1)$$

and

$$B = P^{-1}(1)F'(1)P_u(1)$$

and

$$C = P^{-1}(1)P_v(1)G[P(0) + G'P_v(1)G]^{-1}P(0)m(0).$$

Also, $E[Y(1)]$ is a linear function of $X(0)$, thus let

$$E[Y(1)] = DX(0) + E, \qquad (6.24)$$

where

$$D = Q^{-1}(1)P_u(1)F'(1)P^{-1}(1)[GP^{-1}(0)G' + P_v^{-1}(1)]^{-1}H(1)$$

and

$$E = Q^{-1}(1)P_u(1)F'(1)P^{-1}(1)P_v(1)G[P(0) + G'P_v(1)G]^{-1}P(0)m(0)$$

Now from (6.15),

$$T(1) = (A + BD)X(0) + BE + C$$

and the control vector $X(0)$ should be set at

$$\hat{X}(0) = \{(A + BD)'(A + BD)\}^{-1}(A + BD)'[T(1) - BE - C],$$

$$(6.25)$$

which solves the problem at the first stage, but how does the process continue?

At the second stage one must choose $X(1)$ so that $T(2)$ is close to $\theta(2)$ in the sense of (6.16). Thus we need to know the conditional predictive distribution of $Y(2)$ given $Y(1) = y(1)$ and the posterior mean of $\theta(2)$, then solve equation (6.16) for $X(1)$. In general, referring to equation (6.17), we need to know the predictive conditional distribution of $Y(i)$ given $Y(j) = y(j)$, $j = 1, 2, \ldots, i-1$, and the posterior mean of $\theta(i)$, then solve equation (6.17) for $X(i-1)$.

The general solution is given by the following theorem.

Theorem 6.4. Consider the linear dynamic system (6.1) and (6.14), where $\theta(0) \sim N[m(0), P^{-1}(0)]$, $U(t) \sim N[0, P_u^{-1}(t)]$, and $V(t) \sim N[0, P_v^{-1}(t)]$. Assume for $t = 1, 2, \ldots$, that G, $F(t)$, $H(t)$, $P_u(t)$, $P_v(t)$, $m(0)$, and $P(0)$ are known and that $\theta(0)$; $U(1), U(2), \ldots$; $V(1), V(2), \ldots$, are independent, then at the i-th stage of control, the value $X(i-1)$ which satisfies

$$T(i) = \underset{i}{EE}[\theta(i) | y(1), y(2), \ldots, y(i-1), Y(i), \hat{X}(0), \hat{X}(1), \ldots,$$

CONTROL STRATEGIES

$$\hat{X}(i-2), X(i-1)] \qquad (6.17)$$

is given by

$$\hat{X}(i-1) = \{[A(i) + B(i)D(i)]'[A(i) + B(i)D(i)]\}^{-1}$$
$$\times [A(i) + B(i)D(i)]'[T(i) - B(i)E(i) - C(i)] \qquad (6.26)$$

where

$$A(i) = P^{-1}(i)[GP^{-1}(i-1)G' + P_v^{-1}(i)]^{-1}H(i),$$

$$B(i) = P^{-1}(i)F'(i)F'(i)P_u(i),$$

$$C(i) = P^{-1}(i)P_v(i)G[P(i-1) + G'P_v(1)G]^{-1}P(i-1)m(i-1),$$

$$P(i) = F'(i)P_u(i)F(i) + [G'P^{-1}(i-1)G + P_v^{-1}(i)]^{-1},$$

$$D(i) = Q^{-1}(i)P_u(i)F(i)P^{-1}(i)[GP^{-1}(i-1)G' + P_v^{-1}(i)]^{-1}H(i),$$

$$E(i) = Q^{-1}(i)P_u(i)F(i)P^{-1}(i)P_v(i)G[P(i-1) + G'P_v(i)G]^{-1}$$
$$\times P(i-1)m(i-1),$$

$$Q(i) = P_u(i) - P_u(i)F(i)P^{-1}(i)F'(i)P_u(i),$$

$$m(i) = P^{-1}(i)\{[GP^{-1}(i-1)G' + P_v^{-1}(i)]^{-1}H(i)X(i-1) + F'(i)$$
$$\times P_u(i)Y(i) + P_v(i)G[P(i-1) + G'P_v(i)G]^{-1}P(i-1)m(i-1)\}$$

Now, $m(i)$ and $P(i)$ are the mean vector and precision matrix, respectively, of the posterior distribution of $\theta(i)$ and $Q(i)$ is the precision matrix of the conditional predictive distribution of $Y(i)$ given $Y(j) = y(j)$, $j = 1, 2, \ldots, i-1$.

Proof. The posterior distribution of $\theta(i)$ is normal with mean vector

$$m(i) = P^{-1}(i)\{[GP^{-1}(i-1)G' + P_v^{-1}(1)]^{-1}H(i)X(i-1)$$
$$+ F'(i)P_u(i)Y(i) + P_v(i)G[P(i-1)$$
$$+ G'P_v(i)G]^{-1}P(i-1)m(i-1), \qquad (6.27)$$

and the precision matrix

$$P(i) = F'(i)P_u(i)F(i) + [G'P^{-1}(i-1)G + P_v^{-1}(i)]^{-1}, \qquad (6.28)$$

for $i = 1, 2, \ldots$, and is a recursive formula for the posterior mean and precision matrix of $\theta(i)$ and is similar to that given by the Kalman filter of Theorem 6.1; hence the posterior mean of $\theta(i)$ is expressed as a linear function of the current observation $Y(i)$, the known parameters G, $P_u(i)$, and $P_v(i)$ at the i-th stage, and the previous posterior mean $m(i-1)$ and precision matrix $P(i-1)$.

Next, what is the predictive distribution of $Y(i)$ given $Y(j) = y(j)$ for $j = 1, 2, \ldots, i-1$? It can be shown that this conditional predictive distribution is normal with mean $Q(i)R(i)$ and precision matrix $Q(i)$, where

$$Q(i) = P_u(i) - P_u(i)F(i)P^{-1}(i)F'(i)P_u(i), \qquad (6.29)$$

and

$$R(i) = P_u(i)F'(i)P^{-1}(i)\{[GP^{-1}(i-1)G' + P_v^{-1}(i)]^{-1}$$

$$\times H(i)X(i-1) + P_v(i)G[P(i-1) + G'P_v(i)G]^{-1}$$

$$\times P(i-1)m(i-1), \qquad (6.30)$$

and $P(i-1)$ is the posterior precision matrix of $\theta(i-1)$, given by formula (6.28), and (6.27) gives the posterior mean $m(i-1)$ of $\theta(i-1)$.

Since $m(i)$ is a linear function of $Y(i)$ and $X(i-1)$ and because $Q(i)R(i)$ is a linear function of $X(i-1)$, $\hat{X}(i-1)$ is given as above by (6.26).

The above theorem shows one how to compute the vectors $\hat{X}(0)$, $\hat{X}(1), \ldots$, in order to control the states of the system at target values $T(1), T(2), \ldots$, and (6.17) was the criterion used in developing this strategy, however, other criteria could have been employed. For example, one could use a loss function $L_i\{T(i), E[\theta(i)]\}$ at each stage and choose $X(i-1)$ in such a way as to minimize the expected loss with respect to the conditional predictive distribution of $Y(i)$ given $Y(j) = y(j)$, $j = 1, 2, \ldots, i-1$.

Theorem 6.4 is a single period control strategy, that is, only one-step-ahead predictions are involved. If one is interested in multi-period control strategies, there are several ways to devise control algorithms. For example, Zellner (1971) develops "here and now,"

AN EXAMPLE 251

sequential updating, and adaptive control strategies for multi-period situations involving multiple linear regression models. These methods are easily adapted to linear dynamic systems.

AN EXAMPLE

The preceding sections have introduced the reader to the analysis of linear dynamic systems. Estimation, including smoothing, filtering, and prediction was considered and a single-period control algorithm was devised in order to control the states of the system. These Bayesian methodologies will be illustrated with scalar linear dynamic models.

Let

$$Y(t) = 3\theta(t) + U(t), \qquad t = 1, 2, \ldots, 50,$$
$$\theta(t) = .5\theta(t-1) + V(t),$$

(6.31)

where $U(t) \sim n(0,.2)$, $V(t) \sim n(0,1)$, and $\theta(0) \sim n(2,.7)$, and where $\theta(0); U(1), \ldots, U(50); V(1), V(2), \ldots, V(50)$, are independent random variables. In terms of the general model $F(t) = 3$, $G = .5$, $P_u(t) = .5$, $P_v(t) = 1$, $m(0) = .2$ and $P(0) = (.7)^{-1}$.

Fifty observations were generated from the model (6.31) and Table 6.1 consists of six columns: (1) The U column has the fifty values of the observation errors. (2) The fifty system errors are shown in the V column. (3) The Theta column lists the fifty states where the initial state $\theta(0) = 2.9944$ is given in the first row. The Kalman filter estimates, (6.3), are listed in the m column. (5) The Y column gives the fifty observations (beginning in row 2), and (6) The P column lists the precision of the posterior distribution of the current states and is calculated from formula (6.4).

Since the design matrices and the two precision matrices are constants (scalars), the precision of the estimates reaches the level 18.9870 after three steps and stays at that level (to four decimal places). For the most part, the Kalman filter estimates imitate the behavior of the states with respect to sign; however, since the precision of each estimate is relatively low (18.9870), it is difficult to say that $m(k)$ is a "good" estimate of $\theta(k)$.

The next example considers the control problem with the scalar model.

$$Y(t) = 3X(t) + U(t),$$
$$\theta(t) = .5\theta(t-1) + 2X(t-1) + V(t),$$

(6.32)

Table 6.1
Kalman Filter Estimates of θ

Obs.	Row	T	U	V	θ	m	Y	P
1	1	1	0.3826	0.1421	2.9944	0.2000	0.000	0.7000
2	2	2	-1.0744	1.2243	2.7215	2.2744	7.090	18.7368
3	3	3	0.1626	-0.7466	0.6141	0.6927	2.005	18.9868
4	4	4	-0.6353	0.7390	1.0461	0.8089	2.503	18.9870
5	5	5	0.8767	0.3159	0.8390	1.0934	3.394	18.9870
6	6	6	-0.6793	-0.2560	0.1635	-0.0312	-0.189	18.9870
7	7	7	0.8093	-1.0921	-1.0103	-0.7029	-2.222	18.9870
8	8	8	-1.0961	-0.1825	-0.6876	-1.0165	-3.159	18.9870
9	9	9	1.4850	1.6075	1.2637	1.6408	5.276	18.9870
10	10	10	1.0534	-0.4624	0.1694	0.5362	1.562	18.9870
11	11	11	-1.2282	1.9927	2.0774	1.5952	5.004	18.9870
12	12	12	-2.6058	-1.5175	-0.4788	-1.2359	-4.042	18.9870
13	13	13	-0.1488	-3.3583	-3.5977	-3.4899	-10.942	18.9870
14	14	14	-2.0636	0.2301	-1.5687	-2.2300	-6.770	18.9870
15	15	15	-0.8232	2.1326	1.3482	0.9600	3.221	18.9870
16	16	16	1.3763	0.3588	1.0329	1.4391	4.475	18.9870
17	17	17	-0.3125	0.0436	0.5600	0.4696	1.368	18.9870
18	18	18	-2.0752	-1.5534	-1.2734	-1.8508	-5.895	18.9870
19	19	19	1.4938	-1.0473	-1.6840	-1.1725	-3.558	18.9870
20	20	20	1.8346	0.0716	-0.7704	-0.1811	-0.477	18.9870
21	21	21	-0.9506	0.4381	0.0529	-0.2549	-0.792	18.9870
22	22	22	1.9024	-0.1880	-0.1616	0.4414	1.418	18.9870
23	23	23	-1.8785	0.0780	-0.0028	-0.5848	-1.887	18.9870
24	24	24	-1.7613	-0.7542	-0.7556	-1.2881	-4.028	18.9870

AN EXAMPLE

25	25	25	−1.1793	1.4504	1.0726	0.6107	2.038	18.9870
26	26	26	−0.5606	0.9513	1.4876	1.2490	3.902	18.9870
27	27	27	2.6717	1.1334	1.8772	2.6564	8.303	18.9870
28	28	28	−2.8705	0.5065	1.4451	0.5319	1.465	18.9870
29	29	29	−1.5287	1.7053	2.4278	1.8323	5.755	18.9870
30	30	30	1.3191	0.4895	1.7034	2.0793	6.429	18.9870
31	31	31	−0.9422	−0.2560	0.5957	0.3211	0.845	18.9870
32	32	32	−0.6092	0.4785	0.7764	0.5519	1.720	18.9870
33	33	33	0.2991	0.8278	1.2160	1.2617	3.947	18.9870
34	34	34	−0.0597	0.0517	0.6597	0.6393	1.919	18.9870
35	35	35	0.8914	0.5348	0.8647	1.1180	3.485	18.9870
36	36	36	1.6638	−1.2130	−0.7806	−0.1852	−0.678	18.9870
37	37	37	0.9454	−1.5569	−1.9472	−1.5521	−4.896	18.9870
38	38	38	−0.9849	1.2628	0.2892	−0.0775	−0.117	18.9870
39	39	39	0.5713	0.1308	0.2754	0.4396	1.397	18.9870
40	40	40	2.0040	0.0570	0.1947	0.8292	2.588	18.9870
41	41	41	1.2940	0.1760	0.2734	0.6896	2.114	18.9870
42	42	42	−0.5726	−1.2522	−1.1155	−1.2206	−3.919	18.9870
43	43	43	0.3955	0.1569	−0.4008	−0.2867	−0.807	18.9870
44	44	44	2.7095	−0.0867	−0.2871	0.5766	1.848	18.9870
45	45	45	−2.0137	−0.3484	−0.4919	−1.0877	−3.489	18.9870
46	46	46	0.6470	0.1317	−0.1142	0.0679	0.304	18.9870
47	47	47	2.6622	−2.7617	−2.8189	−1.8293	−5.794	18.9870
48	48	48	−0.9927	−1.0189	−2.4283	−2.6633	−8.278	18.9870
49	49	49	0.3914	1.4949	0.2808	0.3206	1.234	18.9870
50	50	50	2.1436	1.0396	1.1800	1.8043	5.683	18.9870
51	51	51	−0.2302	0.9998	1.5897	1.4813	4.539	18.9870

Table 6.2
Controlling the States of a Dynamic Model ($P_u = 1$, $P_v = 1$)

Obs.	Row	Time	P	Q	A	B
1	1	1	0.70000	0.0000000	0.000000	0.000000
2	2	2	9.73684	0.0756757	0.151351	0.308108
3	3	3	9.97497	0.0977414	0.195483	0.300753
4	4	4	9.97555	0.0977941	0.195588	0.300735
5	5	5	9.97555	0.0977942	0.195588	0.300735
6	6	6	9.97555	0.0977942	0.195588	0.300735
7	7	7	9.97555	0.0977942	0.195588	0.300735
8	8	8	9.97555	0.0977942	0.195588	0.300735
9	9	9	9.97555	0.0977942	0.195588	0.300735
10	10	10	9.97555	0.0977942	0.195588	0.300735
11	11	11	9.97555	0.0977942	0.195588	0.300735
12	12	12	9.97555	0.0977942	0.195588	0.300735
13	13	13	9.97555	0.0977942	0.195588	0.300735
14	14	14	9.97555	0.0977942	0.195588	0.300735
15	15	15	9.97555	0.0977942	0.195588	0.300735
16	16	16	9.97555	0.0977942	0.195588	0.300735
17	17	17	9.97555	0.0977942	0.195588	0.300735
18	18	18	9.97555	0.0977942	0.195588	0.300735
19	19	19	9.97555	0.0977942	0.195588	0.300735
20	20	20	9.97555	0.0977942	0.195588	0.300735
21	21	21	9.97555	0.0977942	0.195588	0.300735
22	22	22	9.97555	0.0977942	0.195588	0.300735
23	23	23	9.97555	0.0977942	0.195588	0.300735
24	24	24	9.97555	0.0977942	0.195588	0.300735
25	25	25	9.97555	0.0977942	0.195588	0.300735
26	26	26	9.97555	0.0977942	0.195588	0.300735
27	27	27	9.97555	0.0977942	0.195588	0.300735
28	28	28	9.97555	0.0977942	0.195588	0.300735
29	29	29	9.97555	0.0977942	0.195588	0.300735
30	30	30	9.97555	0.0977942	0.195588	0.300735
31	31	31	9.97555	0.0977942	0.195588	0.300735
32	32	32	9.97555	0.0977942	0.195588	0.300735
33	33	33	9.97555	0.0977942	0.195588	0.300735
34	34	34	9.97555	0.0977942	0.195588	0.300735
35	35	35	9.97555	0.0977942	0.195588	0.300735
36	36	36	9.97555	0.0977942	0.195588	0.300735
37	37	37	9.97555	0.0977942	0.195588	0.300735
38	38	38	9.97555	0.0977942	0.195588	0.300735
39	39	39	9.97555	0.0977942	0.195588	0.300735
40	40	40	9.97555	0.0977942	0.195588	0.300735

AN EXAMPLE

C	D	E	Z	X	Y	m
0.00000	0	0.0000	−0.40000	0.7325	0.000	0.2000
0.00757	6	0.3000	−0.31380	−0.0426	−0.298	−0.1448
−0.00708	6	−0.2172	−0.22689	−0.2333	−1.410	−0.4924
−0.02408	6	−0.7387	−0.12329	−0.3283	−2.788	−0.9068
−0.04434	6	−1.3603	−0.31526	−0.3921	−0.234	−0.1389
−0.00679	6	−0.2084	−0.23583	−0.2221	−1.291	−0.4567
−0.02233	6	−0.6850	−0.01490	−1.2632	−4.229	−1.3404
−0.06554	6	−2.0106	−0.30429	0.0335	−0.380	−0.1828
−0.00894	6	−0.2743	0.05216	−1.3693	−5.121	−1.6086
−0.07866	6	−2.4129	0.28871	−2.4751	−8.268	−2.5548
−0.12492	6	−3.8323	0.15742	−2.6532	−6.521	−2.0297
−0.09925	6	−3.0445	0.15079	−2.2095	−6.433	−2.0031
−0.09795	6	−3.0047	0.03715	−1.4761	−4.922	−1.5486
−0.07572	6	−2.3229	−0.35570	−0.5359	0.303	0.0228
0.00111	6	0.0342	−0.09660	−0.7670	−3.143	−1.0136
−0.04956	6	−1.5204	−0.15437	−0.7780	−2.374	−0.7825
−0.03826	6	−1.1738	−0.47145	0.3691	1.843	0.4858
0.02375	6	0.7287	−0.16861	−0.8516	−2.185	−0.7256
−0.03548	6	−1.0883	−0.30853	−0.0931	−0.324	−0.1659
−0.00811	6	−0.2488	−0.50668	1.2551	2.312	0.6267
0.03064	6	0.9401	−0.20352	−0.2098	−1.721	−0.5859
−0.02865	6	−0.8789	0.06098	−1.8357	−5.239	−1.6439
−0.08038	6	−2.4659	−0.25907	−0.6816	−0.982	−0.3637
−0.01778	6	−0.5456	−0.35562	−0.4626	0.302	0.0225
0.00110	6	0.0337	−0.30277	−0.5534	−0.401	−0.1889
−0.00924	6	−0.2834	0.44102	−3.2116	−10.293	−3.1641
−0.15471	6	−4.7461	0.01417	−1.4440	−4.616	−1.4567
−0.07123	6	−2.1850	−0.44138	0.7725	1.443	0.3655
0.01787	6	0.5483	−0.32200	−0.0560	−0.145	−0.1120
−0.00548	6	−0.1680	−0.30883	−0.5475	−0.320	−0.1647
−0.00805	6	−0.2470	−0.42132	0.0093	1.176	0.2853
0.01395	6	0.4279	−0.19455	−0.4755	−1.840	−0.6218
−0.03040	6	−0.9327	−0.23864	−0.5623	−1.254	−0.4454
−0.02178	6	−0.6682	0.30711	−2.4864	−8.512	−2.6284
−0.12852	6	−3.9427	−0.17066	−0.9176	−2.158	−0.7174
−0.03508	6	−1.0761	−0.13491	−0.4803	−2.633	−0.8604
−0.04207	6	−1.2906	−0.79765	1.4855	6.182	1.7906
0.08756	6	2.6859	−0.06477	−1.5049	−3.566	−1.1409
−0.05579	6	−1.7114	0.57059	−3.4357	−12.017	−3.6824
−0.18006	6	−5.5236	−0.52164	0.9957	2.511	0.6865

Table 6.2 (Continued)

Obs.	Row	Time	P	Q	A	B
41	41	41	9.97555	0.0977942	0.195588	0.300735
42	42	42	9.97555	0.0977942	0.195588	0.300735
43	43	43	9.97555	0.0977942	0.195588	0.300735
44	44	44	9.97555	0.0977942	0.195588	0.300735
45	45	45	9.97555	0.0977942	0.195588	0.300735
46	46	46	9.97555	0.0977942	0.195588	0.300735
47	47	47	9.97555	0.0977942	0.195588	0.300735
48	48	48	9.97555	0.0977942	0.195588	0.300735
49	49	49	9.97555	0.0977942	0.195588	0.300735
50	50	50	9.97555	0.0977942	0.195588	0.300735
51	51	51	9.97555	0.0977942	0.195588	0.300735

Table 6.3
Controlling the States of a Dynamic Model ($P_u = 10$, $P_v = 10$)

Obs.	Row	Time	P	Q	A	B
1	1	1	0.7000	0.000000	0.000000	0.000000
2	2	2	92.1875	0.237288	0.047458	0.325424
3	3	3	99.7360	0.976175	0.195235	0.300794
4	4	4	99.7555	0.977938	0.195588	0.300735
5	5	5	99.7555	0.977942	0.195588	0.300735
6	6	6	99.7555	0.977942	0.195588	0.300735
7	7	7	99.7555	0.977942	0.195588	0.300735
8	8	8	99.7555	0.977942	0.195588	0.300735
9	9	9	99.7555	0.977942	0.195588	0.300735
10	10	10	99.7555	0.977942	0.195588	0.300735

AN EXAMPLE

C	D	E	Z	X	Y	m
0.03357	6	1.0298	0.01298	−1.5775	−4.600	−1.4519
−0.07099	6	−2.1779	−0.22242	−0.1770	−1.469	−0.5103
−0.02495	6	−0.7655	−0.21163	−0.6936	−1.613	−0.5535
−0.02706	6	−0.8302	−0.04881	−1.4155	−3.778	−1.2048
−0.05891	6	−1.8071	−0.07888	−1.0893	−3.378	−1.0845
−0.05303	6	−1.6267	0.07490	−1.4817	−5.424	−1.6996
−0.08311	6	−2.5494	−0.66884	1.0351	4.468	1.2754
0.06236	6	1.9130	−0.52972	1.0808	2.618	0.7189
0.03515	6	1.0783	−0.07701	−1.5447	−3.403	−1.0920
−0.05339	6	−1.6379	−0.20075	−0.4395	−1.757	−0.5970
−0.02919	6	−0.8955	0.00000	−2.0814	−6.034	−1.8832

C	D	E	Z	X	Y	m
0.000000	0	0.0000	−0.40000	−0.5799	0.0000	0.2000
0.002373	6	0.3000	−0.00402	−1.3917	−4.2016	−1.3839
−0.067548	6	−2.0759	−0.13764	−0.8973	−2.5968	−0.8494
−0.041535	6	−1.2742	−0.25436	−0.4308	−1.0445	−0.3826
−0.018706	6	−0.5738	−0.23623	−0.6033	−1.2856	−0.4551
−0.022252	6	−0.6826	−0.23661	−0.5128	−1.2805	−0.4535
−0.022177	6	−0.6803	−0.21117	−0.4343	−1.6189	−0.5553
−0.027153	6	−0.8330	−0.27782	−0.2504	−0.7324	−0.2887
−0.014117	6	−0.4331	−0.20828	−0.5724	−1.6573	−0.5669
−0.027718	6	−0.8503	−0.03031	−1.2561	−4.0245	−1.2788

Table 6.3 (Continued)

Obs.	Row	Time	P	Q	A	B
11	11	11	99.7555	0.977942	0.195588	0.300735
12	12	12	99.7555	0.977942	0.195588	0.300735
13	13	13	99.7555	0.977942	0.195588	0.300735
14	14	14	99.7555	0.977942	0.195588	0.300735
15	15	15	99.7555	0.977942	0.195588	0.300735
16	16	16	99.7555	0.977942	0.195588	0.300735
17	17	17	99.7555	0.977942	0.195588	0.300735
18	18	18	99.7555	0.977942	0.195588	0.300735
19	19	19	99.7555	0.977942	0.195588	0.300735
20	20	20	99.7555	0.977942	0.195588	0.300735
21	21	21	99.7555	0.977942	0.195588	0.300735
22	22	22	99.7555	0.977942	0.195588	0.300735
23	23	23	99.7555	0.977942	0.195588	0.300735
24	24	24	99.7555	0.977942	0.195588	0.300735
25	25	25	99.7555	0.977942	0.195588	0.300735
26	26	26	99.7555	0.977942	0.195588	0.300735
27	27	27	99.7555	0.977942	0.195588	0.300735
28	28	28	99.7555	0.977942	0.195588	0.300735
29	29	29	99.7555	0.977942	0.195588	0.300735
30	30	30	99.7555	0.977942	0.195588	0.300735
31	31	31	99.7555	0.977942	0.195588	0.300735
32	32	32	99.7555	0.977942	0.195588	0.300735
33	33	33	99.7555	0.977942	0.195588	0.300735
34	34	34	99.7555	0.977942	0.195588	0.300735
35	35	35	99.7555	0.977942	0.195588	0.300735
36	36	36	99.7555	0.977942	0.195588	0.300735
37	37	37	99.7555	0.977942	0.195588	0.300735
38	38	38	99.7555	0.977942	0.195588	0.300735
39	39	39	99.7555	0.977942	0.195588	0.300735
40	40	40	99.7555	0.977942	0.195588	0.300735
41	41	41	99.7555	0.977942	0.195588	0.300735
42	42	42	99.7555	0.977942	0.195588	0.300735
43	43	43	99.7555	0.977942	0.195588	0.300735
44	44	44	99.7555	0.977942	0.195588	0.300735
45	45	45	99.7555	0.977942	0.195588	0.300735
46	46	46	99.7555	0.977942	0.195588	0.300735
47	47	47	99.7555	0.977942	0.195588	0.300735
48	48	48	99.7555	0.977942	0.195588	0.300735
49	49	49	99.7555	0.977942	0.195588	0.300735
50	50	50	99.7555	0.977942	0.195588	0.300735
51	51	51	99.7555	0.977942	0.195588	0.300735

AN EXAMPLE

C	D	E	Z	X	Y	m
−0.062527	6	−1.9181	−0.11359	−1.1526	−2.9168	−0.9457
−0.046240	6	−1.4185	−0.06541	−1.2391	−3.5576	−1.1384
−0.055662	6	−1.7075	−0.08836	−1.0500	−3.2524	−1.0466
−0.051174	6	−1.5699	−0.05701	−1.1103	−3.6693	−1.1719
−0.057305	6	−1.7579	−0.11180	−0.9615	−2.9406	−0.9528
−0.046590	6	−1.4292	−0.18574	−0.5482	−1.9572	−0.6571
−0.032128	6	−0.9856	−0.25178	−0.2025	−1.0788	−0.3929
−0.019210	6	−0.5893	−0.11794	−1.0506	−2.8590	−0.9283
−0.045389	6	−1.3924	−0.21076	−0.6179	−1.6244	−0.5570
−0.027235	6	−0.8355	−0.17227	−0.6563	−2.1364	−0.7109
−0.034763	6	−1.0664	−0.17925	−0.5841	−2.0435	−0.6830
−0.033397	6	−1.0245	−0.11321	−0.8759	−2.9219	−0.9472
−0.046314	6	−1.4208	−0.05174	−1.1211	−3.7395	−1.1931
−0.058337	6	−1.7896	−0.05185	−1.2219	−3.7380	−1.1926
−0.058315	6	−1.7889	−0.13639	−0.8477	−2.6135	−0.8544
−0.041780	6	−1.2817	−0.21908	−0.6364	−1.5137	−0.5237
−0.025607	6	−0.7855	−0.24053	−0.3534	−1.2284	−0.4379
−0.021412	6	−0.6568	−0.17192	−0.6932	−2.1410	−0.7123
−0.034831	6	−1.0685	−0.16675	−0.6304	−2.2097	−0.7330
−0.035841	6	−1.0995	−0.24432	−0.4306	−1.1779	−0.4227
−0.020669	6	−0.6341	−0.11674	−0.8623	−2.8749	−0.9330
−0.045623	6	−1.3995	−0.14040	−0.7969	−2.5603	−0.8384
−0.040996	6	−1.2576	−0.15885	−0.6786	−2.3148	−0.7646
−0.037387	6	−1.1469	−0.22788	−0.6014	−1.3966	−0.4885
−0.023885	6	−0.7327	−0.17246	−0.7181	−2.1337	−0.7101
−0.034724	6	−1.0652	−0.16838	−0.7963	−2.1880	−0.7265
−0.035522	6	−1.0897	−0.14956	−0.7645	−2.4383	−0.8017
−0.039203	6	−1.2026	−0.15651	−0.6689	−2.3459	−0.7739
−0.037843	6	−1.1609	−0.16019	−0.6376	−2.2969	−0.7592
−0.037124	6	−1.1388	−0.21685	−0.4315	−1.5433	−0.5326
−0.026042	6	−0.7989	−0.25865	−0.3983	−0.9874	−0.3654
−0.017866	6	−0.5481	−0.08047	−1.1366	−3.3573	−1.0781
−0.052717	6	−1.6172	−0.26349	−0.3753	−0.9231	−0.3461
−0.016921	6	−0.5191	−0.18463	−0.4555	−1.9720	−0.6615
−0.032345	6	−0.9922	−0.23431	−0.4214	−1.3111	−0.4627
−0.022627	6	−0.6941	−0.23391	−0.5331	−1.3164	−0.4643
−0.022705	6	−0.6965	−0.18750	−0.6470	−1.9337	−0.6500
−0.031783	6	−0.9750	−0.14923	−0.8294	−2.4427	−0.8031
−0.039267	6	−1.2046	−0.09501	−1.0038	−3.1639	−1.0200
−0.049873	6	−1.5299	−0.13909	−0.7544	−2.5776	−0.8436
−0.041252	6	−1.2655	0.00000	0.1014	0.7220	0.1487

Table 6.4
Controlling the States of a Dynamic Model ($P_u = 100$, $P_v = 100$)

Obs.	Row	Time	P	Q	A	B
1	1	1	0.700	0.00000	0.000000	0.000000
2	2	2	902.724	0.30172	0.006034	0.332328
3	3	3	997.305	9.75682	0.195136	0.300811
4	4	4	997.555	9.77937	0.195587	0.300735
5	5	5	997.555	9.77942	0.195588	0.300735
6	6	6	997.555	9.77942	0.195588	0.300735
7	7	7	997.555	9.77942	0.195588	0.300735
8	8	8	997.555	9.77942	0.195588	0.300735
9	9	9	997.555	9.77942	0.195588	0.300735
10	10	10	997.555	9.77942	0.195588	0.300735
11	11	11	997.555	9.77942	0.195588	0.300735
12	12	12	997.555	9.77942	0.195588	0.300735
13	13	13	997.555	9.77942	0.195588	0.300735
14	14	14	997.555	9.77942	0.195588	0.300735
15	15	15	997.555	9.77942	0.195588	0.300735
16	16	16	997.555	9.77942	0.195588	0.300735
17	17	17	997.555	9.77942	0.195588	0.300735
18	18	18	997.555	9.77942	0.195588	0.300735
19	19	19	997.555	9.77942	0.195588	0.300735
20	20	20	997.555	9.77942	0.195588	0.300735
21	21	21	997.555	9.77942	0.195588	0.300735
22	22	22	997.555	9.77942	0.195588	0.300735
23	23	23	997.555	9.77942	0.195588	0.300735
24	24	24	997.555	9.77942	0.195588	0.300735
25	25	25	997.555	9.77942	0.195588	0.300735
26	26	26	997.555	9.77942	0.195588	0.300735
27	27	27	997.555	9.77942	0.195588	0.300735
28	28	28	997.555	9.77942	0.195588	0.300735
29	29	29	997.555	9.77942	0.195588	0.300735
30	30	30	997.555	9.77942	0.195588	0.300735
31	31	31	997.555	9.77942	0.195588	0.300735
32	32	32	997.555	9.77942	0.195588	0.300735
33	33	33	997.555	9.77942	0.195588	0.300735
34	34	34	997.555	9.77942	0.195588	0.300735
35	35	35	997.555	9.77942	0.195588	0.300735
36	36	36	997.555	9.77942	0.195588	0.300735
37	37	37	997.555	9.77942	0.195588	0.300735
38	38	38	997.555	9.77942	0.195588	0.300735
39	39	39	997.555	9.77942	0.195588	0.300735
40	40	40	997.555	9.77942	0.195588	0.300735

AN EXAMPLE

C	D	E	Z	X	Y	m
0.000000	0	0.0000	−0.40000	0.28333	0.0000	0.20000
0.000302	6	0.3000	−0.22431	−0.49200	−1.5064	−0.50275
−0.024526	6	−0.7541	−0.19195	−0.57410	−1.8746	−0.63220
−0.030913	6	−0.9483	−0.16457	−0.71453	−2.2388	−0.74173
−0.036269	6	−1.1126	−0.20299	−0.61187	−1.7278	−0.58805
−0.028754	6	−0.8821	−0.18757	−0.65452	−1.9328	−0.64972
−0.031769	6	−0.9746	−0.18673	−0.60781	−1.9439	−0.65306
−0.031933	6	−0.9796	−0.16810	−0.76964	−2.1918	−0.72761
−0.035578	6	−1.0914	−0.13379	−0.84556	−2.6481	−0.86483
−0.042288	6	−1.2972	−0.14870	−0.73195	−2.4498	−0.80519
−0.039371	6	−1.2078	−0.19428	−0.53905	−1.8435	−0.62286
−0.030456	6	−0.9343	−0.17928	−0.68517	−2.0431	−0.68288
−0.033391	6	−1.0243	−0.17989	−0.74018	−2.0350	−0.68044
−0.033272	6	−1.0207	−0.19120	−0.70264	−1.8846	−0.63522
−0.031060	6	−0.9528	−0.19300	−0.61615	−1.8605	−0.62799
−0.030707	6	−0.9420	−0.13333	−0.88427	−2.6543	−0.86668
−0.042378	6	−1.3000	−0.18604	−0.66993	−1.9532	−0.65585
−0.032069	6	−0.9838	−0.22117	−0.58540	−1.4859	−0.51531
−0.025197	6	−0.7730	−0.19163	−0.65295	−1.8787	−0.63346
−0.030974	6	−0.9502	−0.18874	−0.63355	−1.9173	−0.64505
−0.031541	6	−0.9676	−0.13197	−0.87727	−2.6724	−0.87213
−0.042645	6	−1.3082	−0.17391	−0.70311	−2.1145	−0.70435
−0.034441	6	−1.0565	−0.18155	−0.67364	−2.0129	−0.67379
−0.032947	6	−1.0107	−0.17985	−0.67002	−2.0355	−0.68059
−0.033279	6	−1.0209	−0.20335	−0.59161	−1.7229	−0.58658
−0.028682	6	−0.8799	−0.17897	−0.58401	−2.0473	−0.68414
−0.033452	6	−1.0262	−0.18618	−0.59363	−1.9513	−0.65527
−0.032041	6	−0.9829	−0.17190	−0.74043	−2.1412	−0.71239
−0.034834	6	−1.0686	−0.15734	−0.80895	−2.3349	−0.77064
−0.037682	6	−1.1560	−0.16448	−0.78257	−2.2399	−0.74206
−0.036285	6	−1.1131	−0.18151	−0.73864	−2.0134	−0.67396
−0.032955	6	−1.0109	−0.12254	−0.90182	−2.7978	−0.90984
−0.044489	6	−1.3648	−0.18562	−0.66531	−1.9587	−0.65751
−0.032150	6	−0.9863	−0.15506	−0.79239	−2.3652	−0.77977
−0.038129	6	−1.1697	−0.19711	−0.59683	−1.8059	−0.61155
−0.029903	6	−0.9173	−0.18751	−0.59959	−1.9336	−0.64995
−0.031781	6	−0.9749	−0.18134	−0.68708	−2.0156	−0.67462
−0.032987	6	−1.0119	−0.13812	−0.80512	−2.5906	−0.84754
−0.041442	6	−1.2713	−0.17804	−0.73109	−2.0596	−0.68784
−0.033633	6	−1.0318	−0.16883	−0.74748	−2.1820	−0.72467

LINEAR DYNAMIC SYSTEMS

Table 6.4 (Continued)

Obs.	Row	Time	P	Q	A	B
41	41	41	997.555	9.77942	0.195588	0.300735
42	42	42	997.555	9.77942	0.195588	0.300735
43	43	43	997.555	9.77942	0.195588	0.300735
44	44	44	997.555	9.77942	0.195588	0.300735
45	45	45	997.555	9.77942	0.195588	0.300735
46	46	46	997.555	9.77942	0.195588	0.300735
47	47	47	997.555	9.77942	0.195588	0.300735
48	48	48	997.555	9.77942	0.195588	0.300735
49	49	49	997.555	9.77942	0.195588	0.300735
50	50	50	997.555	9.77942	0.195588	0.300735
51	51	51	997.555	9.77942	0.195588	0.300735

where $t = 1, 2, \ldots, 50$, where $Y(t)$, $\theta(t)$, $U(t)$, $X(t)$ and $V(t)$ are scalar and $U(t) \sim n(0, P_u^{-1})$, $V(t) \sim n(0, P_v^{-1})$, and $\theta(0) \sim n(.2, .7^{-1})$.
Suppose the states of the system are to be controlled at the target values $T(t) = -.7$, $t = 1, 2, \ldots, 50$, in such a way that the future mean of each state is equal to the target value. This is the criterion (6.17) of the preceding section. Now using the formulas (6.26) of Theorem 6.4, fifty observations were generated from the model (6.32) under three conditions:

Experiment One: $P_u = P_v = 1$ \hfill (6.33)

Experiment Two: $P_u = P_v = 10$ \hfill (6.34)

Experiment Three: $P_u = P_v = 100$ \hfill (6.35)

The observations of (6.32) were generated as follows:

(i) $\theta(0)$ is selected from a $n(.2, .7^{-1})$ population,
(ii) $X(0)$ is computed from (6.26),
(iii) $V(1)$ is selected from a $n(0, P_u^{-1})$ population,
(iv) $\theta(1)$ is computed from (6.32),
(v) $U(1)$ is selected from a $n(0, P_v^{-1})$ population,
(vi) and $Y(1)$ is computed from (6.32).
(vii) The first six steps are repeated 49 times to give fifty observations and values of the control variable.

AN EXAMPLE

C	D	E	Z	X	Y	m
−0.035434	6	−1.0870	−0.16695	−0.72732	−2.2071	−0.73220
−0.035803	6	−1.0983	−0.17884	−0.63440	−2.0489	−0.68464
−0.033477	6	−1.0270	−0.18968	−0.67078	−1.9047	−0.64126
−0.031356	6	−0.9619	−0.18641	−0.68313	−1.9483	−0.65437
−0.031997	6	−0.9816	−0.17667	−0.68688	−2.0777	−0.69331
−0.033901	6	−1.0400	−0.15879	−0.74127	−2.3156	−0.76485
−0.037399	6	−1.1473	−0.18696	−0.70172	−1.9410	−0.65218
−0.031889	6	−0.9783	−0.18934	−0.60147	−1.9093	−0.64263
−0.031423	6	−0.9640	−0.18001	−0.65361	−2.0334	−0.67996
−0.033248	6	−1.0199	−0.17062	−0.75086	−2.1583	−0.71752
−0.035085	6	−1.0763	0.00000	−0.71399	−2.1808	−0.72430

Tables 6.2, 6.3, and 6.4 give the results of the three experiments. These tables are organized into several columns. The X, A, B, C, D, E, P, and Q columns are the values needed to compute the control variable X, according to formula (6.26), while the m column is the Kalman filter estimate of the θ column, which lists the fifty states (beginning in row 2) of the system. All the relevant formulas are found in Theorem 6.4.

For example from Table 6.2

$$m(1) = -0.1448 \qquad (6.36)$$

is the Kalman filter estimate of $\theta(1) = -0.0426$ and the precision of the estimate is $P(1) = 9.73684$.

From Table 6.3 ($P_u = P_v = 10$), row 2,

$$m(1) = -1.3839 \qquad (6.37)$$

is the Kalman filter estimate of $\theta(1) = -1.3917$ and the precision of the estimate is $P(1) = 92.1875$. The precision of the Kalman filter estimates increases by a factor of about ten because the observation and system precisions have been increased by a factor of ten, namely from one to ten.

When the precision is increased to $P_u = P_v = 100$,

$$m(1) = -0.50275 \qquad (6.38)$$

is the Kalman filter estimate of $\theta(1) = -0.492$, found in row 2 of Table 6.4, and the precision of the estimate is 902.724.

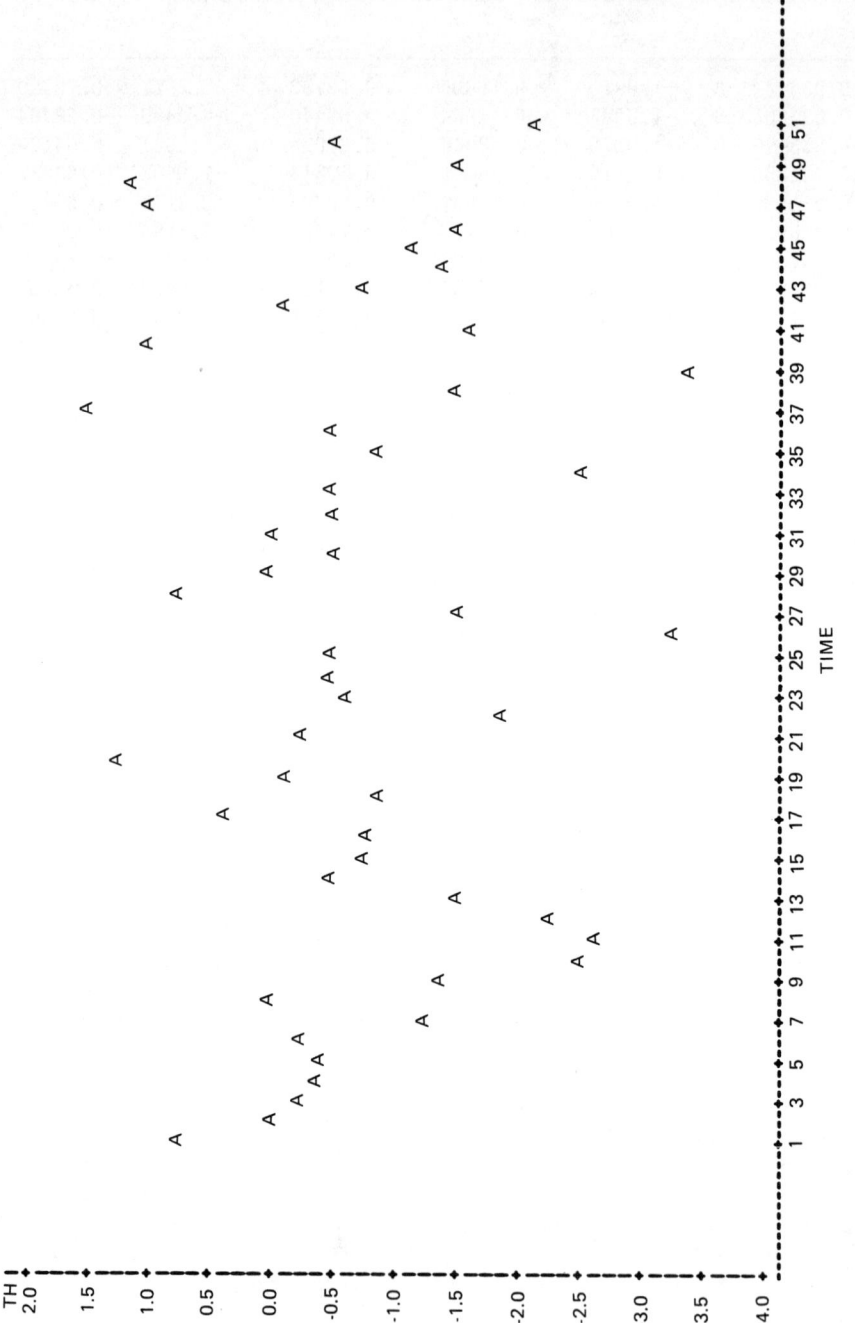

Figure 6.1. Controlling the states of a dynamic model ($P_u = 1$, $P_v = 1$).

AN EXAMPLE

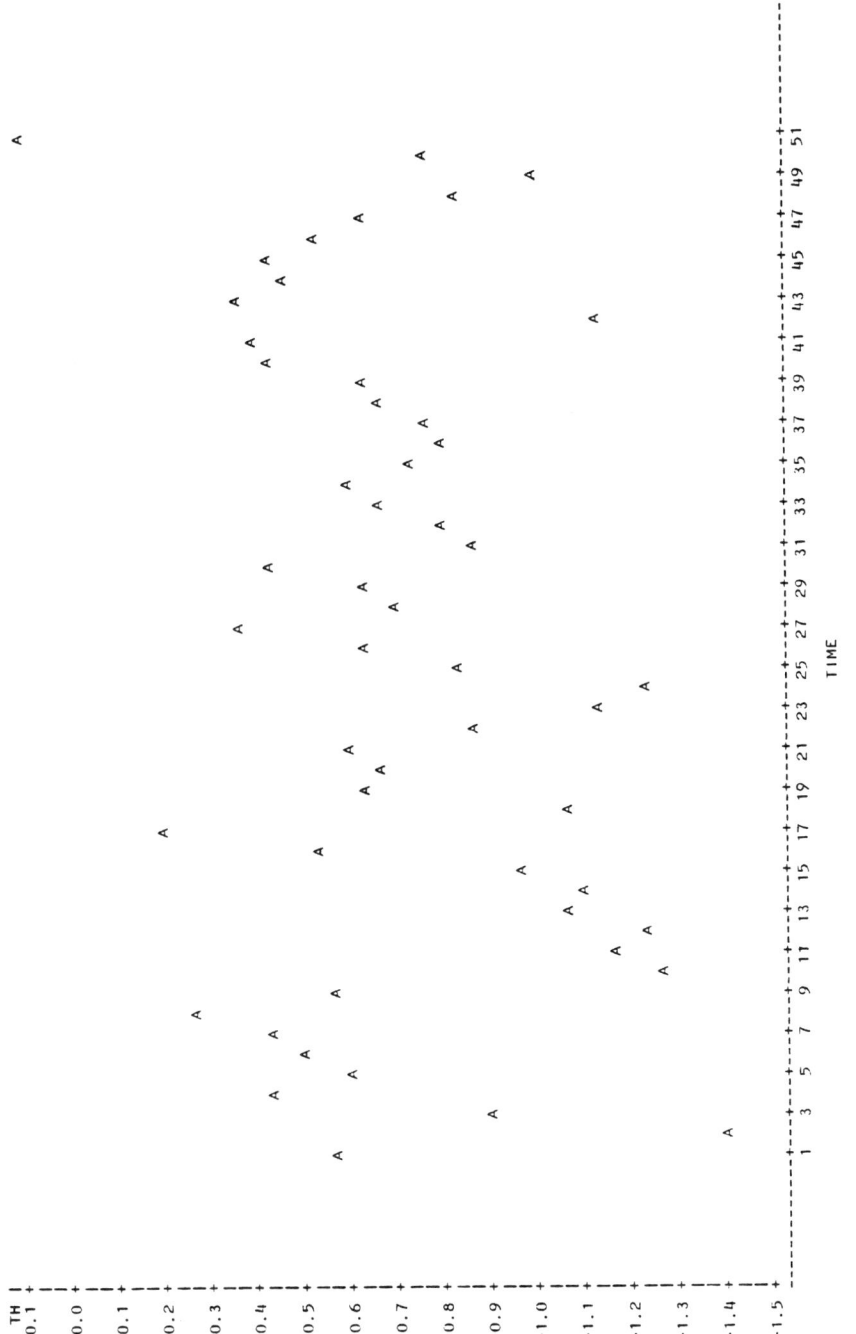

Figure 6.2. Controlling the states of a dynamic model ($P_u = 10$, $P_v = 10$).

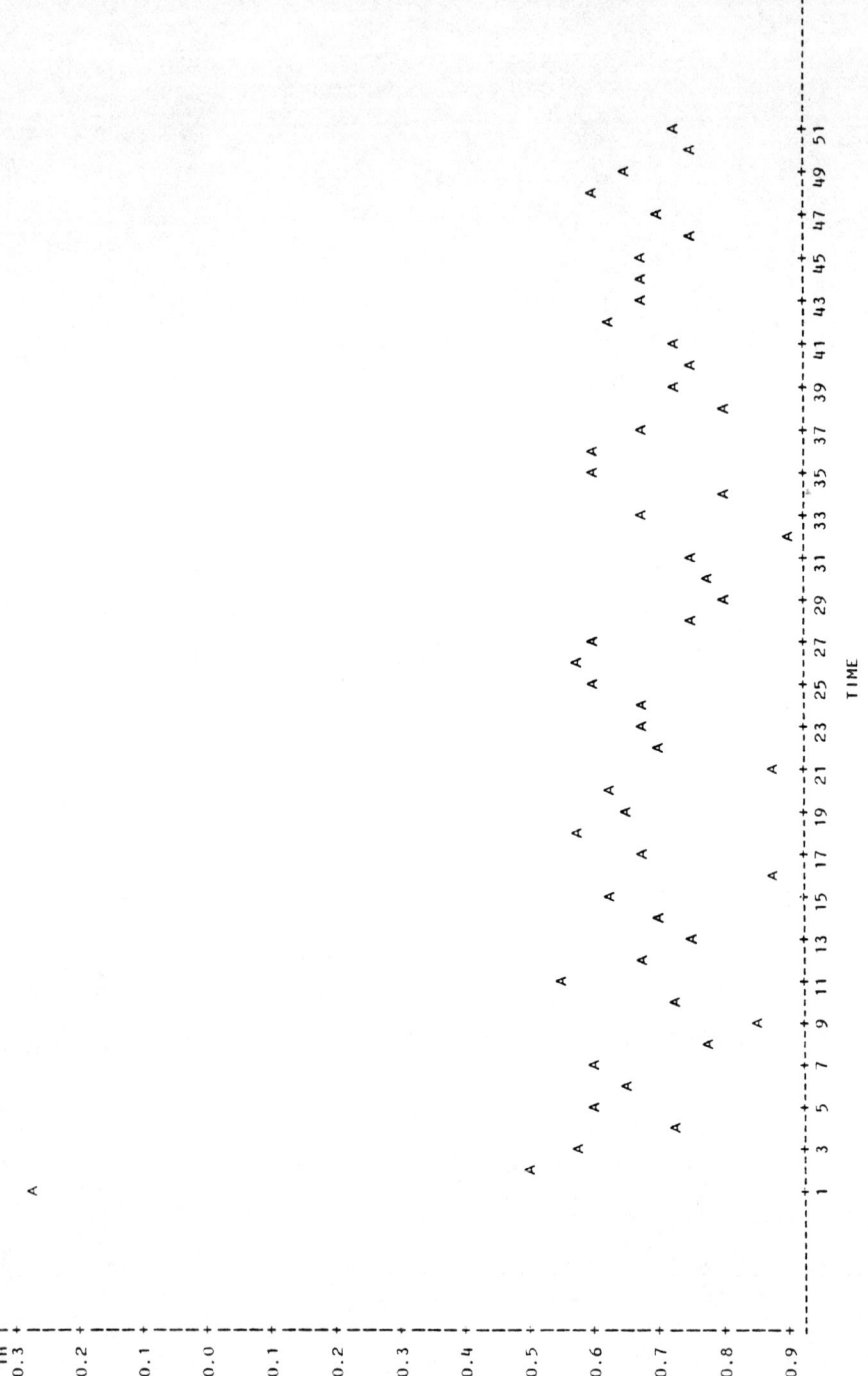

Figure 6.3. Controlling the states of a dynamic model ($P_u = 100$, $P_v = 100$).

NONLINEAR DYNAMIC SYSTEMS

The objective of this study was to see the effect of the observation and system precision on the control of the state vectors, which were supposed to be close to -0.7. From the θ column of the tables and from Figure 6.3 we see the θ values become close to the target value -0.7 when P_u and P_v are one hundred. That this should be the case is evident from the formulas of Theorem 6.4. Figures 6.1 and 6.2 demonstrate that θ is much more difficult to control when the system and observation errors are small as compared to Figure 6.3, when the precisions are each one hundred.

NONLINEAR DYNAMIC SYSTEMS

Consider the dynamic system

$$Y(t) = F_t[\theta(t)] + U(t) \qquad (6.39)$$

$$t = 1, 2, \ldots$$

$$\theta(t) = G_t[\theta(t-1)] + V(t), \qquad (6.40)$$

where F_t is some function of the state vector $\theta(t)$ and G_t is some function of $\theta(t-1)$, the previous state. As before assume F_t, G_t, $P_u(t)$, $P_v(t)$, $M(0)$, and $P(0)$ are known, where $P_u(t)$ is the precision matrix of $U(t)$, $P_v(t)$ is the precision matrix of $V(t)$, $M(0)$ the mean of $\theta(0)$ and $P(0)$ the precision matrix of $\theta(0)$. Suppose $Y(t)$ is the n × 1 observation vector, $\theta(t)$ a real p × 1 state vector, $F_t: R^p \to R^n$, $G_t: R^p \to R^p$, $U(t) \sim N[0, P_u^{-1}(t)]$, $V(t) \sim N[0, P_v^{-1}(t)]$, and that $\theta(0)$; $U(1), U(2), \ldots$; $V(1), V(2), \ldots$, are independent.

When $F_t[\theta(t)] = F(t)\theta(t)$ and $G_t[\theta(t)] = G\theta(t-1)$, we have the usual linear dynamic system with (6.39) as the observation equation with design matrix $F(t)$ and G is the transition matrix of the system equation (6.40). Obviously, the system equation (6.40) can be amended to include a nonlinear function of control variables. There are many versions of the nonlinear dynamic system, and one should refer to Jazwinski (1971) for a detailed account of nonlinear filters.

If F_t and/or G_t are nonlinear functions of $\theta(t)$ and $\theta(t-1)$, respectively, difficulties are encountered with determining the posterior distribution of $\theta(k)$, given k observations, $Y(1), Y(2), \ldots, Y(k)$ for $k = 1, 2, \ldots$. These troubles are not new to statisticians who have worked with nonlinear regression problems. Except for some special nonlinear functions F_t and G_t, the Kalman filter formulas are no longer applicable and one must devise some new methodology for the nonlinear case.

Suppose the function G_t of the system equation is linear in $\theta(t-1)$, that is $G_t[\theta(t)] = G\theta(t-1)$ and G is a $p \times p$ transition matrix, then

$$\theta(t) = G\theta(t-1) + V(t) \tag{6.41}$$

is the system equation.

Consider estimation of $\theta(1)$, now the prior distribution of $\theta(1)$ is induced by the prior distribution of $\theta(0)$, which is $N[M(0), P(0)]$. It is obvious the prior density of $\theta(1)$ is normal with mean $M(1)$ and precision $P(1)$, where

$$M(1) = [P_v^{-1}(1) + GP^{-1}(0)G']P_v(1)G[P(0) + G'P_v(1)G]^{-1}P(0)M(0) \tag{6.42}$$

and

$$P(1) = [GP^{-1}(0)G' + P_v^{-1}(1)]^{-1}. \tag{6.43}$$

What is the posterior distribution of $\theta(1)$? From the observation equation with $t = 1$, the conditional distribution of $Y(1)$ given $\theta(1)$ and $P_u(1)$ is normal with mean $F_1[\theta(1)]$ and precision $P_u(1)$, thus the posterior distribution of $\theta(1)$ is

$$\pi[\theta(1)|Y(1)] \propto \exp -\frac{1}{2}\{Y(1) - F_1[\theta(1)]\}'P_u(1)$$

$$\times \{Y(1) - F_1[\theta(1)]\}$$

$$\times \exp -\frac{1}{2}[\theta(1) - M(1)]'P(1)[\theta(1) - M(1)],$$

$$\theta(1) \in R^p \tag{6.44}$$

but if F_1 is nonlinear in $\theta(1)$, the distribution of $\theta(1)$ given $Y(1)$ is not normal, hence the recursive estimates of $\theta(t)$ given by the Kalman filter (6.3) do not hold. How should one estimate $\theta(1)$?

NONLINEAR DYNAMIC SYSTEMS 269

If p = 1, 2, 3, numerical methods may be used to estimate $\theta(1)$. For example, one could plot the one and two dimensional marginal posterior densities of the components of $\theta(1)$ and compute posterior joint and marginal moments of $\theta(1)$.

Now consider the next stage for estimating $\theta(2)$, where the prior distribution of $\theta(2)$ is induced by the posterior distribution of $\theta(1)$, where

$$\theta(2) = G\theta(1) + V(2).$$

Since the posterior density of $\theta(1)$ is not normal, the prior distribution of $\theta(2)$ is also non-normal, and the posterior density of $\theta(2)$ is

$$\pi[\theta(2)|Y(1),Y(2)] \propto \exp -\frac{1}{2} \{Y(2) - F_2[\theta(2)]\}'P_u(2)$$

$$\times \{Y(2) - F_2[\theta(2)]\}$$

$$\times \int_{R^p} \pi[\theta(1)|Y(1)]\exp -\frac{1}{2}[\theta(2) - G\theta(1)]'$$

$$\times P_v(2)[\theta(2) - G\theta(1)]d\theta(1), \qquad (6.45)$$

where $\theta(2) \in R^p$ and $\pi[\theta(1)|Y(1)]$ is given by (6.44). This integration must be done numerically since the integrand is not in the form of a normal density on $\theta(1)$. In general one would have

$$\pi[\theta(k)|Y(1), \ldots, Y(k)] \propto \exp -\frac{1}{2} \{Y(k) - F_k\theta(k)]\}'P_u(k)$$

$$\times \{Y(k) - F_k[\theta(k)]\}$$

$$\times \int_{R^p} \pi[\theta(k-1)|Y(1), \ldots, Y(k-1)]$$

$$\times \exp -\frac{1}{2}[\theta(k) - G\theta(k-1)]'P_v(k)$$

$$\times [\theta(k) - G\theta(k-1)]d\theta(k-1),$$

$$(6.46)$$

$k = 1, 2, \ldots$, for the posterior density of $\theta(k)$. This is a recursive formula for the posterior density of $\theta(k)$, as a function of the posterior density of $\theta(k - 1)$, and will give good operational results if p is "small," so that the numerical integration of (6.46) can be done quickly. Thus (6.46) yields a generalized Kalman filter, where after (6.46) is determined numerically, the posterior mean $M(k)$ of $\theta(k)$ is given by

$$M_i(k) = \int_{R^p} \pi[\theta(k) | Y(1), Y(2), \ldots, Y(k)] \theta_i(k) d\theta(k),$$

$$i = 1, 2, \ldots, p \quad (6.47)$$

where the integration must be done numerically, and $M_i(k)$ is the i-th component of $M(k)$ and $\theta_i(k)$ the i-th component of $\theta(k)$.

To summarize the above, one may say that if a nonlinear function F_t is included in the observation equation, the computational efficiency of the Kalman filter is lost.

One might try a linear approximation to F_1, as a function of $\theta(1)$ and obtain a normal approximation to the posterior density of $\theta(1)$. Since the prior density of $\theta(1)$ is normal with mean $M(1)$ and precision matrix $P(1)$, replace $F_1[\theta(1)]$ by its first order Taylor series about $M(1) \in R^p$, i.e., let

$$F_1[\theta(1)] \doteq F_1[M(1)] + R_1'[\theta(1) - M(1)], \quad (6.48)$$

where R_1 is the p × 1 vector with i-th component

$$\frac{\partial F_1}{\partial \theta_i(1)} [\theta(1)] \Big|_{\theta(1) = M(1)},$$ where $\theta_i(1)$ is the i-th component of $\theta(1)$.

The approximate posterior density of $\theta(1)$ is, from (6.44),

$$\hat{\pi}[\theta(1) | Y(1)] \propto \exp -\frac{1}{2} \{Y(1) - F_1[M(1)] - R_1'[\theta(1) - M(1)]\}'$$

$$\times P_u(1)\{Y(1) - F_1[M(1)]$$

$$- R_1'[\theta(1) - M(1)]\} \exp -\frac{1}{2}[\theta(1) - M(1)]'P(1)$$

NONLINEAR DYNAMIC SYSTEMS   271

$$\times [\theta(1) - M(1)], \quad \theta(1) \in R^p \tag{6.49}$$

and it can be shown that $\hat{\pi}$ is a normal density with mean

$$\hat{M}(1) = \hat{P}^{-1}(1) R_1 P_u(1) \{Y(1) + R_1' M(1) - F_1[M(1)]\}, \tag{6.50}$$

and precision

$$\hat{P}(1) = R_1 P_u(1) R_1' + P(1), \tag{6.51}$$

where $P(1)$ is the prior precision of $\theta(1)$ and $M(1)$ the mean of the prior distribution of $\theta(1)$. Recall that

$$M(1) = P^{-1}(1) P_v(1) G [P(0) + G' P_v(1) G]^{-1} P(0) M(0), \tag{6.52}$$

where

$$P(1) = [P_v^{-1}(1) + G P^{-1}(0) G']^{-1}. \tag{6.53}$$

Let us consider the second stage where

$$Y(2) = F_2[\theta(2)] + U(2) \tag{6.54}$$

and

$$\theta(2) = G\theta(1) + V(2), \tag{6.55}$$

then the prior distribution of $\theta(2)$ is induced by the $N[\hat{M}(1), \hat{P}^{-1}(1)]$ posterior distribution of $\theta(1)$. The prior (prior to $Y(2)$) distribution of $\theta(2)$ is normal with mean $M(2)$ and precision $P(2)$, where

$$M(2) = P^{-1}(2) \{P_v(2) - P_v(2) G [G' P_v(2) G + \hat{P}(1)]^{-1} \hat{P}(1) \hat{M}(1)\} \tag{6.56}$$

and

$$P(2) = [G \hat{P}^{-1}(1) G' + P_v^{-1}(2)]^{-1}. \tag{6.57}$$

As was done at the first stage, suppose $F_2[\theta(2)]$ is replaced by its first order Taylor's series, then

$$F_2[\dot{\theta}(2)] \equiv F_2[M(2)] + R_2'[\theta(2) - M(2)], \qquad (6.58)$$

where R_2 is the p × 1 vector with i-th component

$$\left. \frac{\partial F_2}{\partial \theta_i(2)} [\theta(2)] \right|_{\theta(2) = M(2)}, \quad \text{where } \theta_i(2) \text{ is the i-th component of } \theta(2);$$

thus the observation equation (6.54) is approximately

$$Y(2) = F_2[M(2)] + R_2'[\theta(2) - M(2)] + U(2) \qquad (6.59)$$

and the posterior distribution of $\theta(2)$ is approximately normal with mean

$$\hat{M}(2) = \hat{P}^{-1}(2) R_2 P_u(2) \{Y(2) + R_2'M(2) - F_2[M(2)]\}, \qquad (6.60)$$

and precision matrix

$$\hat{P}(2) = P_{12} P_u(2) P_{12}' + P(2), \qquad (6.61)$$

where $M(2)$ and $P(2)$ are given by (6.56) and (6.57), respectively. Since $M(2)$ and $P(2)$ are functions of $\hat{M}(1)$ and $\hat{P}(1)$, $\hat{M}(2)$ and $\hat{P}(2)$ are functions of the posterior mean and precision of $\theta(1)$.

Suppose k observations $Y(1), Y(2), \ldots, Y(k)$ are available and one wants to estimate $\theta(k)$ and suppose at each state prior to and including the k-th,

$$F_t[\theta(t)] = F_t[M(t)] + R_t'[\theta(t) - M(t)] \qquad (6.62)$$

for $t = 1, 2, \ldots, k - 1$, where $M(t)$ is the prior mean of $\theta(t)$ and R_t is p × 1 with i-th component

$$\left. \frac{\partial F_t}{\partial \theta_i(t)} [\theta(t)] \right|_{\theta(t) = M(t)},$$

then the posterior distribution of $\theta(k)$ is approximately normal with mean

$$\hat{M}(k) = \hat{P}^{-1}(k) R_k P_u(k) \{Y(k) + R_k'M(k) - F_k[M(k)]\}, \qquad (6.63)$$

NONLINEAR DYNAMIC SYSTEMS 273

and precision matrix

$$\hat{P}(k) = R_k P_u(k) R_k' + P(k), \tag{6.64}$$

where

$$M(k) = P^{-1}(k)\{P_v(k) - P_v(k)G[G'P_v(k)G + \hat{P}(k-1)]^{-1}$$

$$\times \hat{P}(k-1)\hat{M}(k-1)\}, \tag{6.65}$$

and

$$P(k) = [G\hat{P}^{-1}(k-1)G' + P_v^{-1}(k)]^{-1}. \tag{6.66}$$

Thus, $\hat{M}(k)$ the mean of the approximate posterior distribution of $\theta(k)$ acts like a Kalman filter because $\hat{M}(k)$ is a function of $\hat{M}(k-1)$ and $\hat{P}(k-1)$, the previous estimates of the state; however, since the nonlinear functions $F_t[\theta(t)]$ are approximated by linear functions of $\theta(t)$ in a neighborhood of $M(t)$, $\hat{M}(k)$ is only an approximation to the posterior mean of $\theta(k)$, and the accuracy of the approximation needs to be investigated.

We have seen that if the observation equation (6.39) of a dynamic system is nonlinear, then the exact Bayesian solution, i.e., the posterior mean of $\theta(k)$ is given by (6.47), and this integration must be done numerically, but the posterior distribution of $\theta(k)$ can be expressed recursively as a function of the posterior distribution of $\theta(k-1)$, for $k = 1, 2, \ldots$. If p is small, say p = 1, 2, 3, the numerical integration of (6.47) may be quickly done and the on-line filtering of $\theta(k)$ is possible. On the other hand, if p is too large, approximation techniques may be necessary in order for on-line filtering to be feasible.

An approximation method for nonlinear filtering was presented whereby each of the nonlinear functions $F_t[\theta(t)]$ of the observation equation was replaced by its first-order Taylor series about the prior mean $M(t)$, (6.65), of $\theta(t)$, thus allowing one to develop the usual Kalman filter $\hat{M}(k)$, (6.63), as an approximation to the actual mean of the posterior distribution of $\theta(k)$. Needless to say, the error of approximation of $F_t[\theta(t)]$ by its Taylor series induces an error of approximation of $E[\theta(k)]$ by the Kalman filter $\hat{M}(k)$, and this error has not been investigated.

Only one type of nonlinearity was considered, namely when the observation equation (6.39) contained a nonlinear function F_t of $\theta(t)$

and the observation equation (6.40) had a linear transition matrix G_t of $\theta(t-1)$. If, in addition, G_t is nonlinear the posterior analysis is more complex than what was done here; but by replacing $G_t[\theta(t-1)]$ by the Taylor series, one may derive a Kalman filter which is an approximation to the posterior mean of $\theta(k)$.

There are many ways to approach the problem of nonlinear filtering and one should read Jazwinski (1971) or Maybeck (1979) for the traditional solutions to the problem.

ADAPTIVE ESTIMATION

Introduction

The Kalman filter was designed for linear dynamic systems where all the parameters, except the state vectors, are known. The parameters of the system are $m(0)$, $p(0)$, $P_u(t)$ and $P_v(t)$, where $m(0)$ and $p(0)$ are the mean and precision, respectively, of the initial state $\theta(0)$, $P_u(t)$ is the precision matrix of $u(t)$, the observation error, and the precision matrix of $v(t)$, the systems error, is $P_v(t)$.

Often one or more of the parameters are unknown and adaptive estimation is the term which is applied to estimation problems when some of the parameters are unknown. Adaptive estimation is an active area of research and Jazwinski (1970), Mehira (1970,1972), Lainiotis (1971), Kailath (1968), and Maybeck (1979) will give the reader a good introduction to adaptive filtering, smoothing, and prediction.

Adaptive filtering, based on Bayesian methods, is given by Alspach (1974) and Hawkes (1973); however their methods, although related, are not the same as those which are to be presented. Other methods of adaptive filtering depend on covariance-matching, correlation and maximum-likelihood techniques of estimating parameters.

Adaptive Filtering with Unknown Precisions

Our approach to adaptive estimation is begun with the special situation where the observation and systems errors $u(t)$ and $v(t)$ are scalar and have unknown precision parameters τ_u and τ_v, respectively, where $\tau_u = P_u(t)$ and $\tau_v = P_v(t)$ for $t = 1, 2, \ldots$. The observation precision τ_u is common to all observations and there is only one system error precision τ_v. The other parameters are $m(0)$ and $p(0)$, but for the moment, suppose $\theta(0)$ is a known initial state vec-

ADAPTIVE ESTIMATION

tor, that is $m(0) = \theta(0)$ and $p(0) = \infty$. Thus the linear dynamic model is

$$\left.\begin{array}{l} y(t) = F(t)\theta(t) + u(t), \\ \theta(t) = G\theta(t-1) + v(t), \end{array}\right\} \quad t = 1, 2, \ldots \quad (6.67)$$

where $u(t) \sim n(0, \tau_u^{-1})$, $v(t) \sim n(0, \tau_v^{-1})$, $t = 1, 2, \ldots$, and $u(1), u(2), \ldots$ and $v(1), v(2), \ldots$ are independent sequences. The unknown parameters are $\tau_u > 0$ and $\tau_v > 0$, while $\theta(0)$, the initial state, is known.

The problem is to estimate the successive states $\theta(1), \theta(2), \ldots$ of the system as the observations $y(1), y(2), \ldots$ become available. The primary parameters are the states of the system and the precision components τ_u and τ_v are regarded as nuisance parameters, but since they are unknown they must be assigned a prior distribution, thus suppose τ_u and τ_v are independent and that the marginal prior density of τ_u is

$$\pi(\tau_u) \propto \tau_u^{\alpha_u - 1} e^{-\tau_u \beta_u}, \quad \tau_u > 0$$

and that

$$\pi(\tau_v) \propto \tau_v^{\alpha_v - 1} e^{-\beta_v \tau_v}, \quad \tau_v > 0,$$

where α_u, α_v, β_u, and β_v are known positive constants.

First, consider estimating $\theta(1)$, then it may be shown that the posterior density of $\theta(1)$ is

$$\pi[\theta(1)|y(1)] \propto \{2\beta_u + f^2(1)[y(1)/f(1) - \theta(1)]^2\}^{-(2\alpha_u+1)/2}$$

$$\times \{2\beta_u + [\theta(1) - g\theta(0)]^2\}^{-(2\alpha_v+1)/2},$$

$$\theta(1) \in R \quad (6.68)$$

which is a 2/0 univariate poly-t density and is common distribution which is encountered in econometrics. See Dreze (1977), for example, for a review of the literature on this topic. Now in general, it can be shown that $\theta^{(k)} = [\theta(1), \theta(2), \ldots, \theta(k)]'$ has joint marginal density function

$$\pi[\theta^{(k)}|y(1), \ldots, y(k)] \propto \left\{ 2\beta_u + \sum_1^k f^2(t)[y(t)/f(t) - \theta(t)]^2 \right\}^{-(k+2\alpha_u)/2} \left\{ 2\beta_v + \sum_1^k [\theta(t) - g\theta(t-1)]^2 \right\}^{-(k+2\alpha_v)/2}, \quad (6.69)$$

for $\theta^{(k)} \in R^k$, which is a 2/0 k variable poly-t density and is very difficult to work with. For example, the normalizing constant is unknown and the moments must be evaluated numerically. Various approximations have been proposed in working with this distribution. Box and Tiao (1973) and Zellner (1971) have proposed infinite series expansions, while Dreze (1977) has developed a numerical integration algorithm to find the moments and other characteristics of the density.

An interesting feature of this distribution is that the degrees of freedom for the two t densities of (6.69) always have $2\alpha_u$ and $2\alpha_v$ degrees of freedom for all k, k = 1, 2, ... Thus as more observations are obtained, the precision matrices of the two factors do not "decrease" with k! The precision matrices of each factor depend on β_u and β_v, respectively, but β_u and β_v are constant, i.e., they do not depend on k. The two parameters β_u and β_v control the precision of the precision parameters τ_u and τ_v. For example

$$\text{Var}(\tau_u) = \frac{\beta_u^2}{(\alpha_u - 1)^2(\alpha_u - 2)}, \quad \alpha_u > 2,$$

thus small values of β_u (and β_v) decrease the prior variance of τ_u and increase the posterior precision of $\theta(1), \theta(2), \ldots, \theta(k)$. To obtain precise estimates of the states, one must have "small" values of β_u and β_v, but small values of β_u and β_v imply one's prior knowledge of τ_u and τ_v is precise.

The difficulty in working with the posterior density (6.69) is that the marginal density of $\theta(k)$ is difficult to determine, and a recursive formula for $F[\theta(k)|y(1), \ldots, y(k)]$ in terms of $F[\theta(k-1)|y(1), \ldots, y(k-1)]$, $y(k)$, $f(k)$, and G is all but impossible to derive, and a formula analogous to the Kalman filter is difficult to obtain.

ADAPTIVE ESTIMATION

What should be done in order to find a recursive estimate of $\theta(k)$? One could approximate each of the two t-density factors of (6.69) by a normal density (with the same mean vector and precision matrix), then the posterior density of $\theta^{(k)}$ is normal and it would be possible to find a Kalman-type expression for the posterior mean of $\theta(k)$; however, the normal approximation to a poly-t density may not be very good, so this approach, though interesting, will be abandoned.

Inspecting (6.69), one sees that the conditional posterior density of $\theta(k)$, given $\theta(t) = \hat{\theta}(t)$, $t = 1, 2, \ldots, k-1$, is a 2/0 univariate poly-t density, namely

$$\pi[\theta(k)|\theta(t) = \hat{\theta}(t), t = 1, 2, \ldots, k-1] \propto \pi_1[\theta(k)|\theta(t) = \hat{\theta}(t),$$

$$t = 1, 2, \ldots, k-1] \cdot \pi_2[\theta(k)|\theta(t) = \hat{\theta}(t),$$

$$t = 1, 2, \ldots, k-1], \qquad (6.70)$$

where

$$\pi_1[\theta(k)|\theta(t) = \hat{\theta}(t), t = 1, 2, \ldots, k-1]$$

$$\propto \left\{ 1 + \frac{f^2(k)[\theta(k) - y(k)/f(k)]^2}{2\beta_u + \sum_1^{k-1} f^2(t)[y(t)/f(t) - \hat{\theta}(t)]^2} \right\}^{-(k+2\alpha_u)/2} \qquad (6.71)$$

and

$$\pi_2[\theta(k)|\theta(t) = \hat{\theta}(t), t = 1, 2, \ldots, k-1]$$

$$\propto \left\{ 1 + \frac{[\theta(k) - g\hat{\theta}(k-1)]^2}{2\beta_v + \sum_1^{k-1} [\hat{\theta}(t) - g\hat{\theta}(t-1)]^2} \right\}^{-(k+2\alpha_v)/2},$$

where $\theta(k) \in R$.

The first factor π_1 is a t density with $k + 2\alpha_u - 1$ degrees of freedom, location $y(k)/f(k)$ and precision $P_1(k)$, where

$$P_1(k) = \frac{(k + 2\alpha_u - 1)f^2(k)}{2\beta_u + \sum_{1}^{k-1} f^2(t)[y(t)/f(t) - \hat{\theta}(t)]^2}.$$

The degrees of freedom of the second factor is $k + 2\alpha_v - 1$, location $g\hat{\theta}(k-1)$ and precision $P_2(k)$, where

$$P_2(k) = \frac{(k + 2\alpha_v - 1)}{2\beta_v + \sum_{1}^{k-1} [\hat{\theta}(t) - g\hat{\theta}(t-1)]^2}.$$

Since the conditional density of $\theta(k)$ is univariate it is easy to numerically determine the conditional mean, mode, variance and other characteristics of the distribution. A recursive estimate of $\theta(k)$ is found as follows: Let $k = 1$, then the conditional density of $\theta(1)$ given $\theta(0)$ is the 2/0 univariate poly-t distribution given by (6.70) with $k = 1$ and the mode of this density is easily found to be, say $\hat{\theta}(1)$. At the second stage, the conditional density of $\theta(2)$ given $\hat{\theta}(1)$ is given by (6.70) with $k = 2$, and suppose its mode is $\hat{\theta}(2)$. Note $\hat{\theta}(1)$ depends on $y(1)$ and $\hat{\theta}(2)$ depends on $y(1)$ and $y(2)$. Continuing in this manner, one may show the mode $\hat{\theta}(k)$ of the conditional posterior distribution of $\theta(k)$, given $\theta(t) = \hat{\theta}(t)$, $t = 1, 2, \ldots, k - 1$, depends on $y(1), y(2), \ldots, y(k)$ for all k, $k = 1, 2, \ldots$. Thus, a recursive estimate of $\theta(k)$ is easily obtained by the above sequence of conditional distributions (6.71).

In lieu of conditioning on the conditional modes $\hat{\theta}(t)$, one could use the sequence of conditional means $\theta^*(t)$, where

$$\theta^*(t) = E[\theta(t) | \theta^*(1), \ldots, \theta^*(t-1)]$$

and $\theta^*(1)$ is the conditional mean of $\theta(1)$ given $\theta(0)$ and $y(1)$. Of course, one disadvantage of using a sequence of conditional estimates is what characteristic (mean, mode, median) of the sequence of conditional distributions should one use as conditioning values? If one uses a sequence of conditional means, one has different estimates for the successive states than if one uses the sequence of conditional modes!

Now suppose $p > 1$, then what would happen if one adopts the above strategy of using conditional estimates for the state vectors? Can the above procedure be extended to the general case? If $p > 1$, the systems errors $v(t)$ have an unknown common precision matrix say P_v, and if $n > 1$, the observations errors have, say, a common unknown precision matrix P_u.

Thus consider the linear dynamic system

$$y(t) = F(t)\theta(t) + u(t),$$
$$\theta(t) = G\theta(t-1) + v(t), \quad t = 1, 2, \ldots$$

where $\theta(0)$ is known, $u(t) \sim N(0, P_u^{-1})$, $v(t) \sim N(0, P_v^{-1})$, and P_u and P_v are unknown positive definite matrices. If one assumes $u(1), u(2), \ldots$, and $v(1), v(2), \ldots$ are independent sequences and that, a priori, P_u and P_v are independent Wishart (see the appendix) matrices, then it may be shown that one may devise a recursive filtering formula for the states $\theta(1), \theta(2), \ldots$, which is analogous to (6.70).

Only one situation of adaptive estimation has been considered, when the precision parameters of the observation and system errors are unknown, but other cases could have been examined. For example either the design matrices $F(t)$ of the observation equation and/or the transition matrix G of the systems equation may be unknown. For the case when the $F(t)$ are unknown, and one is interested in predicting future observations, see Harrison and Stevens (1976) for a Bayesian solution.

It has been shown that it is indeed difficult to develop a recursive filter for the present state $\theta(k)$ of the system, because the posterior distribution of $\theta(k)$, given by the density (6.69) is a poly-t density and very difficult to work with.

As a consequence, a compromise algorithm (6.70) was derived. By conditioning on previous conditional posterior means or modes of the states, a recursive filter of the present state can be found.

Adaptive Control Problems with Unknown Precision Matrices

Consider a linear dynamic system

$$y(t) = F(t)\theta(t) + u(t),$$
$$\theta(t) = G\theta(t-1) + H(t)x(t-1) + v(t), \quad t = 1, 2, \ldots \quad (6.72)$$

where $H(t)$ is a known $p \times q$ matrix and $x(t)$ is a $q \times 1$ vector of control variables. Suppose the $u(t)$ have a common unknown precision matrix P_u and that P_v is the common unknown precision matrix of the $v(t)$, that $\theta(0)$ is known, and $u(1), u(2), \ldots$ and $v(1), v(2), \ldots$ are independent sequences.

If $n = p = q = 1$, so that the observations, states, errors, and control variables are scalar, the marginal posterior density of $\theta^{(k)} = [\theta(1), \ldots, \theta(k)]$ is

$$\pi[\theta^{(k)} | y(1), y(2), \ldots, y(k)]$$

$$\propto \frac{1}{\left\{2\beta_u + \sum_1^k [y(t) - F(t)\theta(t)]^2\right\}^{(k+2\alpha_u)/2}}$$

$$\times \left\{2\beta_v + \sum_1^k [\theta(t) - G\theta(t-1) + H(t)x(t-1)]^2\right\}^{-(k+2\alpha_v)/2}$$

$$\theta^{(k)} \in R^k$$

where the prior density of $\tau_u = P_u$ and $\tau_v = P_v$ is

$$\pi_0(\tau_u, \tau_v) \propto \tau_u^{\alpha_u - 1} e^{-\beta_u \tau_u} \tau_v^{\alpha_v - 1} e^{-\beta_v \tau_v}, \quad \tau_u > 0, \; \tau_v > 0.$$

(6.74)

This formula is quite similar to (6.69), the only difference being the term $H(t)x(t-1)$, which was included in the system equation, and as before the distribution of $\theta^{(k)}$ is a 2/0 k-variate poly-t distribution. Also, the conditional distribution of $\theta(k)$ given $\theta(t) = \hat{\theta}(t)$, $t = 1, 2, \ldots, k-1$ ($k \geq 2$) is a 2/0 univariate poly-t density

$$\pi[\theta(k) | \theta(t) = \hat{\theta}(t), t = 1, 2, \ldots, k-1]$$

$$\propto \pi_1[\theta(k) | \theta(t) = \hat{\theta}(t), t = 1, 2, \ldots, k-1]$$

$$\times \pi_2[\theta(k) | \theta(t) = \hat{\theta}(t), t = 1, 2, \ldots, k-1],$$

(6.75)

where

$$\pi_1[\theta(k) | \theta(t) = \hat{\theta}(t), t - 1, 2, \ldots, k-1]$$

ADAPTIVE ESTIMATION

$$= \left\{ 1 + \frac{[y(k) - F(k)\theta(k)]^2}{2\beta_u + \sum_{1}^{k-1} [y(t) - F(t)\hat{\theta}(t)]^2} \right\}^{-(k+2\alpha_u)/2},$$

$$\pi_2[\theta(k) | \theta(t) = \hat{\theta}(t), t = 1, 2, \ldots, k-1]$$

$$= \left\{ 1 + \frac{[\theta(k) - G\hat{\theta}(k-1) - H(k)x(k-1)]^2}{2\beta_v + \sum_{1}^{k-1} [\hat{\theta}(t) - G\hat{\theta}(t-1) - H(t)x(t-1)]^2} \right\}^{-(k+2\alpha_v)/2},$$

$$\theta(k) \in R. \quad (6.76)$$

Recall from the section on control strategies (pp. 244-251) that the control problem is to choose $x(0), x(1), \ldots$ in such a way that the posterior mean $E[\theta(k) | y(1), \ldots, y(k)]$ is "close" to some target value $T(k)$, where $k = 1, 2, \ldots$. Also recall what was used as a criterion for close, namely (6.17), that is $x(0)$ is chosen so that the average value of $E[\theta(1) | y(1)]$ with respect to the predictive distribution of $y(1)$ is $T(1)$ and so on. Since the $E[\theta(k) | y^{(k)}]$, $y^{(k)} = [y(1), y(2), \ldots, y(k)]'$ must be found from (6.73), a poly-t distribution, it is difficult to find an explicit expression for the posterior mean of $\theta^{(k)}$ as well as the predictive distribution of $y(k)$ given $y(i)$, $i = 1, 2, \ldots, k-1$, thus one must either find approximate expressions for (6.73) or change the criterion of closeness, or both. One possibility is to approximate the two t-densities of (6.73) by normal densities, then one may find explicit formulas for the posterior mean of $\theta(k)$ and the predictive distribution of $y(k)$, given $y(i)$, $i = 1, 2, \ldots, k-1$. Still another approach is to change the criterion of closeness so that

$$T(1) = E_1 E[\theta(1) | y(1), \theta(0), x(0)]$$

$$T(2) = E_2 E[\theta(2) | \theta(1) = \hat{\theta}(1), y(1), x(0), x(1)]$$

$$\vdots$$

$$T(k) = E_k E[\theta(k) | \theta(t) = \hat{\theta}(t), x(t-1), t = 1, 2, \ldots, k-1,$$

$$y(1), y(2), \ldots, y(k)], \quad (6.77)$$

where E_i is the expectation with respect to the predictive distribution of $y(i)$, given $y(1), y(2), \ldots, y(i-1)$, then the control variables are chosen to satisfy (6.77). Thus instead of averaging the posterior mean of $\theta(k)$, as with (6.17), the conditional posterior means are averaged with respect to the appropriate conditional predictive distribution, as in (6.77). But, if one uses this criterion, one must work with the 2/0 univariate poly-t density (6.75), which is the conditional posterior density of $\theta(k)$, given $\theta(1), \ldots, \theta(k-1)$. Finding the predictive distribution of $y(k)$, given $y(i)$, $i = 1, 2, \ldots, k-1$, as well as the conditional posterior mean must be done numerically, and the only way to avoid this is to approximate the poly-t density by a normal distribution, or some other way. Using a normal approximation, π_1 and π_2 of (6.76) are each approximated by a normal distribution with the same mean and precision of each of these t densities. The result is that the approximate conditional posterior distribution of $\theta(k)$, given $\hat{\theta}(1), \hat{\theta}(2), \ldots, \hat{\theta}(k-1)$, is normal with mean

$$\mu(k) = P^{-1}(k)\{P_1(k)\mu_1(k) + P_2(k)\mu_2(k)\} \tag{6.78}$$

and precision

$$P(k) = P_1(k) + P_2(k), \tag{6.79}$$

where if $k \geq 2$

$$P_1(k) = \frac{(k + 2\alpha_u - 3)F^2(k)}{2\beta_u + \sum_1^{k-1} [y(t) - F(t)\hat{\theta}(t)]^2}, \tag{6.80}$$

$$P_2(k) = \frac{(k + 2\alpha_v - 3)}{2\beta_v + \sum_1^{k-1} [\hat{\theta}(t) - G\hat{\theta}(t-1) - H(t)x(t-1)]^2}, \tag{6.81}$$

$$\mu_1(k) = y(k)/F(k), \quad F(k) \neq 0, \tag{6.82}$$

and

$$\mu_2(k) = G\hat{\theta}(k-1) + H(k)x(k-1). \tag{6.83}$$

If k = 1,

$$\mu_1(1) = y(1)/F(1), \tag{6.84}$$

$$\mu_2(1) = G\theta(0) + H(1)x(0), \tag{6.85}$$

$$P_1(1) = \frac{(\alpha_u - 1)F^2(1)}{\beta_u} \tag{6.86}$$

and

$$P_2(1) = (\alpha_v - 1)/\beta_v. \tag{6.87}$$

What is the predictive distribution of y(k) given y(i), i = 1, 2, ..., k − 1? If k ⩾ 2, it can be shown that this distribution is approximately normal with mean

$$\mu^* = F(k)\mu_2(k) \tag{6.88}$$

and precision

$$P^*(k) = \frac{P_1(k)P_2(k)}{F^2(k)[P_1(k) + P_2(k)]}. \tag{6.89}$$

Of course, this distribution is only approximate because (6.75) is only approximately normal.

When k = 1, the predictive distribution of y(1) is approximately normal with mean

$$\mu^*(1) = F(1)\mu_2(1)$$

and precision

$$P^*(1) = \frac{P_1(1)P_2(1)}{F^2(1)[P_1(1) + P_2(1)]},$$

where $\mu_2(1)$, $P_1(1)$, and $P_2(1)$ are given by (6.85), (6.86) and (6.87), respectively.

The above results for the conditional posterior distribution of $\theta(k)$ and the predictive distribution of y(k) for k = 1, 2, ... are based on the normal approximation to the 2/0 poly-t density (6.75), where each factor which is a t density is replaced by a normal density with the same mean and precision. A 2/0 poly-t density may be asymmetric and bimodal, in which case the normal distribution may not be a good approximation; however, the approximate conditional posterior mean $\mu_1(k)$ of $\theta(k)$, given by (6.78) is usually quite accurate, and the same is true for the conditional mean $\mu^*(k)$ of $y(k)$, given by (6.88). For additional details of the approximation see Press (1972), pages 131-132, and for further information about the analytical properties of the 2/0 univariate poly-t density see Box and Tiao (1973), pages 481-485.

Now, using these normal approximations, one is able to consider the control problem of choosing the control variables $x(0), x(1), \ldots$ in such a way that (6.77) is satisfied. At the first stage $x(0)$ is chosen so that the average value (averaged with respect to $y(1)$) of $\theta(1)$ is $T(1)$. Since

$$E[\theta(1) | y(1), x(0)] = \mu(1)$$

$$= \frac{F^{-1}(1)y(1)P_1(1) + \mu_2(1)P_2(1)}{P_1(1) + P_2(1)},$$

the average value of $\mu(1)$ with respect to $y(1)$ is

$$E_1[\mu(1)] = \frac{\mu_2(1)P_1(1) + \mu_2(1)P_2(1)}{P_1(1) + P_2(1)}$$

$$= G\theta(0) + H(1)x(0)$$

and if this is set equal to $T(1)$, the control variable $x(0)$ must be

$$\hat{x}(0) = \frac{T(1) - G\theta(0)}{H(1)}, \quad H(1) \neq 0. \tag{6.90}$$

$\hat{x}(0)$ is chosen before $y(1)$ is observed and $x(1)$ must be selected before $y(2)$, but after $y(1)$ is observed. It can be shown that

$$\hat{x}(1) = \frac{T(2) - G\hat{\theta}(1)}{H(2)}, \quad H(2) \neq 0, \tag{6.91}$$

where $\hat{\theta}(1)$ is the conditional value of $\theta(1)$ in the conditional distribution of $\theta(1)$. Suppose if $\hat{\theta}(1)$ is chosen as the mean of $\theta(1)$, which from (6.78) is

ADAPTIVE ESTIMATION 285

$$\mu_1[\theta(1)|x(0)] = \frac{P_1(1)y(1)F^{-1}(1) + P_2(1)[G\theta(0) + H(1)x(0)]}{P_1(1) + P_2(1)},$$

(6.92)

then the conditioning value depends on $x(0)$, thus replace $x(0)$ by $\hat{x}(0)$ in (6.92), where $\hat{x}(0)$ is given by (6.90) and use $\mu_1[\theta(1)|\hat{x}(0)]$ for $\hat{\theta}(1)$ in (6.91). In this way, the control variable is determined, and $\hat{x}(1)$ depends on $y(1)$ and $\hat{x}(0)$, the present observation and the past value of the control variable.

The general scheme is as follows: Suppose $y(1), y(2), \ldots$, $y(k-1)$ have been observed and $\hat{x}(0), \hat{x}(1), \ldots, \hat{x}(k-2)$ have been selected as the control variables, then in order to satisfy (6.77), $x(k-1)$ must be chosen as

$$\hat{x}(k-1) = \frac{T(k) - G\hat{\theta}(k-1)}{H(k)}, \quad H(k) \neq 0, \quad k \geq 2 \qquad (6.93)$$

where $\hat{\theta}(k-1)$ is the conditioning value of $\theta(k-1)$ in the conditional posterior distribution of $\theta(k-1)$, given $\theta(1), \ldots, \theta(k-2)$, which has mean

$$\mu[\theta(k-1)|x(k-2)] = \frac{P_1(k-1)y(k-1)F^{-1}(k-1) + P_2(k-1)[G\hat{\theta}(k-2) + H(k-1)x(k-2)]}{P_1(k-1) + P_2(k-1)}$$

(6.94)

where $\hat{\theta}(k-2)$ is the conditioning value of $\theta(k-2)$. Replacing $x(k-2)$ by $\hat{x}(k-2)$ in (6.94) gives a conditioning value for $\theta(k-1)$, and $\hat{x}(k-1)$ depends on the past observations $y(1), \ldots, y(k-1)$ as well as the past values of the control variables $\hat{x}(0), \hat{x}(1), \ldots, \hat{x}(k-2)$. Formula (6.93) is for $k \geq 2$, which gives a way to choose $x(1), x(2), \ldots$, but to choose $x(0)$ at the initial stage, use (6.90).

The General Case of Adaptive Estimation and Control

Adaptive estimation is the general theme of this section and so far only the scalar case has been pursued. We have seen that with regard to filtering the present state, one can easily determine the posterior mean of the state by numerical techniques; however, when one wants to control the states at a particular level, it was necessary to use a normal approximation to the marginal posterior distribution of the states.

In this part of the chapter, adaptive estimation and adaptive control procedures are to be developed for the case when the state vectors are not scalar and the precision matrices $U = P_u(t)$ and $V = P_v(t)$, $t = 1, 2, \ldots$ of the observations and states, respectively, are symmetric, positive definite, and unknown with independent Wishart distributions.

Adaptive Filtering, the Vector Case

Consider the linear dynamic model (6.1) and (6.2), where $\theta(0)$ is known, and $u(1), u(2), \ldots$ and $v(1), v(2), \ldots$ are independent sequences of random vectors. Suppose

$$u(t) \sim N[0, U^{-1}],$$

$$v(t) \sim N[0, V^{-1}],$$

where U is $n \times n$, symmetric, positive definite and unknown and V is $p \times p$, symmetric, positive definite and unknown.

The main goal is to develop a filter for the present state of the system, say $\theta(k)$ and this will be achieved by developing a recursive formula for the posterior mean of $\theta(k)$, given $y(1), y(2), \ldots, y(k)$, for $k = 1, 2, \ldots$. As will be shown, the exact distribution of $\theta(k)$ given $y(1), y(2), \ldots, y(k)$ is a 2/0 p-variate poly-t distribution and is to be approximated by a normal distribution.

Suppose U and V are independent, a priori, and that U has a Wishart distribution with v_1 degrees of freedom and precision matrix T_u, which is positive definite and symmetric, then U has density

$$\pi_0(U | v_1, T_u) \propto |U|^{(v_1 - n - 1)/2} \exp - \frac{1}{2} T_r(T_u U), \quad U > 0,$$

where $U > 0$ means the $n(n+1)/2$ distinct elements of U form a positive definite matrix U.

In a similar way let V have density

$$\pi_0(V | v_2, T_v) \propto |V|^{(v_2 - p - 1)/2} \exp - \frac{1}{2} T_r(T_v V), \quad V > 0,$$

i.e., V is Wishart with v_2 degrees of freedom and precision matrix T_v, which is known, symmetric and positive definite.

The joint density of $y(1)$, $\theta(1)$, U, and V is

ADAPTIVE ESTIMATION

$$g[y(1),\theta(1),U,V] \propto |U|^{(v_1+1-n-1)/2} \exp -\frac{1}{2}T_r\{[y(1) - F(1)$$

$$\times \theta(1)][y(1) - F(1)\theta(1)]' + T_u\}$$

$$\times U|V|^{(v_2+1-p-1)/2} \exp -\frac{1}{2}T_r\{[\theta(1)$$

$$- G\theta(0)][\theta(1) - G\theta(0)]' + T_v\}V,$$

where $U > 0$, $V > 0$, $y(1) \in R^n$ and $\theta(1) \in R^p$, and eliminating U and V by the properties of the Wishart density, gives

$$g[y(1),\theta(1)] \propto \frac{1}{|T_u + [y(1) - F(1)\theta(1)][y(1) - F(1)\theta(1)]'|^{(v_1+1)/2}}$$

$$\times \frac{1}{|T_v + [\theta(1) - G\theta(0)][\theta(1) - G\theta(0)]'|^{(v_1+1)/2}}$$

and upon completing the square with respect to $\theta(1)$ in each factor, the conditional distribution of $\theta(1)$ given $y(1)$ is

$$g[\theta(1)|y(1)] \propto \frac{1}{\{1 + [\theta(1) - \mu_{11}]'P_{11}[\theta(1) - \mu_{11}]\}^{(v_1+1)/2}}$$

$$\times \frac{1}{\{1 + [\theta(1) - \mu_{12}]'P_{12}[\theta(1) - \mu_{12}]\}^{(v_2+1)/2}},$$

$$\theta(1) \in R^p \quad (6.95)$$

Therefore, the marginal posterior density of $\theta(1)$ is a p variable 2/0 poly-t distribution, where

$$\mu_{11} = Q_{11}^{-1}F'(1)T_u^{-1}y(1),$$

$$Q_{11} = F'(1)T_u^{-1}F(1),$$

$$P_{11} = \frac{Q_{11}}{1 + y'(1)T_u^{-1}y(1) - \mu'_{11}F'(1)T_u^{-1}y(1)},$$

$$\mu_{12} = G\theta(0),$$

and

$$P_{12} = T_v^{-1}.$$

If one wants to filter $\theta(1)$, the mean of $\theta(1)$ needs to be found from (6.95), which is unknown, thus it must be found numerically, but this is indeed difficult if $p > 1$.

Suppose each of the two t densities of (6.95) are approximated by normal densities with the same mean vectors and precision matrices, then (6.95) will be approximately normal with mean

$$m(1) = (P^*_{11} + P^*_{12})^{-1}(P^*_{11}\mu_{11} + P^*_{12}\mu_{12}), \tag{6.96}$$

and precision

$$P(1) = (P^*_{11} + P^*_{12}), \tag{6.97}$$

where

$$P^*_{11} = (v_1 - p - 1)P_{11}$$

and

$$P^*_{12} = (v_2 - p - 1)P_{12}.$$

In general, for $k \geq 2$, the conditional posterior distribution of $\theta(k)$, given $\theta(1), \theta(2), \ldots, \theta(k-1)$ is normal with mean

$$m(k) = P^{-1}(k)(P^*_{k1}\mu_{k1} + P^*_{k2}\mu_{k2}), \tag{6.98}$$

and precision matrix

$$P(k) = (P^*_{k1} + P^*_{k2}), \tag{6.99}$$

$$P^*_{k1} = (v_1 + k - p - 2)P_{k1},$$

$$P_{k2}^* = (v_2 + k - p - 2)P_{k2},$$

$$P_{k1} = \frac{F'(k)S_1^{-1}(k-1)F(k)}{1 + y'(k)S_1^{-1}(k-1)y(k) - \mu_{k1}'F'(k)S_1^{-1}(k-1)y(k)},$$

$$\mu_{k1} = [F'(k)S_1^{-1}(k-1)F(k)]^{-1}F'(k)S^{-1}(k-1)y(k),$$

$$S_1(k-1) = T_u + \sum_1^{k-1} [y(t) - F(t)\theta(t)][y(t) - F(t)\theta(t)]',$$

$$\mu_{k2} = G\theta(k-1),$$

$$P_{k2} = S_2(k-1),$$

and

$$S_2(k-1) = T_v + \sum_1^{k-1} [\theta(t) - G\theta(t-1)][\theta(t) - G\theta(t-1)]'.$$

To reiterate, the conditional distribution of $\theta(k)$ given $\theta(1), \ldots, \theta(k-1)$ is normal with mean $m(k)$ and precision $P(k)$, where for $k = 1$, $m(1)$ and $P(1)$ are given by (6.96) and (6.97), respectively, and for $k \geq 2$, $m(k)$ and $P(k)$ by (6.98) and (6.99), respectively.

To filter $\theta(k)$, one may use either the mean of the marginal posterior distribution of $\theta(k)$ or the mean $m(k)$ of the conditional posterior distribution of $\theta(k)$ given $\theta(1), \theta(2), \ldots, \theta(k-1)$ and condition on estimates of the previous states. We choose the latter because it is difficult to express the marginal posterior mean of $\theta(k)$ in a recursive fashion. By using the conditional mean $m(k)$ of $\theta(k)$, given the conditional means of the previous states, $\theta(t) = m(t)$, $t = 1, 2, \ldots, k-1$, one may obtain a recursive filter to $\theta(k)$.

Consider the first state $\theta(1)$, then using the approximate mean $m(1)$ of $\theta(1)$, $\theta(1)$ is easily estimated. Now consider the second state $\theta(2)$, which is estimated by $m(2)$, but $m(2)$ depends on $\theta(1)$ because $S_1(1)$ and μ_{22} depend on $\theta(1)$, thus let $\theta(1) = m(1)$ and $m(2)$ is determined. In the same way, $m(3)$ depends on $\theta(1)$ and $\theta(2)$ because $S_1(2)$ and μ_{32} are functions of $\theta(1)$ and $\theta(2)$, thus letting $\theta(1) = m(1)$ and $\theta(2) = m(2)$, where $m(2)$ is given at the second stage, determines m(3). This process is continued and a recursive filter for θ(k) is easily constructed; however, it depends on the normal approximation to a poly-t distribution and the error of approximation needs to be studied.

A numerical study at the end of the chapter will reveal how reasonable the filter m(k) is in estimating the states.

Adaptive Control, the Vector Case

Dynamic linear models are often used to model the control mechanism of dynamic systems and in a previous section of this chapter a control strategy was developed for a scalar version of the dynamic model (6.72).

The scalar version of the model is now generalized to the case when the states are vectors, and as in the previous section, the conditional posterior density of the present state given the previous states is approximated by a normal distribution.

To recapitulate, again consider the model

$$y(t) = F(t)\theta(t) + u(t),$$
$$\theta(t) = G\theta(t-1) + H(t)x(t-1) + v(t),$$

(6.72)

where $t = 1, 2, \ldots$, and the terms in the model are as those of (6.1) except, $H(t)x(t-1)$ has been added to the system equation, where $H(t)$ is a known $p \times q$ matrix and $x(t-1)$ is a $q \times 1$ vector of control variables, which are to be chosen in such a way that $\theta(t)$ is "close" to $T(t)$, $t = 1, 2, \ldots$.

Suppose $\theta(0)$ is known and that $u(1), u(2), \ldots$ and $v(1), v(2), \ldots$ are independent vectors and that $u(t) \sim N(0, U^{-1})$ and $v(t) \sim N(0, V^{-1})$, $t = 1, 2, \ldots$, where U and V are unknown positive definite symmetric precision matrices of orders n and p, respectively.

First the exact posterior distribution of $\theta(1), \theta(2), \ldots, \theta(k)$ given $y(1), y(2), \ldots, y(k)$, $k = 1, 2, \ldots$ will be derived, then the exact conditional distribution of $\theta(k)$ given $\theta(1), \theta(2), \ldots, \theta(k-1)$ will be identified and finally this will be approximated by a normal distribution, from which a control strategy will be developed.

The main parameters of interest are the states, thus the precision matrices U and V will be regarded as nuisance parameters and independent with prior Wishart distributions. Let U be Wishart with $v_1 > n - 1$ degrees of freedom and precision matrix T_u, which is known, and let $v_2 > p - 1$ be the degrees of freedom and T_v the precision matrix of V.

Under these assumptions, the marginal posterior density of $\theta(1)$ is the 2/0 poly-t density.

ADAPTIVE ESTIMATION

$$g[\theta(1)|y(1)] \propto \prod_{i=1}^{i=2} \frac{1}{\{1 + [\theta(1) - \mu_{1i}]'P_{1i}[\theta(1) - \mu_{1i}]\}^{(v_i+1)/2}},$$

$$\theta(1) \in R^p \quad (6.100)$$

where

$$\mu_{11} = Q_{11}^{-1} F'(1) T_u^{-1} y(1),$$

$$Q_{11} = F'(1) T_u^{-1} F(1),$$

$$P_{11} = \frac{Q_{11}}{1 + y'(1) T_u^{-1} y(1) - \mu_{11}' F'(1) T_u^{-1} y(1)},$$

$$\mu_{12} = G\theta(0) + H(1)x(0),$$

and

$$P_{12} = T_v^{-1}.$$

Also, in general it can be shown that the conditional posterior density of $\theta(k)$ given $\theta(1), \theta(2), \ldots, \theta(k-1)$ is for $k \geq 2$,

$$g[\theta(k)|\theta(1), \ldots, \theta(k-1)] \propto \prod_{i=1}^{2} \{1 + [\theta(k) - \mu_{ki}]'P_{ki} \times [\theta(k) - \mu_{ki}]\}^{-(v_i+k)/2},$$

$$\theta(k) \in R^p \quad (6.101)$$

which is a 2/0 poly-t density and

$$\mu_{k1} = [F'(k)S_1(k-1)F(k)]^{-1} F'(k) S_1^{-1}(k-1) y(k),$$

$$S_1(k - 1) = T_u + \sum_{t=1}^{k-1} [y(t) - F(t)\theta(t)][y(t) - F(t)\theta(t)]',$$

$$P_{k1} = \frac{F'(k)S_1(k-1)F(k)}{1 + y'(k)S_1^{-1}(k-1)y(k) - \mu'_{k1}F'(k)S_1^{-1}(k-1)y(k)},$$

$$\mu_{k2} = G\theta(k-1) + H(k)x(k-1),$$

and

$$S_2(k-1) = T_v + \sum_{t=1}^{k-1} [\theta(t) - G\theta(t-1) - H(t)x(t-1)]$$

$$\times [\theta(t) - G\theta(t-1) - H(t)x(t-1)]'.$$

The marginal posterior density of $\theta(1)$ is given by (6.100) and is to be approximated by a normal distribution with mean

$$m(1) = P^{-1}(1)(P^*_{11}\mu_{11} + P^*_{12}\mu_{12}), \quad (6.102)$$

and precision matrix

$$P(1) = P^*_{11} + P^*_{12}, \quad (6.103)$$

where

$$P^*_{1i} = (v_i - p - 1)P_{1i}, \quad i = 1, 2$$

and P_{11} and P_{12} are given by (6.95). This normal approximation is based on approximating each of the two factors of (6.100) by a normal density with the same mean vector and precision matrix as the factor.

In the same way, the conditional posterior density of $\theta(k)$ given the previous states is approximated by a normal density with mean

$$m(k) = P^{-1}(k)(P^*_{k1}\mu_{k1} + P^*_{k1}\mu_{k2}) \quad (6.104)$$

and precision matrix

$$P(k) = P^*_{k1} + P^*_{k2}, \quad (6.105)$$

ADAPTIVE ESTIMATION 293

where

$$P^*_{ki} = (v_i + k - p - 2)P_{ki}, \quad i = 1,2, \quad k = 2,3, \ldots$$

and μ_{k1}, μ_{k2}, P_{k1}, P_{k2} are given by (6.101).

How does one develop a control strategy based on the normal approximation? First, one must choose a control criterion which tells one how close the preassigned target value T(i) is to be to the state θ(i), i = 1,2, ... Up to this point the criterion was based on the one-step-ahead forecasts of the states, i.e., for example x(0) was chosen so that

$$T(1) = E_1[\theta(1) | x(0), y(1)]$$

$$= E_1[m(1)] \qquad (6.106)$$

$$= E_1[P^{-1}(1)[P^*_{11}\mu_{11} + P^*_{12}\mu_{12})],$$

where E_1 denotes expectation with respect to the distribution (future) of y(1). But, μ_{11} and P^*_{11} both are functions of y(1) and the expectation of (6.106) appears to be intractable, therefore consider an alternative criterion of control. In lieu of choosing the control vector x(0) so that the target value T(1) is the average value of the future state θ(1) via (6.106), choose x(0) where T(1) is the average prior state of θ(1), i.e., let

$$T(1) = \underset{v(1)}{E}[\theta(1)]$$

$$= G\theta(0) + H(1)x(0), \qquad (6.107)$$

and the average is taken with respect to v(1). Solving (6.107) with respect to x(0) gives

$$\hat{x}(0) = [H'(1)H(1)]^{-1}H'(1)[T(1) - G\theta(0)]. \qquad (6.108)$$

if p ⩾ q and H(1) is of full rank.

At the second stage of the control algorithm, y(1) has been observed, x(0) has been chosen according to (6.108) and x(1) must be chosen before y(2) is observed, in such a way that

$$T(2) = \underset{v(2)}{E}[\theta(2) | x(1), \hat{x}(0)]$$

$$= G\hat{\theta}(1) + H(2)x(1)$$

where

$$\hat{\theta}(1) = m(1)|_{\hat{x}(0)},$$

where $m(1)$ is evaluated at $x(0) = \hat{x}(0)$ according to (6.102), the posterior mean of $\theta(1)$, which depends on $x(0) \cdot \mu_{12}$ of (6.100) is a function of $x(0)$, thus $x(1)$ is chosen as

$$\hat{x}(1) = [H'(2)H(2)]^{-1}H'(2)[T(2) - G\hat{\theta}(1)].$$

Thus in general,

$$\hat{x}(k-1) = [H'(k)H(k)]^{-1}H'(k)[T(k) - G\hat{\theta}(k-1)], \qquad (6.109)$$

where

$$\hat{\theta}(k-1) = \begin{cases} \theta(0), & k = 1 \\ m(k-1)|_{\hat{x}(k-2)}, & k \geq 2 \end{cases}$$

where $m(k-1)|_{\hat{x}(k-2)}$ is evaluated according to (6.104), and (6.109) gives a recursive formula for controlling the states at preassigned values $T(1), T(2), \ldots$.

Of course other control criteria could have been used and the reader is referred to Aoki (1967), Sawaragi et al. (1967), and Zellner (1971) for other ways to choose a control strategy.

AN EXAMPLE OF ADAPTIVE ESTIMATION

This example will illustrate adaptive filtering of the bivariate states of a linear dynamic model which has bivariate observations. Consider the model

$$\left. \begin{array}{l} y(t) = F\theta(t) + u(t), \\ \theta(t) = G\theta(t-1) + v(t), \end{array} \right\} \quad t = 1, 2, \ldots, 50 \qquad (6.110)$$

where

$$F = \begin{pmatrix} 3.5 & 3 \\ 3 & 3 \end{pmatrix},$$

AN EXAMPLE OF ADAPTIVE ESTIMATION

$$G = \begin{pmatrix} .5 & .5 \\ .5 & .5 \end{pmatrix},$$

$$u(t) \sim n(0, u^{-1}),$$

$$v(t) \sim n(0, v^{-1}),$$

and $\theta(0)$ is known.

Also assume $u(1), \ldots, u(5), v(1), \ldots, v(50)$ are independent bivariate normal random vectors. Fifty bivariate observations were generated according to (6.110), i.e., $y(1)$ was generated by selecting $v(1)$ from a $N(0,v)$ population and $\theta(1)$ was generated according to the system equation $\theta(1) = G\theta(0) + v(1)$, where $\theta(0)$ is known, then $u(1)$ was selected from a $N(0,u)$ population and $y(1)$ was computed from the observation equation $y(1) = F\theta(1) + u(1)$, and this was repeated fifty times.

Table 6.5 has four columns and the first two consist of the two components of $\theta(1), \theta(2), \ldots, \theta(50)$, and the last two columns list the two components of the filter $m(1), m(2), \ldots, m(50)$, where $m(1)$ is computed by formula (6.96), while the remaining filters $m(t)$, $t = 2, \ldots, 50$ are given by (6.98).

The observation and system precision matrices used in generating the observations and states were

$$u = \begin{pmatrix} .6 & -.4 \\ -.4 & .6 \end{pmatrix} \tag{6.111}$$

and

$$v = \begin{pmatrix} 555.5556 & -444.4445 \\ -444.4445 & 555.5556 \end{pmatrix} \tag{6.112}$$

and the initial state was

$$\theta(0) = \begin{pmatrix} 1 \\ 1 \end{pmatrix}.$$

With regard to prior information, u and v are independent Wishart matrices, where u has $v_1 = 4$ degrees of freedom and precision matrix T_u and $v_2 = 4$ and T_v are the parameters of the prior distribution of v, and

Table 6.5
Estimates of Bivariate States, First Model

1	0.8767608	0.9436042	0.9921297	0.9920480
2	1.0138780	1.0086370	0.9956300	0.9949626
3	1.0157220	0.9275690	0.9968629	0.9964153
4	0.8869861	0.8818690	0.9951978	0.9950728
5	0.9979184	0.9660156	0.9960202	0.9958527
6	1.0177950	0.9694926	0.9971470	0.9968274
7	0.8915292	0.9432775	0.9970379	0.9969366
8	0.9890499	1.0239550	0.9964256	0.9968209
9	1.0458780	1.0531150	0.9965436	0.9969116
10	0.8683957	0.8866324	0.9977383	0.9972972
11	1.0401780	1.1292990	0.9984227	0.9980738
12	0.9417663	0.9990182	0.9983795	0.9984148
13	1.0145750	0.9442801	0.9992406	0.9993359
14	0.9947690	1.0194400	0.9993104	0.9996912
15	1.0400200	1.0639820	0.9997271	0.9999698
16	1.0981140	1.0044710	0.9995579	0.9992192
17	1.1150730	1.0627030	0.9989278	0.9997275
18	0.8976504	0.7979441	0.9993744	0.9995686
19	0.9583021	0.9701377	0.9992608	0.9994334
20	0.9606810	0.8540279	0.9992424	0.9991309
21	1.0641170	1.0544220	0.9989031	0.9992933
22	1.0035550	1.0304920	0.9984049	0.9988465
23	0.9376380	1.0469480	0.9988129	0.9983499
24	0.8141437	0.8606508	0.9981064	0.9984301
25	1.0743160	1.0896180	0.9981948	0.9981479
26	1.0292130	0.9388443	0.9982485	0.9979065
27	0.9791433	0.9985471	0.9979768	0.9982306
28	1.0833180	1.0673180	0.9978260	0.9981133
29	1.0762960	1.0282070	0.9979589	0.9977634
30	0.9052299	0.9344323	0.9978279	0.9978729
31	0.8707097	0.9098074	0.9979017	0.9977839
32	1.0610300	1.0355550	0.9979369	0.9979443
33	0.9989973	0.9829768	0.9981555	0.9976802
34	1.0556150	1.1018600	0.9978477	0.9975105
35	0.9656247	1.0633060	0.9976317	0.9973399
36	0.8883852	0.8781859	0.9972868	0.9973239
37	1.0536730	1.1227540	0.9969144	0.9974268
38	0.9135162	0.9449835	0.9971144	0.9969935
39	1.0512460	1.0342730	0.9970482	0.9971711
40	1.0604350	0.9912644	0.9972612	0.9967551

AN EXAMPLE OF ADAPTIVE ESTIMATION 297

Table 6.5 (Continued)

41	1.0291380	0.9817958	0.9966899	0.9971806
42	0.9045110	0.9742339	0.9968914	0.9967545
43	1.0577140	1.0639700	0.9966074	0.9970306
44	0.9408895	0.8657800	0.9966914	0.9968976
45	0.9251381	1.0265350	0.9967267	0.9972832
46	0.8869721	0.9065623	0.9968241	0.9969598
47	0.9849322	1.0250230	0.9967179	0.9969848
48	0.8835386	0.9120919	0.9967266	0.9969863
49	0.9539953	0.9177560	0.9966724	0.9970572
50	1.0018050	1.0242950	0.9968926	0.9970560

$$T_u = v_1 u^{-1}$$

and

$$T_v = v_2 v^{-1},$$

where u and v are given by (6.111) and (6.112), respectively.

Now, referring to Table 6.5, the last two columns estimate the first two, thus m(1) = (.9921297, .9920480) estimates θ(1) = (.8767608, .9436042) and m(50) = (.9968926, .9970560) estimates θ(50) = (1.0018050, 1.0242950) and one sees there is very little variation in the states and estimates. That there is little variation is due to the "large" system precision matrix V, which allows for small variation in the state vectors. Thus 555.5556 is the precision in the first component of the state vector.

Table 6.6 lists the fifty precision matrices of the approximate conditional normal distribution of θ(k) given θ(t) = m(t), t = 1, 2, ..., k − 1 and k = 1, 2, ..., 50. Remember m(k) is the mean of the conditional distribution of θ(k), given θ(1), θ(2), ..., θ(k − 1), and Table 6.6 gives the precision of the estimates of this distribution for k = 1, ..., 50. We see as the number of observations increases, the precision of the first of the two components of m(k) increases from 139.9763 to 6936.667.

If the systems precision matrix is changed to

$$V = \begin{pmatrix} .6 & -0.4 \\ -0.4 & .6 \end{pmatrix} \qquad (6.113)$$

and the observation precision matrix to

Table 6.6
Estimates of Precision Matrix, First Model

1	139.9763000	−110.1362000	−110.1362000	139.7888000
2	278.1474000	−221.9297000	−221.9297000	278.0527000
3	416.8376000	−333.1662000	−333.1662000	416.8354000
4	555.5969000	−444.3525000	−444.3525000	555.6538000
5	694.3935000	−555.5400000	−555.5400000	694.4687000
6	833.1870000	−666.7160000	−666.7160000	833.3015000
7	971.9570000	−777.8828000	−777.8828000	972.1335000
8	1110.7590000	−889.0532000	−889.0532000	1110.9590000
9	1249.5300000	−1000.2010000	−1000.2010000	1249.7910000
10	1388.3120000	−1111.3460000	−1111.3460000	1388.6030000
11	1527.0440000	−1222.4810000	−1222.4810000	1527.4240000
12	1665.7910000	−1333.6250000	−1333.6250000	1666.2520000
13	1804.5810000	−1444.7840000	−1444.7840000	1805.0830000
14	1943.3680000	−1555.9500000	−1555.9500000	1943.8970000
15	2082.1280000	−1667.0730000	−1667.0730000	2082.6820000
16	2220.9120000	−1778.2190000	−1778.2190000	2221.4880000
17	2359.6900000	−1889.3500000	−1889.3500000	2360.2680000
18	2498.2870000	−2000.3210000	−2000.3210000	2498.9140000
19	2637.0580000	−2111.4570000	−2111.4570000	2637.7170000
20	2775.8220000	−2222.5910000	−2222.5910000	2776.5240000
21	2914.6010000	−2333.7330000	−2333.7330000	2915.3340000
22	3053.3150000	−2444.8210000	−2444.8210000	3054.0960000
23	3191.9850000	−2555.8980000	−2555.8980000	3192.8710000
24	3330.6710000	−2666.9500000	−2666.9500000	3331.5910000
25	3469.3740000	−2778.0430000	−2778.0430000	3470.3690000
26	3608.1450000	−2889.1820000	−2889.1820000	3609.1800000
27	3746.8620000	−3000.2620000	−3000.2620000	3747.9270000
28	3885.5970000	−3111.3650000	−3111.3650000	3886.7010000
29	4024.3060000	−3222.4520000	−3222.4520000	4025.4670000
30	4163.0580000	−3333.5700000	−3333.5700000	4164.2500000
31	4301.8120000	−3444.6980000	−3444.6980000	4303.0500000
32	4440.5620000	−3555.8140000	−3555.8140000	4441.8390000
33	4579.3240000	−3666.9430000	−3666.9430000	4580.6400000
34	4717.9600000	−3777.9470000	−3777.9470000	4719.3120000
35	4856.6790000	−3889.0160000	−3889.0160000	4858.0350000
36	4995.3940000	−4000.0910000	−4000.0910000	4996.7650000
37	5134.1400000	−4111.2030000	−4111.2030000	5135.5540000
38	5272.7260000	−4222.1750000	−4222.1750000	5274.2100000
39	5411.4680000	−4333.2810000	−4333.2810000	5412.9840000
40	5550.2100000	−4444.3900000	−4444.3900000	5551.7650000

AN EXAMPLE OF ADAPTIVE ESTIMATION

Table 6.6 (Continued)

41	5688.8000000	−4555.3350000	−4555.3350000	5690.3710000
42	5827.3780000	−4666.2920000	−4666.2920000	5829.0030000
43	5966.1090000	−4777.3860000	−4777.3860000	5967.7650000
44	6104.7340000	−4888.3820000	−4888.3820000	6106.4290000
45	6243.4450000	−4999.4600000	−4999.4600000	6245.1830000
46	6381.9880000	−5110.3350000	−5110.3350000	6383.6950000
47	6520.6990000	−5221.4210000	−5221.4210000	6522.4570000
48	6659.3710000	−5332.4680000	−5332.4680000	6661.1750000
49	6798.0460000	−5443.5070000	−5443.5070000	6799.8820000
50	6936.6670000	−5554.4960000	−5554.4960000	6938.5390000

Table 6.7
Estimates of Bivariate States, Second Model

1	−2.0187330	0.2714834	−1.7885740	0.0073242
2	−0.3568782	−0.5616732	−0.1711372	1.0364430
3	−0.1034499	−2.8283520	−0.4052987	−2.7328530
4	−4.2550140	−4.1827350	−3.8115030	−3.5973520
5	−3.6362530	−4.6128130	−3.4780980	4.2119970
6	−3.3096080	−4.8248720	−3.1248320	−4.3479190
7	−6.3195590	−4.5256390	−5.0647860	−3.8704520
8	−4.6620440	−3.5833760	−5.0297330	−4.0316240
9	−3.3241550	−3.2063650	−4.5411190	−4.1322040
10	−7.4801380	−6.6583880	−4.8145800	4.7252330
11	−3.7249180	−1.1017790	−5.1964110	−3.9505730
12	−5.9570100	−4.0974540	−5.2468970	−4.1224350
13	−4.2883490	−6.4610460	−4.1150940	5.3106630
14	−4.8190750	−4.0589170	−5.1326100	−4.3283750
15	−3.7379370	−3.0930690	−4.9735230	−4.3302470
16	−2.2448510	−5.2986450	−3.4472240	−5.5369760
17	−1.6583950	−3.4925020	−4.1790720	−4.4818610
18	−6.8210420	−9.6436080	−3.7749360	5.1844730
19	−5.4881420	−5.0424010	−4.7625970	−4.3596320
20	−5.5082360	−8.6731960	−3.6413460	−5.4908290
21	−2.9756040	−3.4056260	−4.5341880	−4.4563310
22	−4.3860670	−3.5756530	−4.9361630	−4.1328890
23	−6.0284100	−2.5753710	−5.3793240	−3.8960780
24	−9.1554870	−7.3568390	−5.5199390	4.2653220
25	−3.0298220	−2.7223140	−4.8775220	−4.7453790

Table 6.7 (Continued)

26	−4.0510470	−6.8654740	−3.7927980	−5.6437160
27	−5.1820460	−4.5522410	−5.0457170	−4.4836900
28	−2.6773540	−3.3413930	−4.6268300	−4.7481430
29	−2.7688590	−4.3950170	−4.0422420	5.0999170
30	−6.8400740	−5.7588750	−4.9973520	−4.3651140
31	−7.7955340	−6.3414780	−5.1617560	−4.4704800
32	−3.2683430	−4.1748700	−4.5055150	4.9825980
33	−4.7181680	−5.2078570	−4.3622970	−5.0384390
34	−3.2870640	−1.9998420	−4.8541070	−4.4801670
35	−5.4523020	−2.4132670	−5.4350430	−4.0666450
36	−7.4232510	−7.5067710	−4.8126640	−4.8422420
37	−3.4467090	−1.4618470	−5.6200360	−4.1190240
38	−6.9186350	−5.7871800	−5.1857880	−4.7001310
39	−3.6154960	−4.2446610	−4.7909640	5.0114490
40	−3.3500490	−5.5865210	−3.9588040	−5.6019570
41	−3.9933370	−5.5007500	−4.5554680	−4.9468580
42	−7.0150830	−4.7013150	−5.4134860	−4.3033640
43	−3.3668910	−3.3031310	−5.0504150	−4.6506340
44	−6.2205110	−8.3897360	−4.3390080	−5.3339380
45	−6.5916890	−3.3572670	−6.1055210	−3.8858910
46	−7.6909470	−6.8660590	−5.3444560	−4.8167030
47	−5.3735310	−4.1286990	−5.5777810	−4.6881830
48	−7.9085680	−6.8041210	−5.6229320	−4.8358590
49	−6.2792730	−7.2928890	−5.1296930	−5.3598010
50	−5.1237140	−4.4496240	−5.5265420	−5.0118690

Table 6.8
Estimates of Precision Matrix, Second Model

1	758.8698000	664.1608000	564.1604000	585.1857000
2	0.8138938	0.4229654	0.4229654	6.8305330
3	1.1235380	−0.1353001	−0.1353030	2.5455650
4	1.4086680	−0.2876567	−0.2876605	0.7875984
5	0.9312764	−0.6302108	−0.6302108	0.8866733
6	1.0907000	−0.7812513	−0.7812513	0.9866041
7	1.2407220	−0.8738025	−0.8737891	1.0532150
8	1.0807830	−0.8331743	−0.8331743	1.1221170
9	1.0940570	−0.8739011	−0.8739011	1.2227610
10	1.1929490	−0.9670018	−0.9670018	1.3511640

AN EXAMPLE OF ADAPTIVE ESTIMATION 301

Table 6.8 (Continued)

11	1.3056440	−1.0688860	−1.0688860	1.4792010
12	1.3159190	−1.0775630	−1.0775630	1.5309000
13	1.3486230	−1.1105050	−1.1105360	1.6132620
14	1.3669840	−1.0775690	−1.0776020	1.5714780
15	1.4258790	−1.1277940	−1.1278000	1.6615020
16	1.5057740	−1.1944110	−1.1944110	1.7638520
17	1.3708450	−1.0020730	−1.0020570	1.5602130
18	1.4474810	−1.0563220	−1.0563470	1.6408530
19	1.4582770	−1.0213700	−1.0213440	1.6057760
20	1.5198770	−1.0675250	−1.0675060	1.6855690
21	1.4835190	−0.9850793	−0.9849992	1.6061190
22	1.5480850	−1.0297110	−1.0297320	1.6797060
23	1.5788550	−1.0468060	−1.0468280	1.7324680
24	1.5478350	−1.0077560	−1.0077670	1.7345170
25	1.5434750	−0.9992738	−0.9992023	1.7701100
26	1.6102210	−1.0458600	−1.0458350	1.8458860
27	1.5587260	−0.9563593	−0.9563465	1.7667630
28	1.6061540	−0.9865548	−0.9865147	1.8299230
29	1.6609260	−1.0198510	−1.0198790	1.8910060
30	1.6834240	−1.0155720	−1.0156010	1.9096880
31	1.7227500	−1.0394010	−1.0393260	1.9677200
32	1.7573640	−1.0596560	−1.0596250	2.0228830
33	1.8067600	−1.0860420	−1.0860420	2.0751410
34	1.8513450	−1.1049870	−1.1049220	2.1187620
35	1.8974430	−1.1316270	−1.1316100	2.1756700
36	1.8817690	−1.1020990	−1.1020650	2.1818270
37	1.9336460	−1.1328220	−1.1328580	2.2407580
38	1.9107270	−1.0931890	−1.0930800	2.2340560
39	1.9533640	−1.1179410	−1.1179230	2.2908620
40	1.9965790	−1.1390550	−1.1390170	2.3391290
41	1.9646730	−1.0760300	−1.0759400	2.2950500
42	2.0108160	−1.0993710	−1.0992560	2.3447920
43	2.0133390	−1.0870810	−1.0869860	2.3677530
44	2.0570000	−1.1111050	−1.1111490	2.4217210
45	2.0767010	−1.1023650	−1.1021750	2.4339880
46	1.9753590	−0.9883214	−0.9883955	2.3575520
47	2.0107060	−1.0061040	−1.0061070	2.4068760
48	2.0310540	−1.0085630	−1.0085220	2.4413430
49	2.0556070	−1.0163300	−1.0163910	2.4821100
50	2.0949360	−1.0328120	−1.0328550	2.5257250

$$u = \begin{pmatrix} 211.2676 & -78.169 \\ -78.169 & 70.47254 \end{pmatrix},$$

Table 6.7 lists m(k) versus θ(k) and Table 6.8 gives the precision matrix P(k), which was computed according to (6.99).

The prior information about u and v is such that the prior parameters are $T_u = v_1 u^{-1}$, $T_v = v_2 v^{-1}$, $v_1 = v_2 = 4$ and θ(0) = (1,1)'.

An examination of Table 6.8 reveals the precision of the two components of m(k) slowly increases to the value 2.094936 of the first component and 2.525725 for the second component and this causes more variation in the θ(k) and m(k), k = 1, 2, ..., 50 of Table 6.7. More variation is present in θ(k) and m(k) (as compared to the first example) because of the "small" system precision matrix V, given by (6.113).

SUMMARY

In this chapter, a strictly Bayesian approach to the analysis of linear dynamic systems was taken. The usual approach (see Maybeck, 1979) is to use both Bayesian and non-Bayesian methods of analysis. For example, the Kalman filter is usually derived from the Bayesian approach, but sometimes the Kalman filter is based on sampling theory where it is derived as a minimum mean squared error estimator.

The analysis on estimation examined all phases of the estimation procedure, i.e., of filtering, smoothing, and prediction, where the appropriate posterior distribution was derived. With filtering it was the posterior distribution of the present state and the Kalman filter was given by Theorem 6.1, Theorems 6.1 and 6.2 specify the posterior distribution of the past and future states, thus giving a solution to the smoothing and prediction problems.

A Bayesian control strategy was introduced (pp. 244-251), where the average value of the posterior mean of the one-step-ahead future states was equal to the predetermined target values. A recursive algorithm, similar to the Kalman filter for finding the values of the control variables, was given and shown to produce an operational on-line way to control the linear dynamic system.

The analysis of dynamic systems was extended to the case where the design function of the observation equation was nonlinear and the transition function of the systems equation was linear. It was shown that nonlinearities of this type produce filters which do not possess the advantages of the Kalman filter. The mean of the posterior distribution can be determined in a recursive fashion but must be calculated by numerical integration.

EXERCISES

To avoid the difficulty, a linear approximation (based on a first-order Taylor's series) to the design function was taken and the result was a recursive filter which can be easily computed, however, the effect of the error of approximation on the properties of the filter was not studied.

The last part of the chapter examined a particular case of adaptive estimation, namely when the precision matrices of the observation and system errors are unknown. Unfortunately, the posterior distribution of the present state could not be expressed as a recursive relation, because the posterior distribution of the past and present states is a 2/0 multivariate poly-t distribution; therefore an approximate filter was devised. The approximation was based on the conditional posterior distribution of the present state given the past states. In this way, one may derive a recursive filter, which is similar to the Kalman filter; however, a normal approximation to the poly-t distribution was also employed. In addition, a control strategy based on the above approximation was developed.

Linear dynamic systems are an important class of linear models. These models are usually not studied by statisticians, but as has been shown, present one with challenging problems in the Bayesian approach to inference, and this chapter is only a very brief introduction to the subject. Many interesting problems are waiting to be solved.

EXERCISES

1. Let

    (i) $Y(t) = F(t)\theta(t) + u(t)$
    (ii) $\theta(t) = G\theta(t-1) + v(t)$

    where $t = 1, 2, \ldots$, that is, consider the linear dynamic model (6.1) and (6.2), and that the initial state $\theta(0)$ is known and also that $F(t)$ and G are known, where all quantities above in the model are scalar. Let $u(t)$; $t = 1, 2, \ldots$ be a sequence of n.i.d. $(0, \sigma_u^2)$ variables and suppose $v(t)$, $t = 1, 2, \ldots$ is a sequence of n.i.d. $(0, \sigma_v^2)$ variables. Given k observations $y(1), y(2), \ldots y(k)$, find the predictive distribution of $y(k+1)$ and explain how you would forecast a future observation. Assume $\sigma_u^2 = \sigma_v^2$ and that they are known.

2. Referring to the above system but now letting

(iii) $Y(t) = \theta(t) Y(t-1) + u(t)$
(iv) $\theta(t) = G\theta(t-1) + v(t)$,

that is above let $F(t) = Y(t-1)$. This is a dynamic autoregressive model of order one. Assuming $u(1), u(2), \ldots$ and $v(1)$, $v(2), \ldots$ are independent random vectors estimate (filter) the current state $\theta(k)$ using the first k observations $Y(1), Y(2), \ldots, Y(k)$.

3. Theorem 6.4 gives a solution to the control problem using a one-step-ahead strategy. Now use a two-step-ahead strategy and find values for the control variables $X(0)$ and $X(1)$ such that $\theta(1)$ and $\theta(2)$ are close to $T(1)$ and $T(2)$ respectively, where $Y(1)$ and $Y(2)$ have yet to be observed. Hint: Use the joint predictive density of $Y(1)$ and $Y(2)$. Assume what is given in Theorem 6.4.

4. Write a short essay on the difficulties of a Bayesian approach to nonlinear filtering and review the nonlinear regression section of Chapter 3.

5. Propose alternative ways to filter the states of a dynamic linear system for the general case of adaptive estimation. See the section entitled "Adaptive Filtering, the Vector Case" (pp. 286-290).

6. Consider the linear dynamic model of (6.1) and (6.2):

   (a) Explain how the usual general linear model (1.1) is a special case of the dynamic linear model.
   (b) Define the mixed model (4.1) as a special case of the dynamic model.
   (c) Is the autoregressive process AR(1) a special case of the dynamic linear model? If so, explain in what sense.

7. Consider a dynamic autoregressive model

$$Y(t) = \theta_1(t) Y(t-1) + \theta_2(t) Y(t-1) + u(t), \quad t = 1, 2, \ldots$$

with autoregressive parameters $\theta_1(t)$ and $\theta_2(t)$ which follow a bivariate AR(1) process

$$\theta(t) = G\theta(t-1) + V(t)$$

where $\theta(0)$ is known, G is a known 2×2 transition matrix and

$$\theta(t) = [\theta_1(t), \theta_2(t)]', \quad t = 1, 2, \ldots$$

Let $u(t) \sim N_1(0, P_u^{-1})$, $V(t) \sim N_2(0, P_v^{-1})$, where P_u and P_v are known, $P_u > 0$ and P_v is a 2nd order positive definite precision matrix.

Suppose $u(1), u(2), \ldots; V(1), V(2), \ldots$ are independent and develop a test of the null hypothesis $\theta_2(t) = 0$, $t = 1, 2, \ldots, k$, where k is a fixed positive integer.

REFERENCES

Alspach, D.L. (1974). "A parallel filtering algorithm for linear systems with unknown time varying noise statistics," IEEE Transactions on Automatic Control, Vol. AC-19, 555-556.

Aoki, Masanao (1967). *Optimization of Stochastic Systems, Topics in Discrete Time Series,* Academic Press, New York, London.

Box, G. E. P. and G. C. Tiao (1973). *Bayesian Inference in Statistical Analysis,* Addison-Wesley, Reading, Mass.

Downing, D.J., D. H. Pike, and G. W. Morrison (1980). "Application of the Kalman filter to inventory control," Technometrics, Vol. 22, No. 1, pp. 17-22.

Dreze, Jacques H. (1977). "Bayesian regression analysis using poly-t densities," in *New Developments in the Application of Bayesian Methods,* edited by Ahmet Aykac and Carlo Bremat, North-Holland Publishing Company, Amsterdam.

Harrison, P. J. and C. F. Stevens (1976). "Bayesian forecasting" (with discussion), Journal of the Royal Statistical Society, Series B, Vol. 38, 205-247.

Hawkes, R. M. (1973). *Demodulation of Pulse Modulated Signals Using Adaptive Estimation,* unpublished master's thesis, University of Newcastle.

Jazwinski, Andrew H. (1970). *Stochastic Processes and Filtering Theory,* Academic Press, New York.

Kailath, T. (1968). "An innovations approach to least squares estimation—Part I: Linear filtering in addative white noise," IEEE Transactions on Automatic Control, Vol. AC-13, 639-645.

Kalman, R. E. (1960). "A new approach to linear filtering and prediction problems," Transactions, ASME, J. of Basic Engineering, Vol. 82-D, pp. 34-45.

Lainiotis, D. G. (1971). "Optimal adaptive estimation: structure and parameter adaptation," IEEE Transactions on Automatic Control, Vol. AC-16, pp. 160-170.

Maybeck, Peter S. (1979). *Stochastic Models, Estimation, and Control,* Volumes I and II, Academic Press, New York.

Mehira, R. A. (1970). "Approaches to adaptive filtering," IEEE Transactions on Automatic Control, Vol. AC-17, pp. 693-698.

Mehira, Raman K. (1970). "On the identification of variances and adaptive Kalman filtering," IEEE Transactions on Automatic Control, Vol. AC-15, pp. 175-184.

Plackett, R. L. (1950). "Some theorems in least squares," Biometrika, Vol. 37, pp. 149-157.
Press, James S. (1972). *Applied Multivariate Analysis*, Holt, Rinehart, and Winston, Inc., New York.
Sallas, William M. and David A. Harville (1981). "Best linear recursive estimation for mixed linear models," Journal of American Statistical Association, Vol. 76, pp. 860-869.
Sawaragi, Yashikazu, Yoshifumi Sunaharo, and Takayoshi Nakamizo (1967). *Statistical Decision Theory in Adaptive Control Systems*, Academic Press, New York, London.
Shumway, R. H., D. E. Olsen and L. J. Levy (1981). "Estimation and tests of hypotheses for the initial mean and covariance in the Kalman filter model," Communications in Statistics, Part A—Theory and Methods, Vol. A10, No. 16, pp. 1625-1642.
Zellner, A. (1971). *An Introduction to Bayesian Inference in Econometrics*, John Wiley and Sons, Inc., New York.

7
STRUCTURAL CHANGE IN LINEAR MODELS

INTRODUCTION

This chapter will give a brief introduction to linear models which exhibit a change in some or all of the parameters during the period of observing the data. These models are quite useful for some economic, social, physical, and biological processes.

Since 1954, when Page discussed the problem of detecting a parameter change in the context of quality control, statistical techniques for examining structural change have been rapidly developing.

Page (1954, 1955, 1957) continued to develop nonparametric procedures for detecting shifts in univariate sequences and is well-known for his cusums (cumulative sums). After that, the next development was by Chernoff and Zacks (1964) and Kander and Zacks (1966), who use Bayesian methods to estimate the current mean and to test for a shift in the mean ($H_0: \mu_1 = \mu_2$ versus $H_a: \mu_1 \neq \mu_2$) of a sequence of independent random variables. Bhattacharyya and Johnson (1968) study optimality criteria related to testing hypotheses of shifting parameters, while Bacon and Watts (1970) and Broemeling (1972) give a Bayesian analysis of shifting normal sequences and switching regression models and explain estimation and detection techniques.

Much of the work in structural change is in the field of economics and the pioneering work of Quandt (1958, 1960, 1972, 1974). Goldfield and Quandt (1973a, 1973b, 1974) introduced techniques to analyze switching regression models.

Other contributions to this area of structural change are Hinkley (1969, 1971), Farley, Hinich, and McGuire (1971, 1975), Chow (1960), Brown and Durbin (1968), who use recursive residuals to detect changes in regression parameters, Madala and Nelson (1974), and Hartley and Mallela (1975). For a review of estimation methods for structural change, Poirier (1976) gives a thorough survey.

From the Bayesian viewpoint, there have also been many contributions. Bacon and Watts (1970) appears to be the first Bayesian study of switching regression models, while Broemeling (1972), as mentioned previously, studied sequences of normal random variables. The Bacon and Watts method employs a transition function to model possibly smooth changes from one regression line to the other, while Broemeling uses a shift point, which allows for more general changes in the mean.

Other contributions to the Bayesian methods of analyzing structural change are Holbert (1973), Holbert and Broemeling (1977), Chin Choy (1977), Chin Choy and Broemeling (1980, 1981), Chi (1979), Salazar (1980), Salazar, Broemeling and Chi (1981), Smith (1975, 1977, 1980), Ferreira (1975), and Tsurumi (1977, 1978).

This brief review fails to mention many contributions to structural change, but one must not fail to mention the many articles by Sen and Srivastava (1973, 1975a, 1975b, 1975c), who studied changes in the mean vector of multivariate normal sequences.

SHIFTING NORMAL SEQUENCES

A simple example of structural change is a normal sequence $X_1, X_2, \ldots, X_m, X_{m+1}, \ldots, X_n$ of independent random variables, where the first m observations have a $n(\theta_1, \tau)$ distribution and the last $(n - m)$ have a $n(\theta_2, \tau)$ distribution, where $\theta_1 \in R$, $\theta_2 \in R$, $\tau > 0$, $\theta_1 \neq \theta_2$, and the shift point $m = 1, 2, \ldots, n - 1$, where $n \geq 3$. The primary interest here is in m, the shift point and the pre- and postchange values of the mean.

The likelihood function of the parameters is

$$L(\theta_1, \theta_2, \tau, m | s) \propto \tau^{n/2} \exp{-\frac{\tau}{2}\left\{\sum_{i=1}^{m}(x_i - \theta_1)^2 + \sum_{i=m+1}^{n}(x_i - \theta_2)^2\right\}}, \quad (7.1)$$

for $(\theta_1, \theta_2) \in R^2$ but $\theta_1 \neq \theta_2$, $\tau > 0$, and $m = 1, 2, \ldots, n - 1$. How should one analyze structural change? The posterior analysis will consist of deriving the marginal posterior mass function of m, and the marginal posterior densities of θ_1, θ_2, and τ, but first one must select prior information for the parameters.

It is difficult to find a family of conjugate densities for the parameters, however the form of the likelihood function suggests a quasi-conjugate type prior density, namely, let (θ_1, θ_2) and τ have a normal-gamma distribution, and suppose m is independent of θ_1, θ_2, and τ and has a uniform distribution over $\{1, 2, \ldots, n-1\}$, i.e., let $\theta = (\theta_1, \theta_2)'$, and

$$p(\theta|\tau) \propto \tau^{2/2} \exp -\frac{\tau}{2}(\theta - \mu)' P(\theta - \mu), \quad \theta \in R^2 \quad (7.2)$$

be the prior density of θ, given τ, where P is a 2×2 positive definite matrix, and $\mu \in R^2$ are known hyperparameters. Now suppose the marginal prior density of τ is

$$p(\tau) \propto \tau^{\alpha-1} e^{-\tau\beta}, \quad \tau > 0 \quad (\alpha > 0, \beta > 0) \quad (7.3)$$

and

$$p(m) = (n-1)^{-1}, \quad m = 1, 2, \ldots, n-1, \quad (7.4)$$

the marginal posterior mass function of m. It seems imperative that the conditional prior density of θ depend on m, namely through the hyperparameters, μ and P, but it is difficult to devise such a density which conveniently combines with the likelihood function.

Combining the likelihood function with the prior density gives the joint posterior density of the parameters.

$$p(\theta, \tau, m | s) \propto \tau^{(n+2+2\alpha)/2-1} \exp -\frac{\tau}{2}\{[\theta - A^{-1}(m)B(m)]'A(m)$$

$$\times [\theta - A^{-1}(m)B(m)] + C(m) - B'(m)A^{-1}(m)B(m)\},$$

$$(7.5)$$

where $s = (x_1, x_2, \ldots, x_n)$ is the sample, $\theta \in R^2$, $\theta_1 \neq \theta_2$, $\tau > 0$, $m = 1, 2, \ldots, n-1$, and

$$A(m) = P + \begin{pmatrix} m & 0 \\ 0 & n-m \end{pmatrix},$$

$$B(m) = P\mu + \begin{pmatrix} \sum_{1}^{n} x_i \\ \sum_{m+1}^{n} x_i \end{pmatrix},$$

and

$$C(m) = 2\beta + \sum_{1}^{n} x_i^2 + \mu'P\mu.$$

Now θ and τ are easily eliminated from the joint density (7.5), to give

$$p(m|s) \propto \frac{1}{|A(m)|^{1/2}[C(m) - B'(m)A^{-1}(m)B(m)]^{(n+2\alpha)/2}},$$

$$m = 1, 2, \ldots, n-1 \qquad (7.6)$$

for the marginal posterior mass function of the shift point.

Upon inspection of (7.5), one sees the conditional posterior density of θ, given m, after eliminating τ, is a bivariate t distribution with density

$$p(\theta|m,s) \propto \frac{1}{\{[\theta - A^{-1}(m)B(m)]'A(m)[\theta - A^{-1}(m)B(m)] + C(m) - B'(m)A^{-1}(m)B(m)\}^{(n+2+2\alpha)/2}}$$

$$(7.7)$$

for $\theta \in R^2$. The location parameter is $A^{-1}(m)B(m)$, the precision matrix is

$$T(m) = \frac{(n+2\alpha)A(m)}{C(m) - B'(m)A^{-1}(m)B(m)},$$

and the degrees of freedom are $n + 2\alpha$.

SHIFTING NORMAL SEQUENCES

In the same way, one may show the conditional posterior distribution of τ, given m, is a gamma with parameters $(n + 2\alpha)/2$ and $[C(m) - B'(m)A^{-1}(m)B(m)]/2$.

To find the marginal distributions of θ and τ, one may average the conditional distributions of θ and τ with respect to the marginal posterior mass function of m, thus for θ

$$p(\theta|s) = \sum_{m=1}^{n-1} p(m|s)p(\theta|m,s), \quad \theta \in R^2 \qquad (7.8)$$

where $p(m|s)$ and $p(\theta|m,s)$ are given by (7.6) and (7.7), respectively, is the marginal posterior density of θ. This is a mixture of $n - 1$ conditional t densities and the mixing distribution is the marginal distribution of the shift point.

As for τ, one has

$$p(\tau|s) = \sum_{m=1}^{n-1} p(m|s)p(\tau|m,s), \quad \tau > 0, \qquad (7.9)$$

where $p(\tau|m,s)$ is a gamma density with parameters $\alpha(m) = (n + 2\alpha)/2$, and $\beta(m) = [C(m) - B'(m)A^{-1}(m)B(m)]/2$.

The above may be summarized by the following theorem.

Theorem 7.1. Consider a sequence $X_1, X_2, \ldots, X_n$ of independent normal random variables, where

$$\begin{aligned} X_i &\sim n(\theta_1, \tau^{-1}), \quad i = 1, 2, \ldots, m \text{ and} \\ X_i &\sim n(\theta_2, \tau^{-1}), \quad i = m + 1, \ldots, n \end{aligned} \qquad (7.10)$$

where θ_1, θ_2, τ, and m are unknown, and $\theta_1 \in R$, $\theta_2 \in R$, $\theta_1 \neq \theta_2$, $\tau > 0$, and $m = 1, 2, \ldots, n - 1$.

Suppose a priori that $\theta = (\theta_1, \theta_2)'$ and τ is a normal gamma with parameters μ, P, α and β, where $\mu \in R^2$, P is a 2×2 positive definite symmetric matrix and that $\alpha > 0$ and $\beta > 0$ and suppose m is uniform over $\{1, 2, \ldots, n - 1\}$ and independent of (θ, τ), then the joint posterior density of all the parameters is given by (7.5), the marginal posterior mass function of the shift point by (7.6), the marginal posterior mass function of θ is a mixture of conditional t densities, given by (7.8), and (7.9) expresses the marginal posterior density of τ as a mixture of $(n - 1)$ conditional gamma densities.

We will see that these results extend in a straightforward way to two-phase regression and more general linear models, including time series processes.

How does one represent vague prior information when one has a shifting sequence of random variables? If one lets $P \to 0(2 \times 2)$, $\beta \to 0$, and $\alpha \to -1$ in the joint posterior distribution (7.5), of all the parameters, then the joint posterior density of the parameters is the one that would have been obtained if one had used

$$p(\theta,\tau,m) \propto (n-1)^{-1}\tau^{-1}, \qquad \tau > 0, \; \theta \in R^2,$$

$$m = 1, 2 \ldots, n-1 \qquad (7.11)$$

for the prior density of the parameters. Under this assumption, one may prove the following

Corollary 7.1. Consider a shifting sequence (7.10) of normal random variables and suppose the prior distribution of θ_1, θ_2, τ, and m is given by the density (7.11), then the marginal posterior mass function of m is

$$p^*(m|s) \propto \frac{1}{\sqrt{m(n-m)} \, [C^*(m) - B^{*\prime}(m)A^{*-1}(m)B^*(m)]^{(n-2)/2}},$$

$$m = 1, 2 \ldots, n-1 \qquad (7.12)$$

where $n \geq 3$ and

$$A^*(m) = \begin{pmatrix} m & 0 \\ 0 & n-m \end{pmatrix},$$

$$B^*(m) = \begin{pmatrix} \sum_{1}^{m} x_i \\ \sum_{m+1}^{n} x_i \end{pmatrix},$$

and

$$C^*(m) = \sum_{1}^{n} x_i^2.$$

Also, the marginal posterior density of θ is the mixture

$$p(\theta|s) \propto \sum_{m=1}^{n-1} p^*(m|s)p^*(\theta|m,s), \quad \theta \in R^2 \qquad (7.13)$$

of $n - 1$ conditional t densities $p^*(\theta|m,s)$, which have $n - 2$ degrees of freedom, location vector $A^{*-1}(m)B^*(m)$ and precision

$$T^*(m) = \frac{(n-2)A^*(m)}{C^*(m) - B^{*\prime}(m)A^{*-1}(m)B^*(m)}.$$

As for τ, its marginal posterior distribution is a mixture of $n - 1$ conditional gamma densities $p^*(\tau|m,s)$, which have parameters

$$(n-2)/2 \text{ and } \beta(m) = [C^*(m) - B^{*\prime}(m)A^{*-1}(m)B^*(m)]/2,$$

and the mixing distribution is the posterior mass function $p^*(m|s)$ of m.

The above theorem and corollary provide a way to make inferences about the parameters of the model and this will be given in detail for the two-phase regression model, but for now let us study the forecasting problem for a shifting normal sequence.

Suppose $S = (x_1, x_2, \ldots, x_n)$ represents the observed values of a shifting normal sequence (7.10), and that X_{n+1} is a future value one wants to forecast, then what is the Bayesian predictive density of the future observation? To find this distribution the joint density of $X_1, X_2, \ldots, X_m, X_{m+1}, \ldots, X_n, X_{n+1}$ given θ, τ, and m will be multiplied by the proper prior density (7.2) of the parameters, then the product integrated with respect to θ, τ, and m, and the result will be the marginal joint density of the observations and the future value X_{n+1}, namely,

$$g(x_1, \ldots, x_n, x_{n+1}) = \sum_{m=1}^{n-1} p(m|s)g(x_1, \ldots, x_n, x_{n+1}|m),$$

$$(7.14)$$

for $(x_1, x_2, \ldots, x_{n+1}) \in R^{n+1}$, $p(m|x)$ is the posterior mass function of m given by (7.12), and

$$g(x_1, \ldots, x_{n+1}|m) \propto$$

$$\frac{1}{D^{1/2}(m)[F(m) - E'(m)D^{-1}(m)E(m)]^{(n+2\alpha+1)/2}} \qquad (7.15)$$

is the conditional density of the observations and x_{n+1}, given m, where

$$D(m) = \begin{pmatrix} m & 0 \\ 0 & n-m+1 \end{pmatrix} + P,$$

$$E(m) = \begin{pmatrix} \sum_{1}^{m} x_i \\ \sum_{m+1}^{n} x_i + x_{n+1} \end{pmatrix} + P,$$

and

$$F(m) = 2\beta + \mu'P\mu + \sum_{1}^{n} x_i^2 + x_{n+1}^2.$$

Since $F(m) - E'(m)D^{-1}(m)E(m)$ is a quadratic function of x_{n+1}, the predictive density of X_{n+1} is a mixture of $n-1$ conditional t densities (7.15) and the mixing distribution is the marginal posterior distribution of the shift point m with density (7.12). Each conditional density has $n + 2\alpha$ degrees of freedom. What are the other parameters?

To forecast a future value, one averages $n - 1$ conditional forecasts with respect to the marginal posterior probabilities of the possible change points, i.e., one assumes the change occurred at $m = 1$, then forecasts X_{n+1} with the conditional t density (7.15) letting $m = 1$, then one assumes a change occurred at $m = 2$, and forecasts with the conditional distribution of X_{n+1} given s and $m = 2$, and so on until one completes $n - 1$ forecasts. Each forecast is weighted by the appropriate posterior probability of a shift occurring at that point.

For further details about the posterior analysis of shifting normal sequences, the reader is referred to Holbert (1973) and Broemeling (1974), and for Bayesian forecasting methods with normal sequences, Broemeling (1977).

STRUCTURAL CHANGE IN LINEAR MODELS

Introduction

Structural change in linear models is easily generalized from changing normal sequences. A special case of the changing linear model is two-phase regression, where

$$Y_i = \alpha_1 + \beta_1 x_i + e_i, \quad i = 1, 2, \ldots, m \text{ and}$$
$$Y_i = \alpha_2 + \beta_2 x_i + e_i, \quad i = m+1, \ldots, n, \quad (7.16)$$

where the e_i's are i.i.d. $n(0, \tau^{-1})$, the x_i's are the values of a concomitant variable x, the regression coefficients satisfy $(\alpha_1, \beta_1) \neq (\alpha_2, \beta_2)$, and m is the shift point such that if m = n, there is no shift, but when m = 1, 2, ..., n − 1 exactly one shift has occurred. This model has been studied from classical and Bayesian perspectives. For example, Quandt (1958) estimated the switch point m and the regression parameters by maximum likelihood, and Hinkley (1969, 1971), under the assumption that the structural change was continuous, estimated and made other inferences about the intersection $\gamma = (\alpha_2 - \alpha_1)(\beta_1 - \beta_2)^{-1}$ between the two regression lines.

From the Bayesian viewpoint, Holbert (1973) and Holbert and Broemeling (1977) studied two-phase regression by assigning a marginal uniform proper prior to the shift point and an improper prior for the unknown regression parameters. Ferreira (1975) also assigned a vague-type prior distribution to the unknown regression parameters and assigned three prior mass functions to the shift point and developed some sampling properties for the estimated shift point m. Some other studies related to two-phase regression problems are Quandt (1960), Sprent (1961), Hudson (1966), Feder (1975), Farley, Hinick, and McGuire (1975), and Brown, Durbin, and Evans (1975).

From a non-Bayesian viewpoint, the changing linear model presents analytical and numerical difficulties and to avoid such problems, the Bayesian approach, as will be seen, provides one with some advantages.

The earliest Bayesian solutions used improper prior distributions with the result that the marginal posterior mass function of the shift point did not exist at all possible values, thus it is important to re-examine the problem with proper prior distributions.

Essentially, there are two problems associated with changing linear models: detecting the change (what is m?) and making inferences about the shift point and the other parameters, assuming a change has occured, where m = 1, 2, ..., n − 1. In the first problem, one tests the hypothesis m = n versus the alternative m ≠ n, but presently we will deal with only the second problem. For details about the second problem, see Broemeling (1972), Smith (1975), Chin Choy (1977), and Chin Choy and Broemeling (1981). In many practical problems either the data themselves will validate the assumption that there is a change or there will be reasons which make this assumption reasonable. For example in biological systems, the threshold level of a chemical is specific, i.e., the response of the system to the chemical is additive to the threshold level, and after the level is attained, the response stays constant or the chemical becomes toxic, thereby resulting in a decreasing response with increasing concentration. Ohki (1974) found the top growth of cotton increased sharply with a very slight increase of manganese content of the blade tissue. After reaching the inflection point of the nutrient calibration curve, the manganese content of the blade tissue increased sharply with no increase in plant growth. Some other examples of this problem can be seen in the papers of Green and Fekette (1933), Needham (1935), and Pool and Borchgrevinck (1964).

In what is to follow, the posterior analysis of a changing linear model will be presented, then an example will be given in detail which will illustrate how one makes inferences about the parameters of the model.

The Posterior Analysis

Consider a sequence $Y_1, Y_2, \ldots, Y_n$ of normally and independently distributed random variables, where

$$Y_i = x_i \beta_1 + e_i, \quad i = 1, 2, \ldots, m \text{ and}$$
$$Y_i = x_i \beta_2 + e_i, \quad i = m+1, \ldots, n,$$

(7.17)

where x_i, $i = 1, 2, \ldots, n$, is $1 \times p$ vector of observations on p independent variables, β_1 and β_2 are distinct $p \times 1$ vectors of coefficient parameters, the e_i's are i.i.d. $n(0, \tau^{-1})$ unobservable random errors, m is the shift point where $m = 1, 2, \ldots, n-1$, and the precision parameter $\tau > 0$. This model can be written as

$$y_1(m) = x_1(m)\beta_1 + e_1,$$
$$y_2(m) = x_2(m)\beta_2 + e_2,$$

(7.18)

or

STRUCTURAL CHANGE IN LINEAR MODELS

$$y = x(m)\beta + e$$

where $y_i(m)$, $x_i(m)$, β_i, and e_i, $i = 1, 2$, denote the observation vector, design matrix, parameter vector, and error vector of the i-th model. Also,

$$y = [y_1'(m), y_2'(m)]', \quad \beta = (\beta_1', \beta_2')', \quad e = (e_1', e_2')',$$

$$x(m) = \begin{pmatrix} x_1(m) & \phi_{12} \\ \phi_{21} & x_2(m) \end{pmatrix}.$$

Note $y_1(m)$ is $m \times 1$, $y_2(m)$ is $(n - m) \times 1$, $x_1(m)$ is $m \times p$, $x_2(m)$ is $(n - m) \times p$, β_1 is $p \times 1$, β_2 is $p \times 1$, e_1 is $m \times 1$, e_2 is $(n - m) \times 1$, and ϕ_{12} and ϕ_{21} are matrices of zeros.

The probability density function of y given m, β and τ is

$$f(y|m, \beta, \tau) \propto \tau^{n/2} \exp -\frac{\tau}{2}[y - x(m)\beta]'[y - x(m)\beta], \quad (7.19)$$

where $\tau > 0$, $\beta \in R^{2p}$, and $m = 1, 2, \ldots, n - 1$.

In this presentation, only the most general case will be considered, where all the parameters are unknown. For this and special cases, one may refer to Chin Choy (1977). The posterior analysis is given by the following theorem.

Theorem 7.2. If model (7.17) holds and m, β, and τ are unknown and if m is uniformly distributed over $I_{n-1} = \{1, 2, \ldots, n - 1\}$ and the joint marginal prior distribution of β and τ is such that: the conditional distribution of β given τ is normal with mean μ and precision matrix τr ($\tau > 0$) where r is a given $2p \times 2p$ positive definite matrix and μ a $2p \times 1$ constant vector, and the marginal prior distribution of τ is gamma with parameters $a > 0$ and $b > 0$, and m is independent of (β, τ), then

(i) The posterior probability mass function of m is

$$\pi(m|y) \propto D^{-a^*}(m) |x'(m)x(m) + r|^{-1/2}, \quad m = 1, 2, \ldots, n - 1,$$

$$(7.20)$$

where $a^* = a + n/2$,

$$D(m) = b + \frac{1}{2} \{y'y + \mu'r\mu - \beta^{*'}(m)[x'(m)x(m) + r]\beta^*(m)\},$$

and

$$\beta^*(m) = [x'(m)x(m) + r]^{-1}[r\mu + x'(m)y].$$

(ii) The posterior density function of β is

$$\pi(\beta|y) \propto \sum_{m=1}^{n-1} t[\beta; 2p, 2a^*, \beta^*(m), p(m)] \cdot \pi(m|y),$$

$$\beta \in R^{2p} \quad (7.21)$$

where $t[\beta; 2p, 2a^*, \beta^*(m), p(m)]$ is a 2p-dimensional t-density with $2a^*$ degrees of freedom, location $\beta^*(m)$ and precision matrix

$$p(m) = [a^*/p(m)][x'(m)x(m) + r].$$

Partition the vector β, the location $\beta^*(m)$, and the precision matrix $p(m)$ as

$$\beta = \begin{pmatrix} \beta_1 \\ \beta_2 \end{pmatrix}, \quad \beta^*(m) = \begin{pmatrix} \alpha_1^*(m) \\ \alpha_2^*(m) \end{pmatrix}, \quad p(m) = \begin{pmatrix} p_{11}(m) & p_{12}(m) \\ p_{21}(m) & p_{22}(m) \end{pmatrix},$$

where β_i and $\alpha_i^*(m)$, $i = 1, 2$, are each $p \times 1$ and the $p_{ij}(m)$, $i,j = 1,2$, are $p \times p$, then

(iii) The marginal posterior density of β_1 is

$$\pi(\beta_1|y) \propto \sum_{m=1}^{n-1} t[\beta_1; p, 2a^*, \alpha_1^*(m), p_1^*(m)] \pi(m|y),$$

$$\beta_1 \in R^p \quad (7.22)$$

where

$$p_1^*(m) = p_{11}(m) - p_{12}(m) p_{22}^{-1}(m) p_{21}(m).$$

STRUCTURAL CHANGE IN LINEAR MODELS

(iv) In a similar fashion, the marginal posterior density of β_2 is

$$\pi(\beta_2|y) \propto \sum_{m=1}^{n-1} t[\beta_2; p, 2a^*, \alpha_2^*(m), p_2^*(m)] \cdot \pi(m|y), \quad (7.23)$$

where

$$p_2^*(m) = p_{22}(m) - p_{21}(m) p_{11}^{-1}(m) p_{12}(m).$$

(v) The marginal posterior density of τ is

$$\pi(\tau|y) \propto \sum_{m=1}^{n-1} g[\tau; a^*, D(m)] \pi(m|y)], \quad \tau > 0, \quad (7.24)$$

where $g[\tau, a^*, D(m)]$ is a gamma density with parameters a^* and $D(m)$.

Proof. Combining the likelihood function (7.19) with the prior distribution of the parameters gives

$$\pi(m, \beta, \tau|y) \propto \tau^{a^*+p-1} \exp\left\{-\tau\left\{b + \frac{1}{2}(\beta-\mu)'r(\beta-\mu) + \frac{1}{2}[y - x(m)\beta]'[y - x(m)\beta]\right\}\right\}, \quad (7.25)$$

where $m \in I_{n-1}$, $\beta \in R^{2p}$, and $\tau > 0$, for the joint posterior density of the parameters. If (7.25) is integrated with respect to (β, τ), (m, τ); and (m, β), one obtains the marginal posterior p.d.f. of m, β, and τ respectively. Using the properties of the multivariate t distribution, see DeGroot (1970), the posterior p.d.f. of β_i, $i = 1, 2$, is obtained from (7.21).

As a corollary to the theorem, one may show:

Corollary 7.2.

(i) The conditional posterior p.d.f. of β given m is

$$\pi(\beta|y,m) \propto t[\beta; 2p, 2a^*, \beta^*(m), p(m)], \quad \beta \in R^{2p}. \quad (7.26)$$

(ii) The conditional posterior p.d.f. of β_1 given m is

$$\pi(\beta_1|y,m) \propto t[\beta_1; p, 2a^*, \alpha_1^*(m), p_1^*(m)], \quad \beta_1 \in R^p. \quad (7.27)$$

(iii) The conditional posterior density of τ given m is

$$\pi(\tau|y,m) = g[\tau; a^*, D(m)]. \quad (7.28)$$

These formulas for the marginal and conditional posterior distributions of the shift point (7.20), regression coefficients (7.21), (7.26), and precision (7.24), (7.28) will be illustrated with data generated from a model which has known parameters.

Marginal and Conditional Moments

To estimate the parameters one may use several estimators corresponding to different loss functions. It is well known that with a squared error loss function, the Bayes estimator is the mean of the posterior distribution and the Bayes risk is the average variance of the posterior distribution. The mean and dispersion matrix of the conditional posterior distributions of β_1, β_2, and τ given m are

$$E(\beta_1|y,m) = \alpha_1^*(m),$$

$$E(\beta_2|y,m) = \alpha_2^*(m),$$

$$E(\tau|y,m) = a^*/D(m),$$

$$\mathrm{Cov}(\beta_1|y,m) = a^*(a^* - 1)^{-1}[p_1^*(m)]^{-1}, \quad (7.29)$$

$$\mathrm{Cov}(\beta_2|y,m) = a^*(a^* - 1)^{-1}[p_2^*(m)]^{-1},$$

$$\mathrm{Var}(\tau|y,m) = a^*/[D(m)]^2.$$

The mean and dispersion matrix of the marginal posterior distributions of β_1, β_2, and τ are

$$E(\beta_1|y) = \sum_{m=1}^{n-1} \pi(m|y) E(\beta_1|m,y),$$

$$E(\beta_2|y) = \sum_{m=1}^{n-1} \pi(m|y) E(\beta_2|m,y),$$

STRUCTURAL CHANGE IN LINEAR MODELS

$$E(\tau|y) = \sum_{m=1}^{n-1} \pi(m|y)E(\tau|m,y),$$

$$\text{Cov}(\beta_1|y) = \underset{m}{E} \text{Cov}(\beta_1|m,y) + \underset{m}{\text{Cov}} E(\beta_1|m,y), \quad (7.30)$$

$$\text{Cov}(\beta_2|y) = \underset{m}{E} \text{Cov}(\beta_2|m,y) + \underset{m}{\text{Cov}} E(\beta_2|m,y),$$

$$\text{Var}(\tau|y) = \underset{m}{E} \text{Var}(\tau|m,y) + \underset{m}{\text{Var}} E(\tau|m,y),$$

where $\underset{m}{E}$, $\underset{m}{\text{Cov}}$, $\underset{m}{\text{Var}}$ denote the mean, dispersion and variance are taken with respect to the marginal distribution of m.

A Numerical Example

In this section, an example given by Chin Choy (1977) will illustrate the method of estimating the shift point and all unknown regression parameters of a two-phase regression model. This example uses the data generated by Quandt (1958) shown in Table 7.1. These data are twenty observations generated from the model

$$Y_i = 2.5 + 0.7x_i + e_i, \quad i = 1, 2, \ldots, 12,$$
$$Y_i = 5.0 + 0.5x_i + e_i, \quad i = 13, \ldots, 20, \quad (7.31)$$

where the e_i's are i.i.d. $n(0,1)$.

Assume that the two-phase regression model is

$$Y_i = \alpha_1 + \beta_1 x_i + e_i, \quad i = 1, 2, \ldots, m$$
$$Y_i = \alpha_2 + \beta_2 x_i + e_i, \quad i = m+1, \ldots, n, \quad (7.32)$$

where the e_i's are n.i.d. $(0, \tau^{-1})$ where τ and $\beta = (\alpha_1, \beta_1, \alpha_2, \beta_2)$ are unknown.

The first part of the example demonstrates the sensitivity of the marginal posterior distribution of m to different prior distributions, and the second part will consist of providing inferences for all the parameters.

Table 7.1
Quandt's Data Set

Obs. no. (i)	1	2	3	4	5	6	7	8	9	10
x_i	4	13	5	2	6	8	1	12	17	20
y_i	3.473	11.555	5.714	5.710	6.046	7.650	3.140	10.312	13.353	17.197

Obs. no. (i)	11	12	13	14	15	16	17	18	19	20
x_i	15	11	3	14	16	10	7	19	18	9
y_i	13.036	8.264	7.612	11.802	12.551	10.296	10.014	15.472	15.65	9.871

STRUCTURAL CHANGE IN LINEAR MODELS

Sensitivity Analysis

Two sources of prior information are used, where the first is data based, that is, the parameters μ, r, a, and b of the normal-gamma prior density are determined from the data. In order to obtain parameter values, assume the shift is near 12 and group the first nine consecutive observations into three sets and the last six into two sets of three observations per set. Based on the three observations in each set, a regression analysis is performed on each set and the usual least squares estimators $\hat{\alpha}$, $\hat{\beta}$, and $\hat{\sigma}^2$ ($\sigma^2 = \tau^{-1}$) are obtained from the first three sets, the mean and variance of $\hat{\alpha}$ and $\hat{\beta}$ are obtained and the covariance between $\hat{\alpha}$ and $\hat{\beta}$ is calculated. The values are used for the prior parameters of the first regression, and in the same way, using the last two sets, values for the prior parameters of the second regression are found. The numerical results are shown in Table 7.2, and from this table, the prior mean vector of β is μ = (2.7523, 0.5878, 6.1420, 0.4440).

Assuming the coefficients of the first regression are independent of the second, the covariance matrix of β is

$$\text{cov}(\beta) = \begin{pmatrix} 4.2926 & -0.5704 & 0 & 0 \\ -0.5704 & 0.0762 & 0 & 0 \\ 0 & 0 & 4.8878 & -0.4647 \\ 0 & 0 & -0.4647 & 0.0442 \end{pmatrix}$$

Now τ has a gamma distribution with parameters a and b, thus $\sigma^2 = \tau^{-1}$ has an inverse gamma distribution with parameters a and b and by the method of moments, two estimators of a and b are used. One is based on the five values of $\hat{\sigma}^2$ in Table 7.2 and the other on the five values of $\hat{\tau} = 1/\hat{\sigma}^2$ of the same table. For the first one, the estimates are $\hat{a}_1$ = 3.5032 and $\hat{b}_1$ = 1.0550 and for the second method the estimates based on $\hat{\tau}$ are $\hat{a}_2$ = 0.3625 and $\hat{b}_2$ = .0226. Since β has a four variate t distribution with mean μ and precision (a/b)r, it can be shown the dispersion matrix of β is $\text{cov}(\beta) = b(a-1)^{-1}r^{-1}$, when a > 1. If one uses the second set of estimates $\hat{a}_2$ and $\hat{b}_2$, $\text{cov}(\beta)$ does not exist, thus the first set of estimates $\hat{a}_1$ and $\hat{b}_1$ are used, giving

$$\text{cov}(\beta) = \begin{pmatrix} 17.4064 & 130.2590 & 0 & 0 \\ 130.2590 & 980.3119 & 0 & 0 \\ 0 & 0 & 422.1041 & 4439.3427 \\ 0 & 0 & 4439.3427 & 46698.8884 \end{pmatrix}$$

Table 7.2
Least Squares Estimators for Each of the Five Sets

Set No.	Numbers of observations contained in each set	$\hat{\alpha}$	$\hat{\beta}$	$\hat{\sigma}^2$	$\hat{\tau} = 1/\hat{\sigma}^2$
1	(1, 2, 3)	0.8009	0.8336	1.0007	0.9993
2	(4, 5, 6)	4.9266	0.2891	0.5892	1.6973
3	(7, 8, 9)	2.5294	0.6406	0.0144	69.2712
		Mean = 2.7523	0.5878		
		Var. = 4.2926	0.0762		
		Cov $(\hat{\alpha}, \hat{\beta})$ = −0.5704			
4	(15, 16, 17)	7.7053	0.2953	0.2043	4.8960
5	(18, 19, 20)	4.5787	0.5925	0.2993	3.3407
		Mean = 6.1420	0.4440		
		Var. = 4.8878	0.0442		
		Cov$(\hat{\alpha}, \hat{\beta})$ = −0.4647			

$\hat{a}_1 = 3.5027 \qquad \hat{a}_2 = 0.3623$

$\hat{b}_1 = 1.0550 \qquad \hat{b}_2 = 0.0226$

STRUCTURAL CHANGE IN LINEAR MODELS 325

for the marginal prior dispersion matrix of β.

The values a, b, μ, and r complete the specification of the prior normal gamma distribution. Using the values for μ, r, a, and b, the marginal p.m.f. of m is calculated and shown in Table 7.3, and shows the p.m.f. of m at 12 is 0.8728, which is an extremely high probability. It appears unreasonable to believe the shift did not occur at m = 12.

Now consider two experiments in order to see the sensitivity of the marginal posterior mass function of m to the prior parameters: the mean μ of β, the precision matrix $\Gamma(\beta) = (a/b)r$ of β (or the covariance matrix of β, $\text{cov}(\beta) = [b/(a-1)]r^{-1}$), and the parameters a and b of τ, where $E(\tau) = a/b$ and $\text{var}(\tau) = a/b^2$. Also note the variance $\sigma^2 = \tau^{-1}$ has mean and variance $E(\sigma^2) = b(a-1)^{-1}$ and $\text{var}(\sigma^2) = b^2(a-1)^{-2}(a-2)^{-1}$.

The two experiments were as follows:

Experiment One—The values of the parameters are assigned as follows:
1. $\mu = (2.5, 0.7, 5, 0.5)$,
2. $\Gamma(\beta) = \lambda I_4$ and $\lambda = .01, .1, 1, 10, 100$, thus all regression coefficients are uncorrelated, a priori.
3. $E(\tau) = 1$
4. $\text{var}(\tau) = .01, 0.1, 1, 10, 100$.

Once the values of these four parameters are assigned, the values of μ, r, a, b are known and the result is 25 different prior distributions. Based on each prior, the p.m.f. of m is calculated as shown in Tables 7.4-7.8.

Experiment Two—The values of the parameters are
1. $\mu = (2.5, 0.7, 5, 0.5)'$,
2. $\text{cov}(\beta) = \nu I_4$, $\nu = .01, .1, 1, 10, 100$,
3. $E(\sigma^2) = 1$,
4. $\text{var}(\sigma^2) = .01, .1, 1, 10, 100$.

Once the values of μ, $\text{cov}(\beta)$, $E(\sigma^2)$ and $\text{var}(\sigma^2)$ are assigned, one knows the values of μ, r, a, and b, and for each of the 25 sets of these, one has a marginal p.m.f. of m, which is tabulated in Tables 7.9-7.13.

The results of Experiment One demonstrate that

1. The posterior mass function of m is maximum at m = 12 for all values of λ and $\text{var}(\lambda)$.
2. When λ decreases, the probability of the end points m = 1 and m = 19 increases and is very noticeable when $\lambda = .01$ and 0.1. The reason for this is as $\lambda \to 0$, $r = \lambda I$ approaches singularity and

Table 7.3

Posterior Distribution of m for Data Based Prior

m	1	2	3	4	5	6	7	8	9	10
p.m.f.	0.0002	0.0002	0.0004	0.0000	0.0002	0.0006	0.0117	0.0234	0.0279	0.0280

m	11	12	13	14	15	16	17	18	19
p.m.f.	0.0244	0.8728	0.0017	0.0022	0.0039	0.0019	0.0001	0.0001	0.0002

STRUCTURAL CHANGE IN LINEAR MODELS

Table 7.4
Posterior Distribution of m When $\lambda = 0.01$, $E(\tau) = 1$
and $\beta_\mu = (2.5, 0.7, 5, 0.5)'$

m	Var (τ)				
	0.01	0.1	1	10	100
1	0.3505	0.3480	0.3410	0.3390	0.3389
2	0.0159	0.0158	0.0155	0.0154	0.0154
3	0.0132	0.0131	0.0128	0.0128	0.0127
4	0.0010	0.0013	0.0016	0.0017	0.0017
5	0.0021	0.0024	0.0027	0.0027	0.0027
6	0.0034	0.0037	0.0039	0.0039	0.0040
7	0.0186	0.0177	0.0167	0.0165	0.0164
8	0.0274	0.0254	0.0234	0.0230	0.0230
9	0.0279	0.0257	0.0235	0.0230	0.0230
10	0.0362	0.0331	0.0301	0.0295	0.0295
11	0.0347	0.0318	0.0289	0.0283	0.0283
12	0.4077	0.4173	0.4323	0.4360	0.4364
13	0.0054	0.0055	0.0054	0.0054	0.0054
14	0.0078	0.0077	0.0075	0.0074	0.0074
15	0.0169	0.0159	0.0149	0.0146	0.0146
16	0.0146	0.0139	0.0132	0.0130	0.0130
17	0.0015	0.0019	0.0022	0.0023	0.0023
18	0.0022	0.0028	0.0032	0.0033	0.0033
19	0.0129	0.0172	0.0211	0.0219	0.0220

Table 7.5
Posterior Distribution of m When $\lambda = 0.1$, $E(\tau) = 1$ and $\underset{\sim}{\beta}_\mu$ (2.5, 0.7, 5, 0.5)'

m	\multicolumn{5}{c}{Var (τ)}				
	0.01	0.1	1	10	100
1	0.1454	0.1458	0.1436	0.1429	0.1428
2	0.0151	0.0155	0.0155	0.0155	0.0155
3	0.0151	0.0142	0.0150	0.0150	0.0149
4	0.0013	0.0017	0.0021	0.0022	0.0022
5	0.0027	0.0032	0.0036	0.0037	0.0037
6	0.0045	0.0050	0.0053	0.0053	0.0053
7	0.0259	0.0246	0.0232	0.0229	0.0228
8	0.0381	0.0354	0.0326	0.0320	0.0320
9	0.0389	0.0358	0.0328	0.0321	0.0321
10	0.0500	0.0458	0.0417	0.0408	0.0407
11	0.0480	0.0440	0.0400	0.0392	0.0391
12	0.5459	0.5580	0.5745	0.5784	0.5789
13	0.0072	0.0073	0.0072	0.0072	0.0072
14	0.0103	0.0102	0.0099	0.0098	0.0098
15	0.0225	0.0212	0.0198	0.0196	0.0195
16	0.0189	0.0181	0.0171	0.0169	0.0169
17	0.0018	0.0023	0.0027	0.0028	0.0028
18	0.0026	0.0032	0.0038	0.0039	0.0039
19	0.0058	0.0077	0.0095	0.0099	0.0099

Table 7.6
Posterior Distribution of m When $\lambda = 1$, $E(\tau) = 1$
and β_μ (2.5, 0.7, 5, 0.5)'

m	Var (τ)				
	0.01	0.1	1	10	100
1	0.0384	0.0419	0.0444	0.0448	0.0449
2	0.0040	0.0050	0.0057	0.0059	0.0059
3	0.0071	0.0080	0.0087	0.0088	0.0088
4	0.0009	0.0014	0.0019	0.0020	0.0020
5	0.0025	0.0031	0.0037	0.0038	0.0038
6	0.0046	0.0053	0.0059	0.0060	0.0060
7	0.0351	0.0338	0.0324	0.0320	0.0320
8	0.0519	0.0489	0.0457	0.0451	0.0450
9	0.0541	0.0505	0.0469	0.0462	0.0461
10	0.0659	0.0613	0.0566	0.0557	0.0556
11	0.0629	0.0585	0.0541	0.0532	0.0531
12	0.6036	0.6120	0.6233	0.6259	0.6262
13	0.0079	0.0082	0.0084	0.0084	0.0084
14	0.0113	0.0114	0.0113	0.0113	0.0113
15	0.0250	0.0240	0.0229	0.0229	0.0226
16	0.0186	0.0183	0.0177	0.0177	0.0176
17	0.0014	0.0018	0.0023	0.0023	0.0023
18	0.0019	0.0025	0.0030	0.0030	0.0031
19	0.0029	0.0040	0.0051	0.0051	0.0053

Table 7.7
Posterior Distribution of m When $\lambda = 10$, $E(\tau) = 1$ and $\underset{\sim}{\beta}_\mu$ (2.5, 0.7, 5, 0.5)'

m	Var (τ)				
	0.01	0.1	1	10	100
1	0.0026	0.0042	0.0060	0.0063	0.0064
2	0.0005	0.0008	0.0013	0.0014	0.0014
3	0.0015	0.0022	0.0030	0.0031	0.0032
4	0.0003	0.0005	0.0009	0.0010	0.0010
5	0.0013	0.0019	0.0026	0.0027	0.0028
6	0.0033	0.0042	0.0050	0.0052	0.0052
7	0.0431	0.0426	0.0418	0.0416	0.0416
8	0.0647	0.0624	0.0598	0.0593	0.0592
9	0.0702	0.0671	0.0639	0.0632	0.0632
10	0.0817	0.0777	0.0737	0.0728	0.0727
11	0.0776	0.0739	0.0701	0.0693	0.0692
12	0.5888	0.5944	0.6012	0.6027	0.6028
13	0.0081	0.0088	0.0093	0.0094	0.0094
14	0.0114	0.0119	0.0123	0.0124	0.0124
15	0.0247	0.0246	0.0243	0.0243	0.0243
16	0.0164	0.0169	0.0171	0.0171	0.0171
17	0.0008	0.0013	0.0017	0.0019	0.0018
18	0.0011	0.0017	0.0022	0.0024	0.0024
19	0.0018	0.0027	0.0038	0.0040	0.0040

Table 7.8
Posterior Distribution of m When $\lambda = 100$, $E(\tau) = 1$
and $\underset{\sim}{\beta}_\mu = (2.5, 0.7, 5, 0.5)'$

m	Var (τ)				
	0.01	0.1	1	10	100
1	0.0003	0.0006	0.0011	0.0012	0.0013
2	0.0002	0.0004	0.0007	0.0008	0.0008
3	0.0008	0.0013	0.0019	0.0020	0.0020
4	0.0002	0.0004	0.0007	0.0007	0.0007
5	0.0010	0.0015	0.0022	0.0023	0.0023
6	0.0027	0.0036	0.0045	0.0047	0.0047
7	0.0425	0.0426	0.0424	0.0424	0.0424
8	0.0644	0.0629	0.0612	0.0609	0.0608
9	0.0694	0.0673	0.0650	0.0645	0.0645
10	0.0918	0.0882	0.0846	0.0838	0.0838
11	0.0881	0.0847	0.0812	0.0805	0.0804
12	0.5822	0.5850	0.5885	0.5893	0.5894
13	0.0083	0.0092	0.0100	0.0101	0.01011
14	0.0111	0.0120	0.0127	0.0129	0.0129
15	0.0212	0.0218	0.0222	0.0222	0.0223
16	0.0134	0.0143	0.0150	0.0152	0.0152
17	0.0006	0.0010	0.0015	0.0015	0.0016
18	0.0008	0.0012	0.0018	0.0019	0.0019
19	0.0012	0.0019	0.0028	0.0030	0.0030

Table 7.9

Posterior Distribution of m for $v = 0.01$, $E(\sigma^2) = 1$ and $\underset{\sim}{\beta}_\mu = (2.5, 0.7, 5, 0.5)'$

m	Var(σ^2)				
	0.01	0.1	1	10	100
1	0.0003	0.0004	0.0006	0.0006	0.0006
2	0.0002	0.0003	0.0004	0.0004	0.0004
3	0.0007	0.0010	0.0011	0.0012	0.0013
4	0.0001	0.0002	0.0003	0.0004	0.0004
5	0.0009	0.0012	0.0013	0.0014	0.0014
6	0.0026	0.0029	0.0030	0.0031	0.0031
7	0.0417	0.0387	0.0359	0.0354	0.0353
8	0.0636	0.0588	0.0542	0.0534	0.0533
9	0.0686	0.0635	0.0587	0.0578	0.0577
10	0.0911	0.0848	0.0788	0.0776	0.0775
11	0.0874	0.0813	0.0755	0.0744	0.0743
12	0.5876	0.6137	0.6387	0.6436	0.6441
13	0.0080	0.0079	0.0076	0.0075	0.0075
14	0.0108	0.0104	0.0100	0.0098	0.0098
15	0.0208	0.0195	0.0182	0.0180	0.0179
16	0.0130	0.0125	0.0118	0.0117	0.0009
17	0.0006	0.0008	0.0009	0.0009	0.0011
18	0.0007	0.0009	0.0011	0.0011	0.0018
19	0.0011	0.0015	0.0017	0.0018	0.0018

Table 7.10

Posterior Distribution of m for $v = 0.1$, $E(\sigma^2) = 1$ and $\underset{\sim}{\beta}_\mu = (2.5, 0.7, 5, 0.5)'$

m	Var (σ^2)				
	0.01	0.1	1	10	100
1	0.0025	0.0031	0.0036	0.0037	0.0037
2	0.0004	0.0006	0.0007	0.0008	0.0008
3	0.0015	0.0017	0.0019	0.0019	0.0019
4	0.0003	0.0004	0.0005	0.0005	0.0005
5	0.0012	0.0015	0.0016	0.0017	0.0017
6	0.0031	0.0033	0.0034	0.0034	0.0034
7	0.0424	0.0387	0.0352	0.0346	0.0345
8	0.0639	0.0581	0.0529	0.0518	0.0517
9	0.0695	0.0634	0.0578	0.0567	0.0566
10	0.0810	0.0744	0.0681	0.0669	0.0668
11	0.0770	0.0706	0.0646	0.0635	0.0633
12	0.5945	0.6248	0.6541	0.6598	0.6604
13	0.0079	0.0075	0.0071	0.0070	0.0070
14	0.0111	0.0104	0.0097	0.0095	0.0095
15	0.0242	0.0222	0.0203	0.0199	0.0199
16	0.0160	0.0149	0.0137	0.0135	0.0135
17	0.0008	0.0010	0.0011	0.0011	0.0011
18	0.0011	0.0013	0.0014	0.0143	0.0014
19	0.0017	0.0021	0.0023	0.0024	0.0024

Table 7.11

Posterior Distribution of m for $v = 1$, $E(\sigma^2) = 1$
and $\underset{\sim}{\beta}_\mu = (2.5, 0.7, 5, 0.5)'$

	Var (σ^2)				
m	0.01	0.1	1	10	100
1	0.0372	0.0351	0.0325	0.0319	0.0319
2	0.0039	0.0040	0.0039	0.0039	0.0039
3	0.0068	0.0066	0.0062	0.0061	0.0061
4	0.0009	0.0011	0.0011	0.0011	0.0011
5	0.0023	0.0025	0.0025	0.0024	0.0024
6	0.0044	0.0043	0.0041	0.0041	0.0041
7	0.0344	0.0305	0.0269	0.0262	0.0261
8	0.0512	0.0453	0.0398	0.0388	0.0387
9	0.0534	0.0474	0.0418	0.0408	0.0406
10	0.0653	0.0584	0.0519	0.0506	0.0505
11	0.0624	0.0557	0.0495	0.0483	0.0481
12	0.6104	0.6478	0.6844	0.6915	0.6923
13	0.0077	0.0071	0.0064	0.0063	0.0062
14	0.0110	0.0100	0.0089	0.0087	0.0086
15	0.0245	0.0217	0.0191	0.0186	0.0186
16	0.0182	0.0162	0.0144	0.0140	0.0140
17	0.0013	0.0014	0.0015	0.0015	0.0015
18	0.0018	0.0019	0.0020	0.0019	0.0019
19	0.0027	0.0031	0.0032	0.0032	0.0032

Table 7.12

Posterior Distribution of m for $v = 10$, $E(\sigma^2) = 1$
and $\underset{\sim}{\beta}_\mu = (2.5, 0.7, 5, 0.5)'$

m	Var (σ^2)				
	0.01	0.1	1	10	100
1	0.1419	0.1274	0.1128	0.1099	0.1096
2	0.0147	0.0134	0.0119	0.0116	0.0116
3	0.0147	0.0132	0.0117	0.0114	0.0114
4	0.0013	0.0014	0.0014	0.0014	0.0014
5	0.0026	0.0026	0.0025	0.0024	0.0024
6	0.0043	0.0041	0.0038	0.0037	0.0037
7	0.0254	0.0223	0.0194	0.0189	0.0188
8	0.0377	0.0330	0.0286	0.0278	0.0277
9	0.0385	0.0339	0.0295	0.0286	0.0285
10	0.0497	0.0441	0.0387	0.0377	0.0377
11	0.0477	0.0423	0.0371	0.0361	0.0360
12	0.5541	0.6008	0.6475	0.6566	0.6576
13	0.0070	0.0063	0.0056	0.0055	0.0055
14	0.0101	0.0090	0.0079	0.0077	0.0077
15	0.0221	0.0194	0.0169	0.0164	0.0163
16	0.0185	0.0163	0.0142	0.0138	0.0137
17	0.0017	0.0019	0.0019	0.0019	0.0018
18	0.0025	0.0026	0.0025	0.0025	0.0025
19	0.0055	0.0060	0.0061	0.0061	0.0061

Table 7.13

Posterior Distribution of m for $v = 100$, $E(\sigma^2) = 1$ and $\underset{\sim}{\beta}_\mu = (2.5, 0.7, 5, 0.5)'$

m	Var (σ^2)				
	0.01	0.1	1	10	100
1	0.3442	0.3152	0.2846	0.2784	0.2777
2	0.0156	0.0143	0.0129	0.0126	0.0126
3	0.0130	0.0119	0.0107	0.0105	0.0105
4	0.0010	0.0011	0.0011	0.0011	0.0011
5	0.0020	0.0020	0.0020	0.0020	0.0019
6	0.0033	0.0032	0.0030	0.0030	0.0030
7	0.0184	0.0165	0.0147	0.0144	0.0143
8	0.0273	0.0244	0.0217	0.0212	0.0211
9	0.0278	0.0250	0.0223	0.0217	0.0217
10	0.0362	0.0329	0.0296	0.0289	0.0288
11	0.0347	0.0315	0.0283	0.0277	0.0276
12	0.4164	0.4645	0.5143	0.5256	0.5267
13	0.0053	0.0049	0.0045	0.0044	0.0044
14	0.0077	0.0070	0.0063	0.0062	0.0061
15	0.0167	0.0150	0.0133	0.0130	0.0130
16	0.0144	0.0130	0.0116	0.0113	0.0113
17	0.0015	0.0015	0.0016	0.0016	0.0016
18	0.0022	0.0023	0.0023	0.0023	0.0023
19	0.0124	0.0138	0.0144	0.0144	0.0144

STRUCTURAL CHANGE IN LINEAR MODELS   337

   $x'(m)x(m)$ is singular at m = 1 and 19, thus $x'(m)x(m) + r$ approaches singularity.
3. The posterior probability at m = 12 increases with an increase in var(τ).

The results from the second experiment show that:

1. The p.m.f. of m is maximum at m = 12 for all values of ν and var(σ^2) and the probability at m = 12 increases very little as var(σ^2) increases from 1 to 100.
2. As ν increases, the probability at m = 1 and m = 19 increases, especially at m = 1, and is most noticeable when ν = 10 and 100 because $x'(m)x(m) + r$ approaches singularity (as ν increases).
3. The posterior probability at m = 12 increases with an increase in var(σ^2).

A safe conclusion from the above results is that the shift point is at m = 12 for all prior distributions and that the p.m.f. shows a remarkable insensitivity to the different prior distributions.

Inferences for the Parameters

This part of the example will provide inferences for the parameters β, including β_1 and β_2, τ, and m of the model. Together with Quandt's data set, Table 7.1, and a normal-gamma prior distribution with parameters $\mu = (2.5, 0.7, 5, 0.5)'$, $r = I_4$, a = 3, and b = 2, i.e., $E(\sigma^2) = 1$, var(σ^2) = 1, $E(\beta) = \mu$, and var(β) = I_4, the marginal posterior distributions of each parameter will be analyzed.

The posterior p.m.f. of m is given in Table 7.2 and clearly shows a shift occurred at m = 12 with a probability of .6844 and that the mode is at m = 12, the median at 12, and the mean at 11.11.

Inferences about the regression parameters α_1, β_1, α_2, β_2 may be based on either the marginal posterior distribution or the conditional posterior distribution, given, say, m = 12, of the parameter, and these densities were derived in an earlier section of the chapter, for example the conditional densities are give by formulas (7.26) through (7.28), while the marginal densities were given by Theorem 7.2.

Point estimates and regions of highest posterior density will be obtained for each of the parameters using both the marginal and conditional posterior distributions of that parameter. The reader is referred to Box and Tiao (1965) and to Chapter 1 for an explanation of HPD regions.

Inferences about α_1. Let $\pi(\alpha_1|y)$ and $\pi(\alpha_1|y,12)$ denote the marginal and conditional (given m = 12) posterior densities of α_1. Figure 7.1 is a plot of $\pi(\alpha_1|y)$, $\pi(\alpha_1|y,12)$ and the marginal prior density of α_1, $\pi_0(\alpha_1)$, and the following table gives point estimates and HPD regions for α_1.

Point and Interval Estimates of α_1

Estimates	$\pi(\alpha_1|y)$	$\pi(\alpha_1|y, 12)$
mean	2.36	2.29
mode	2.32	2.29
median	2.35	2.29
variance	.2541	.1937
HPD regions		
90%	(1.52, 3.19)	(1.56, 3.01)
95%	(1.34, 3.42)	(1.42, 3.16)
99%	(0.86, 3.95)	(1.11, 3.46)

Point and Interval Estimates of β_1

Estimates	$\pi(\beta_1|y)$	$\pi(\beta_1|y, 12)$
mean	.67	.69
mode	.68	.69
median	.69	.69
variance	.0069	.0017
HPD regions		
90%	(0.58, 0.77)	(0.62, 0.75)
95%	(0.55, 0.80)	(0.61, 0.77)
99%	(0.02, 0.90)	(0.58, 0.80)

The prior density of $\pi_0(\beta_1)$ and the marginal and conditional posterior densities are plotted in Figure 7.2.

STRUCTURAL CHANGE IN LINEAR MODELS

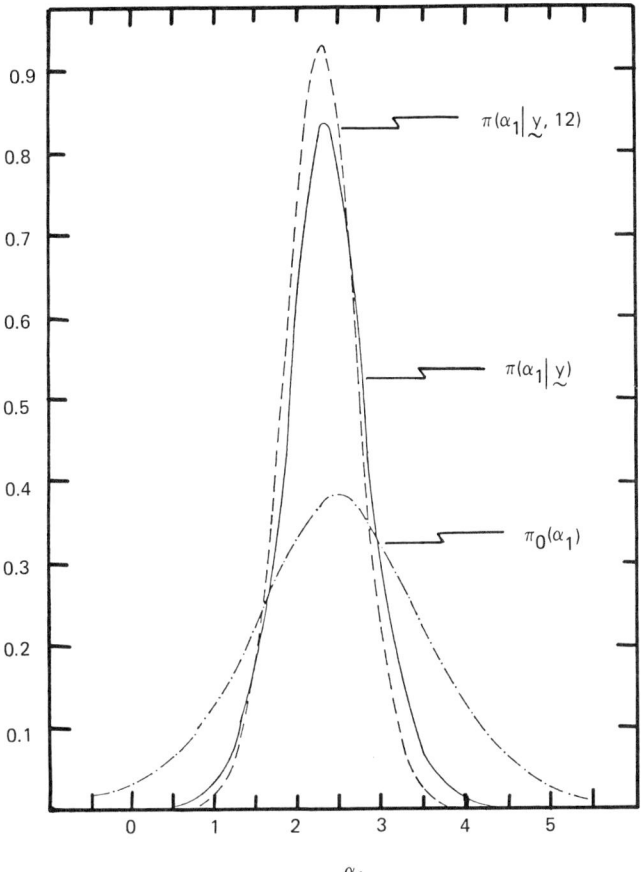

Figure 7.1. Prior, marginal posterior, and conditional posterior densities of α_1.

Point and Interval Estimates of α_2

Estimates	$\pi(\alpha_2|y)$	$\pi(\alpha_2|y, 12)$
mean	5.34	5.52
mode	5.45	5.52
median	5.39	5.52
variance	.3933	.3617

340 STRUCTURAL CHANGE IN LINEAR MODELS

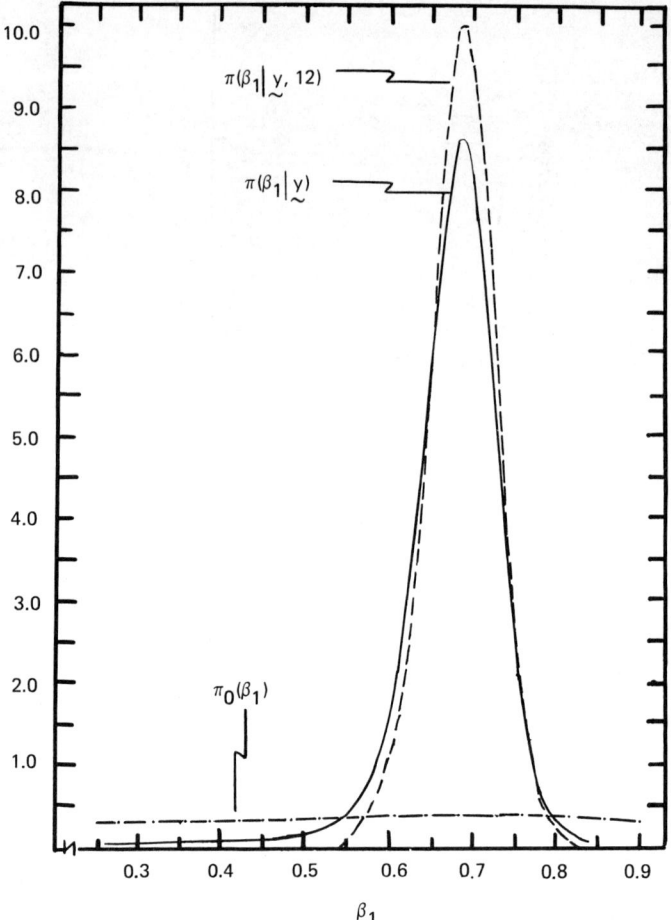

Figure 7.2. Prior, marginal posterior, and conditional posterior densities of β_1.

Point and Interval Estimates of α_2 (Continued)

Estimates	$\pi(\alpha_2|y)$	$\pi(\alpha_2|y, 12)$
HPD regions		
90%	(4.15, 6.54)	(4.50, 6.50)
95%	(3.80, 6.72)	(4.33, 6.71)
99%	(3.20, 7.12)	(3.91, 7.12)

STRUCTURAL CHANGE IN LINEAR MODELS 341

Figure 7.3 is a graph of the marginal prior density $\pi_0(\alpha_2)$ and the marginal and conditional posterior densities of α_2.

Point and Interval Estimates of β_2

Estimates	$\pi(\beta_2\|y)$	$\pi(\beta_2\|y, 12)$
mean	0.52	.51
mode	0.51	.51
median	0.52	.51
variance	.0026	.0024
HPD regions		
90%	(.43, .61)	(.43, .59)
95%	(.41, .63)	(.41, .60)
99%	(.37, .67)	(.38, .64)

Figure 7.4 is a graph of the three densities of the slope parameter of the second regression. Figure 7.5 is a plot of the marginal prior, the marginal posterior, and the conditional posterior density, given m = 12, of τ, and the following gives point and interval estimates of the precision parameter τ.

Point and Interval Estimates of τ

Estimates	$\pi(\tau\|y)$	$\pi(\tau\|y, 12)$
mean	1.20	1.30
mode	1.09	1.20
median	1.17	1.27
variance	.1126	.1297
HPD regions		
90%	(.61, 1.78)	(.71, 1.87)
95%	(.54, 1.93)	(.64, 2.02)
99%	(.43, 2.26)	(.51, 2.33)

Of course, one may also make inferences about the residual variances $\sigma^2 = \tau^{-1}$, which are as follows.

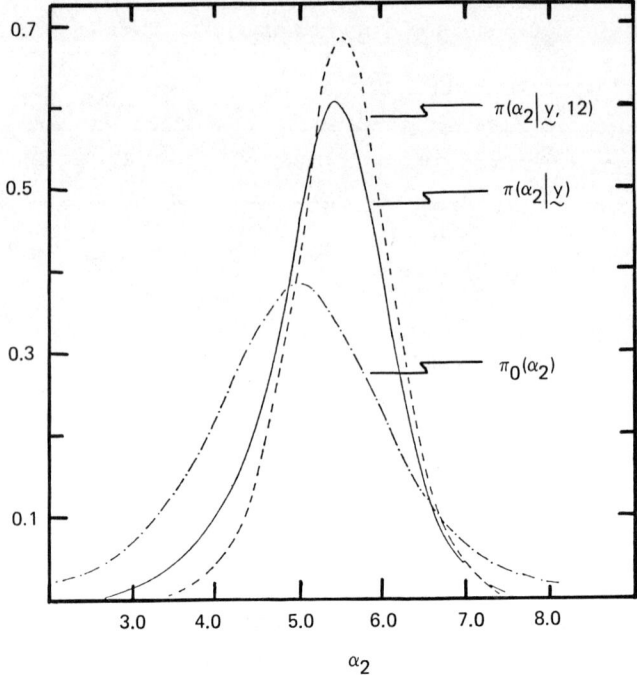

Figure 7.3. Prior, marginal posterior, and conditional posterior densities of α_2.

Point and Interval Estimates of σ^2

Estimates	$\pi(\sigma^2\|y)$	$\pi(\sigma^2\|y, 12)$
mean	.92	.83
mode	.76	.72
median	.86	.80
variance	.078	.0633
HPD regions		
90%	(.47, 1.36)	(.46, 1.20)
95%	(.44, 1.55)	(.43, 1.34)
99%	(.38, 1.97)	(.37, 1.66)

STRUCTURAL CHANGE IN LINEAR MODELS

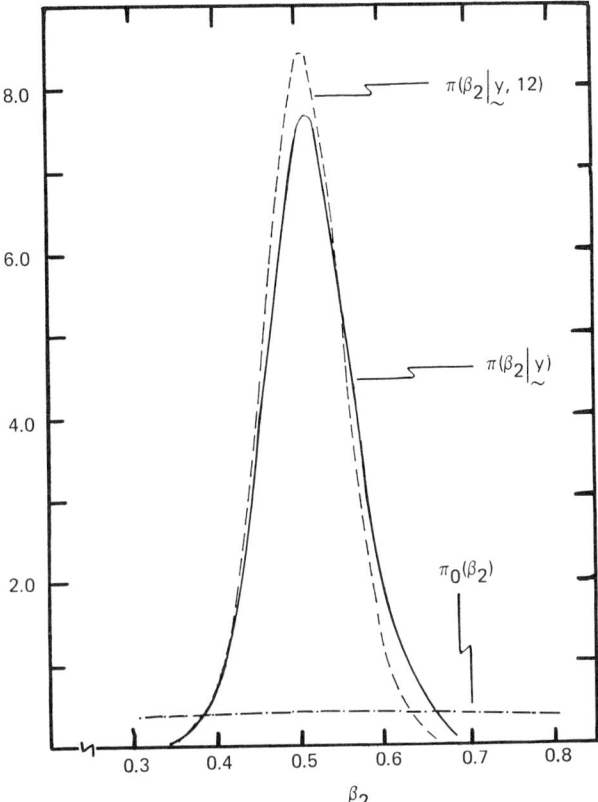

Figure 7.4. Prior, marginal posterior, and conditional posterior densities of β_2.

The prior, marginal and conditional posterior densities of σ^2 are depicted in Figure 7.6.

This example shows that the conditional posterior distribution is a "good" approximation to the corresponding marginal posterior distribution of the parameter. The conditional is easier to use in finding point and interval estimates because they are standard densities, either t, in the case of the regression coefficients, or a gamma in the case of the residual precision. The conditional distribution of a parameter is perhaps a good approximation to the marginal distribution if the conditioning value of m has a relatively large posterior probability mass, as is the case in this example.

It has been shown that the Bayesian analysis of a changing linear model provides a convenient way to make informal inferences about the

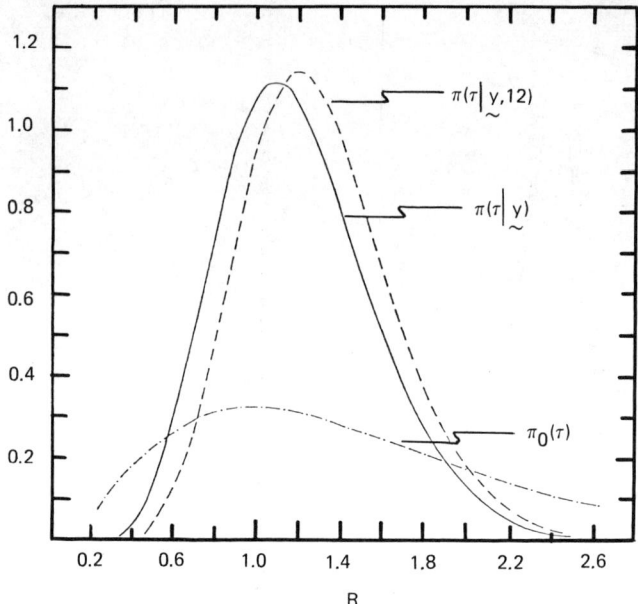

Figure 7.5. Prior, marginal posterior, and conditional posterior densities of τ.

model. It is possible to investigate how the model changes and what parameters account for the change.

First, the posterior p.m.f. of the shift point m is found and the most probable value when the change is likely to have occurred is determined. The posterior distribution of the regression parameters is calculated in order to find which parameters account for the change. Point and interval estimates are easily obtained from the appropriate marginal and conditional distributions, and conditional inferences, when applicable, are much easier to obtain.

An advantage of the Bayesian analysis is that the exact small sample distribution of the parameters can be found in terms of well-known distributions thus avoiding the large-sample methods of maximum likelihood estimation and the likelihood ratio test, which was used by Hinkley (1971) in his analysis of a changing linear model.

The above analysis uses a normal-gamma proper prior distribution for the parameters, but if one wants to employ vague prior information, one could let $r \to 0$, $a \to -p/2$ and $b \to 0$ in the joint posterior distribution with density (8.25), and in this way obtain the joint posterior

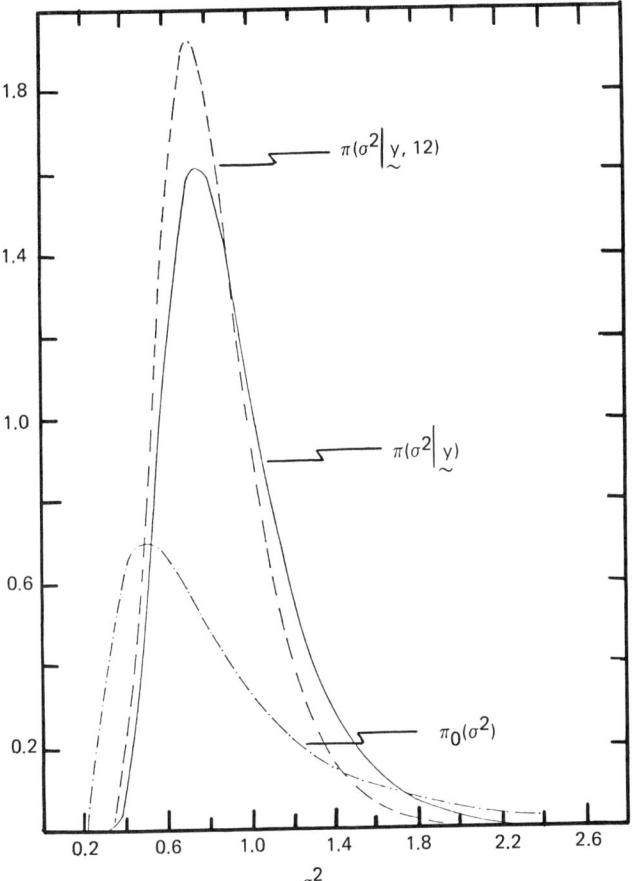

Figure 7.6. Prior, marginal posterior, and conditional posterior densities of σ^2.

density derived by Holbert and Broemeling (1977), who used the Jeffreys' type prior density

$$\pi_0(\beta, \tau, m) \propto \frac{1}{\tau}, \quad \tau > 0, \quad \beta \in R^4, \quad m = 1, 2, \ldots, n-1.$$

The above analysis can be found in Chin Choy and Broemeling (1980) and in a more detailed version in Chin Choy (1977), where she explains the numerical procedures for computing point and interval estimates of the parameters.

DETECTION OF STRUCTURAL CHANGE

Up to this point, one assumed exactly one change in the model, that is one knew a change will occur, however in some situations one is not sure if the model has changed its parameter values, and the model (7.17) should be revised to

$$Y_i = x_i\beta_1 + e_i, \quad i = 1, 2, \ldots, m$$
$$Y_i = x_i\beta_2 + e_i, \quad i = m+1, \ldots, n. \tag{7.33}$$

where $m = 1, 2, \ldots, n-1, n$. By using the additional mass point n, one is allowing for the a priori possibility of no change, since when $m = n$, there is only one relationship between the dependent and independent variables with regression coefficient vector β_1. As before, let the e_i be n.i.d. $(0, \tau^{-1})$ variables $\beta_1 \in R^p$, $\beta_2 \in R^p$, and $\tau > 0$. If $1 \leq m \leq n-1$, then $\beta_1 \neq \beta_2$ and if $m = n$ no change has occurred in the linear relation between independent and dependent variables. If $1 \leq m \leq n-1$, then the first m observations have a linear structure with regression coefficients β_1 and $x_1, x_2, \ldots, x_m$ are each $1 \times p$ known vectors, and the remaining $n - m$ observations have a linear structure with coefficients β_2, where $x_{m+1}, \ldots, x_n$ are known $1 \times p$ vectors.

It is not known if structural stability is present and the objective of this part of the chapter is to test the null hypothesis of no change versus the alternative of exactly one change, i.e., test

$$H_0: m = n \text{ versus } H_1: 1 \leq m \leq n-1.$$

This analysis is a generalization of previous studies of Broemeling (1972) and Smith (1975) and can be found in Broemeling and Chin Choy (1981) and a more complete version by Chin Choy (1977). An informal Bayesian analysis is used where the test is based on the marginal posterior mass function of m. The posterior probability of no change $\pi_m(n)$ is computed and compared to the probability of one change. Zellner (1971), Box and Tiao (1973), and Lindley (1965) give this way of testing hypotheses. This test will be illustrated with Quandt's (1958) data set and the sensitivity of the posterior distribution of m to its prior will be analyzed.

The likelihood function for m, β_1, β_2, and τ is, from (7.33),

DETECTION OF STRUCTURAL CHANGE

$$L(m, \beta_1, \beta_2, \tau | y) \propto \begin{cases} \tau^{n/2} \exp - \frac{\tau}{2} [y_1(n) - x_1(n)\beta_1]'[y_1(n) - x_1(n)\beta_1], \\ \qquad m = n, \; \beta_1 \in R^p, \quad \tau > 0 \\ \tau^{n/2} \exp - \frac{\tau}{2} [y(m) - x(m)\beta]'[y(m) - x(m)\beta], \\ \qquad \beta \in R^{2p}, \quad 1 \leq m \leq n - 1, \quad \tau > 0, \end{cases}$$

(7.34)

where

$$y(m) = \begin{pmatrix} y_1(m) \\ y_2(m) \end{pmatrix},$$

$$x(m) = \begin{pmatrix} x_1(m) & \phi_1 \\ \phi_2 & x_2(m) \end{pmatrix},$$

$$\beta = \begin{pmatrix} \beta_1 \\ \beta_2 \end{pmatrix},$$

where ϕ_1 and ϕ_2 are zero matrices, $y_1(m)$ a $m \times 1$ vector consisting of the first m observations of the sequence, and $y_2(m)$ the vector of the remaining $n - m$ observations. Now $x_1(m)$ is the design matrix corresponding to the first m observations, namely $x_1(m) = (x_1, x_2, \ldots, x_m)'$, and $x_2(m) = (x_{m+1}, \ldots, x_n)'$. Prior information is assigned as follows: first, the marginal p.m.f. of m is

$$\pi_0(m) = \begin{cases} q, & m = n \\ (1-q)/(n-1), & 1 \leq m \leq n - 1, \end{cases} \quad (7.35)$$

where $0 < q < 1$.

Secondly, if $m = n$, β_1 and τ are the parameters and assigned a normal-gamma density, where

$$\pi_0(\beta_1|\tau) \propto \tau^{p/2} \exp\left[-\frac{\tau}{2}(\beta_1 - \beta_{10})'\Lambda_{11}(\beta_1 - \beta_{10})\right],$$

$$\beta_1 \in R^p \quad (7.36)$$

is the conditional prior p.d.f. of β_1 given τ, and the marginal prior density of τ is

$$\pi_0(\tau) \propto \tau^{a-1} e^{-\tau b}, \quad \tau > 0, \quad (7.37)$$

where β_{10} is a given $p \times 1$ vector, $a > 0$, and $b > 0$.

Thirdly, if $1 \leq m \leq n - 1$, $\beta' = (\beta_1', \beta_2')$ and τ are the parameters and assigned a normal-gamma prior density, where

$$\pi_0(\beta_1, \beta_2|\tau) \propto \tau^{2p/2} \exp\left[-\frac{\tau}{2}(\beta - \beta_0)'\Lambda(\beta - \beta_0)\right],$$

$$\beta \in R^{2p} \quad (7.38)$$

is the conditional prior density of β given τ, and the marginal prior density of τ is still (7.37). The mean vector of β is β_0 where $\beta_0 = (\beta_{10}', \beta_{20}')'$ and β_{20} is $1 \times p$, and the precision matrix Λ is partitioned as

$$\Lambda = \begin{pmatrix} \Lambda_{11} & \Lambda_{12} \\ \Lambda_{21} & \Lambda_{22} \end{pmatrix},$$

where Λ_{11}, Λ_{12}, and Λ_{22} are given matrices such that Λ is positive definite. Note that β_{10} and Λ_{11} are also used in the prior density of β_1 given τ when $m = n$.

Combining the likelihood function (7.34) with the prior information gives

$$\pi(m,\beta_1,\beta_2,\tau|y) \propto \begin{cases} |\Lambda_{11}|^{1/2} q\tau^{(n+p+2a)/2-1} \exp-\frac{\tau}{2} \\ \quad \times \{D(n) + (\beta_1 - \beta_1^*)'[x'(n)x(n) + \Lambda_{11}] \\ \quad \times (\beta_1 - \beta_1^*)]\}, \quad m = n, \quad \beta_1 \in R^p, \\ \tau > 0, \\ |\Lambda|^{1/2}(1-q)\tau \exp-\frac{\tau}{2}\{D(m) \\ \quad + [\beta - \beta^*(m)]'[x'(m)x(m) + \Lambda] \\ \quad \times [\beta - \beta^*(m)]\}, \quad 1 \leq m \leq n-1, \\ \beta \in R^{2p}, \quad \tau > 0, \end{cases}$$

(7.39)

where

$$\beta^*(m) = [x'(m)x(m) + \Lambda]^{-1}[\Lambda\beta_0 + x'(m)y(m)],$$

$$D(m) = 2b + [y(m) - x(m)\beta^*(m)]'y(m) + [\beta_0 - \beta^*(m)]'\Lambda\beta_0,$$

$$\beta_1^* = [x'(n)x(n) + \Lambda_{11}]^{-1}[\Lambda_{11}\beta_{10} + x'(n)y(n)],$$

and

$$D(n) = 2b + \{y'(n)y(n) + \beta_{10}'\Lambda_{11}\beta_{10} - \beta_1^{*'}[x'(n)x(n) + \Lambda_{11}]\beta_1^*\}.$$

The shift point m is of primary interest, thus the nuisance parameters β_1, β_2, and τ are integrated from (7.39) giving

$$\pi(m|y) \propto \begin{cases} q|\Lambda_{11}|^{1/2} D(n)^{-(n+2a)/2} |x'(n)x(n) + \Lambda_{11}|^{1/2} \\ \text{if } m = n, \\ \frac{(1-q)}{(n-1)} |\Lambda|^{1/2} D^{-(n+2a)/2}(m) |x'(m)x(m) + \Lambda|^{1/2} \\ \text{if } 1 \leq m \leq n-1 \end{cases}$$

(7.40)

as the marginal posterior mass function of m, which will be used to test for stability.

Suppose the marginal posterior mass function of m is used to test for stability in the Quandt data set, which was used in previous sections to illustrate some Bayesian inference procedures for a two-phase regression model. Recall the observations $Y_1, Y_2, \ldots, Y_n$ were generated as $Y_i \sim$ n.i.d. $(2.5 + 0.7 x_i, 1)$, $i = 1, 2, \ldots, 12$ and $Y_i \sim$ n.i.d. $(5.0 + 0.5 x_i, 1)$, $i = 13, \ldots, 20$, and the (x, Y) pairs are given in Table 7.1. Thus there is a change from one linear regression to the other after the 12th observation. For purposes of illustration let the parameters of the prior distribution be $\beta_0 = (2.5, 0.7, 5.0, 0.5)'$, $\beta_{10} = (2.5, 0.7)'$, $\beta_{20} = (5.0, 0.5)'$, $p = 2$, $\Lambda = I_4$, $\Lambda_{11} = \Lambda_{22} = I_2$, $\Lambda_{12} = \Lambda_{21}' = 0(2 \times 2)$, and $a = b = 1$, thus the prior means of the regression coefficients are set equal to their "true" values and the prior mean of the error precision is set at the "true" precision about the regression line. Table 7.14 gives the posterior mass function of m for $q = .05$, 0.5, .95, and .99, and reveals the posterior mass function of m is sensitive to the parameter q, and that the latter is always greater than the former. This indicates that information from the sample is confirming the model is unstable. If one is not sure the model is unstable, $q = .5$ seems appropriate, in which case the posterior probability of no change is .3855, which is evidence the model is unstable.

Another Bayesian way to detect stability is Smith's (1975) sequential procedure and the best-known classical procedure is Brown, Durbin, and Evans (1975). Poirier (1976) cites many other references for testing structural stability.

With this analysis one is able to test structural stability in a linear model by computing the posterior probability of no change and even though this probability is sensitive to the prior probability of no change, for Quandt's data set, instability is correctly detected.

One should be careful in specifying prior information about the nature of the shift from one linear model to another. If one knew the change from one to the other was a "continuous" function of the regressor variable X, then the use of the shift point parameter to model change is inappropriate. The shift point prior distribution does not take into account additional information about the nature of the structural instability, it only implies exactly one shift will occur, somewhere. If the instability is "smooth," one should employ a transition function, which was how Bacon and Watts (1971) and Tsurumi (1978) approached the problem. For additional information one is referred to Hinkley (1971) and Poirier (1976).

Table 7.14
Posterior Probability Mass Function of m

	Prior Probability of No Change, q.			
m	0.05	0.50	0.95	0.99
1	0.0430	0.0273	0.0034	0.0007
2	0.0056	0.0035	0.0004	0.0001
3	0.0084	0.0053	0.0007	0.0001
4	0.0018	0.0011	0.0001	0.0000
5	0.0036	0.0023	0.0003	0.0001
6	0.0057	0.0036	0.0005	0.0001
7	0.0313	0.0199	0.0025	0.0005
8	0.0443	0.0281	0.0035	0.0007
9	0.0454	0.0288	0.0036	0.0007
10	0.0548	0.0348	0.0044	0.0009
11	0.0523	0.0332	0.0042	0.0009
12	0.6034	0.3830	0.0482	0.0099
13	0.0081	0.0052	0.0006	0.0001
14	0.0110	0.0070	0.0009	0.0001
15	0.0222	0.0141	0.0018	0.0004
16	0.0172	0.0109	0.0014	0.0003
17	0.0022	0.0014	0.0002	0.0000
18	0.0029	0.0018	0.0002	0.0000
19	0.0049	0.0031	0.0004	0.0001
20	0.0320	0.3855	0.9226	0.9842

STRUCTURAL STABILITY IN OTHER MODELS

Our study of structural change, up to this point, has been confined to univariate linear models but is easily extended to multivariate sequences and multivariate linear models. For example Salazar (1980) and Sen and Srivastava (1973, 1980) have developed detection and estimation methods for sequences of random vectors which have multivariate normal distributions. In this section, the Bayesian analysis for structural change of time series models will be presented and is taken from Salazar (1980).

Regression models with autocorrelated errors and autoregressive models will be examined in this section. Recall, these models were introduced in Chapter 5 on time series and will now be studied from a different viewpoint, i.e., these models will incorporate structural change into their structure. Instead of assuming the parameters of the model are static over time, one will allow for one change in some of the parameters over the observation period.

Structural change will be modeled in two ways, first with a shift point m and secondly with a transition function, which was first used by Bacon and Watts (1971), then by Tsurumi (1977) and Salazar (1980).

Structural Change in a Regression Model with Autocorrelated Errors

Assuming only one change in the regression parameters, the model is

$$Y_t = x_t' \beta_1 + a_t, \quad t = 1, 2, \ldots, m$$
$$Y_t = x_t' \beta_2 + a_t, \quad t = m+1, \ldots, T$$
(7.41)

$$a_t = \rho a_{t-1} + e_t, \quad t = 1, 2, \ldots, T \tag{7.42}$$

where Y_t is the t-th observation on the dependent variable, β_1 and β_2 are distinct $k \times 1$ vectors of regression parameters, x_t is the t-th observation on k independent variables, ρ is the unknown autocorrelation parameter, and the e_t's are i.i.d. $n(0, \tau^{-1})$ random variables, where the precision $\tau > 0$, and a_0 a known initial quantity. The shift point m models the time where the change occurs and $m = 1, 2, \ldots, T - 2$.

The model is rewritten as

$$Y_t = \rho Y_{t-1} + (x_t' - \rho x_{t-1}') \beta_1 + e_t, \quad t = 1, 2, \ldots, m$$
(7.43)
$$Y_{m+1} = \rho Y_m + x_{m+1}' \beta_2 - \rho x_m' \beta_1 + e_{m+1},$$

$$Y_t = \rho Y_{t-1} + (x_t' - \rho x_{t-1}') \beta_2 + e_t, \quad t = m+2, \ldots, T$$

STRUCTURAL STABILITY IN OTHER MODELS 353

where x_0 and Y_0 are "known" initial values of the independent and dependent variables, and the errors a_t still satisfy (7.42). The model allows one change in the regression parameter, but the precision parameter τ and autocorrelation parameters do not undergo change.

The following theorem gives the marginal posterior mass function of m and the marginal posterior densities of β and τ.

Theorem 7.3. Given the above model (7.43) and supposing the joint prior distribution of m, β_1, β_2, ρ, and τ is as follows: The marginal prior distribution of m is uniform over 1, 2, ..., T − 2, the conditional prior density of $\beta = (\beta_1', \beta_2')'$ given τ is a 2k-dimensional normal with mean vector $\mu = (\mu_1', \mu_2')'$ and precision matrix P, where $\mu \in R^{2k}$, $\mu_i \in R^k$, i = 1, 2, and P is positive definite, the marginal prior distribution of τ is gamma with positive parameters a and b, the marginal prior density of ρ is constant over R, and m, (β, τ), and ρ are independent, a priori, then:

(i) The marginal posterior mass function of m is

$$\pi(m|y) \propto \int_R |H(\rho)|^{-1/2} c(\rho)^{-(2a+T)/2} d\rho,$$

$$m = 1, 2, \ldots, T - 2, \quad (7.44)$$

where

$$c(\rho) = 2b + \mu'P\mu + \sum_{t=1}^{T}(y_t - \rho y_{t-1})^2 - \hat{\beta}'H(\rho)\hat{\beta},$$

$$H(\rho) = \begin{pmatrix} P_{11} + z_1'z_1 + \rho x_m x_m' & P_{12} - \rho x_m x_{m+1}' \\ P_{21} - \rho x_{m+1} x_m' & P_{22} + z_2'z_2 + x_{m+1} x_{m+1}' \end{pmatrix},$$

$$P = \begin{pmatrix} P_{11} & P_{12} \\ P_{21} & P_{22} \end{pmatrix},$$

where P_{11} is of order k and $H(\rho)$ is partitioned as $H(\rho) = (H_{ij}(\rho))$, where $H_{ij}(\rho)$ is of order k,

$$z_1 = (x_1 - \rho x_0, x_2 - \rho x_1, \ldots, x_m - \rho x_{m-1}),$$

$$z_2 = (x_{m+2} - \rho x_{m+1}, \ldots, x_T - \rho x_{T-1}),$$

$$v_1 = (y_1 - \rho y_0, y_2 - \rho y_1, \ldots, y_m - \rho y_{m-1}),$$

$$v_2 = (y_{m+2} - \rho y_{m+1}, \ldots, y_T - \rho y_{T-1}),$$

$$\hat{\beta}_1 = H_{11.2}^{-1}(\rho)[\alpha_1 - H_{12}(\rho)H_{22}^{-1}(\rho)\alpha_2],$$

$$\hat{\beta}_2 = H_{22.1}^{-1}(\rho)[\alpha_2 - H_{21}(\rho)H_{11}^{-1}(\rho)\alpha_1],$$

where

$$H_{11.2}(\rho) = H_{11}(\rho) - H_{12}(\rho)H_{22}^{-1}(\rho)H_{21}(\rho),$$

$$H_{22.1}(\rho) = H_{22}(\rho) - H_{21}(\rho)H_{11}^{-1}(\rho)H_{12}(\rho),$$

$$\alpha_1 = P_{12}\mu_2 + P_{11}\mu_1 + z_1'v_1 - \rho(y_{m+1} - \rho y_m)x_m,$$

$$\alpha_2 = P_{21}\mu_1 + P_{22}\mu_2 + z_2'v_2 - (y_{m+1} - \rho y_m)x_{m+1}.$$

(ii) The marginal posterior density of β is

$$\pi(\beta|y) \propto \sum_{m=1}^{T-2} \int_R |H(\rho)|^{-1/2} c(\rho)^{-(2a+T)/2}$$

$$\times f(\beta, \rho) \, d\rho, \qquad \beta \in R^{2k} \quad (7.45)$$

where $f(\beta, \rho)$ is the density of a t distribution on β with $2a + T$ degrees of freedom, location $\hat{\beta} = (\hat{\beta}_1', \hat{\beta}_2')'$, and precision matrix $(2a + T)H(\rho)c^{-1}(\rho)$.

(iii) The marginal posterior density of τ is

$$\pi(\tau|y) \propto \sum_{m=1}^{n-1} \int_R \tau^{(2a+T)/2-1} |H(\rho)|^{-1/2} e^{-c(\rho)/2} \, d\rho,$$

$$\tau > 0. \quad (7.46)$$

Proof. The likelihood function of the parameters is

$$L(\beta, \tau, \rho, m|y) \tau^{T/2} \exp -\frac{\tau}{2}[(v_1 - z_1\beta_1)'(v_1 - z_1\beta_1)$$

STRUCTURAL STABILITY IN OTHER MODELS 355

$$+ (v_2 - z_2\beta_2)'(v_2 - z_2\beta_2)$$

$$+ (y_{m+1} - \rho y_m - x'_{m+1}\beta_2 + \rho x'_m\beta_1)^2],$$

where $\beta \in R^{2k}$, $\tau > 0$, $\rho \in R$, and $m = 1, 2, \ldots, T-2$, and the joint prior density of the parameters is

$$\pi(\beta, \tau, \rho, m) \propto \tau^{a+k-1} \exp -\frac{\tau}{2}[2b + (\beta - \mu)'P(\beta - \mu)],$$

where $\beta \in R^{2k}$, $\tau > 0$, $\rho \in R$, and $m = 1, 2, \ldots, T-2$ and combined with the likelihood function gives the joint posterior density of the parameters. The marginal posterior densities (i), (ii), and (iii) of the theorem are easily found by integrating the joint posterior density of the parameters with respect to the appropriate parameters.

This theorem gives the foundation for making inferences about the parameters of a regression model with autocorrelated errors, but, of course, it doesn't directly provide all the information about the parameters. For example, part (ii) of the theorem gives the posterior distribution of β but not the marginal distributions of β_1 and β_2, but by using the properties of the multivariate t, they are easily found.

Consider a numerical example to illustrate inferences for m, the shift point. Suppose the model is

$$Y_t = \begin{cases} 3x_t + a_t, & t = 1, 2, \ldots, m^* \\ (3 + \Delta)x_t + a_t, & t = m^* + 1, \ldots, 15, \end{cases} \quad (7.47)$$

$a_t = \rho a_{t-1} + \varepsilon_t$, the ε_t are i.i.d. $n(0,1)$, $\Delta = .2, .3, .5, .7, .8$, $\rho = .5, 1.25$, and the x_t are rescaled investment expenditures used by Zellner (1971, page 89).

The joint prior density of the parameters is $\pi_0(\beta, \tau, \rho, m) \propto \tau^{a-1} e^{-\tau b}$, $\beta \in R$, $\rho \in R$, $\tau > 0$, $m = 1, 2, \ldots, 13$, where $a = 1$ and $b = 2$. The data were generated using IMSL subroutines and the p.m.f. of m (7.44) was obtained using Simpson's rule for numerical integration in double precision on an IBM 370/168 computer.

Table 7.15 gives the marginal p.m.f. of m* for five values of Δ and m* = 3, 8, 12, where the first entry in each cell corresponds to $\rho = 1.25$, the "explosive" case, and the second entry corresponds to $\rho = .5$.

The results show when $\Delta > .5$, the posterior p.m.f. of m gives a clear indication of the shift, and a shift located near the center

Table 7.15
Posterior Probability of m* for T = 15

m*	Δ				
	.2	.3	.5	.7	.8
3	.07	.06	.89	.99	1.00
	.03	.07	.42	.90	.97
8	.05	.12	.65	.96	1.00
	.09	.03	.77	1.00	1.00
12	.10	.14	.35	.69	.83
	.05	.11	.38	.73	.84

The first value in each cell corresponds to $\rho = 1.25$ and the second to $\rho = .5$.

(m* = 8) is easier to detect than if m* is located at either extreme (m* = 3 or 12).

Table 7.16 is the p.m.f. of m when m* = 8, $\rho = .5$, $\tau = 1$, Δ as before is .2 through .8, and T = 15. The correct value of m* has probability of at least .77 when Δ > .5, but when Δ = .2, m = 5 has the largest posterior mass of .16. If Δ = .2, the distribution of m is rather uniform and one would be hesitant in identifying the shift at any particular mass point. When Δ = .3, one perhaps would identify incorrectly m as 12.

The results of Table 7.17 are similar to those of Table 7.16 in that if Δ > .5, the posterior probability of a shift at the "correct" value of m = 3 becomes easier to detect as Δ increases, but these probabilities are not as large as those in Table 7.16 because the m* in Table 7.16 is set at 8 (which is at the center of the data) while m* is set at one extreme (m* = 3) for Table 7.17. When the value of m* is set at the middle of the observations, the probability of a change at m* is larger than when m* is set equal to one extreme or the other.

Transition Functions and Structural Change in Regression Models with Autocorrelated Errors

When a shift point m is employed to model structural change, one is assuming, a priori, only that a change has occurred, exactly once, somewhere during the period of observation, thus if one has additional

STRUCTURAL STABILITY IN OTHER MODELS

Table 7.16
Posterior Distribution of m for $\rho = .5$ and $m^* = 8$

m	\Delta .2	.3	.5	.7	.8
1	.0923	.0624	.0031	.0001	0.0000
2	.0449	.0191	.0010	0.0000	0.0000
3	.0616	.0155	.0011	0.0000	0.0000
4	.0680	.0139	.0015	.0001	0.0000
5	.1623	.0142	.0014	.0001	0.0000
6	.1042	.0136	.0023	.0001	0.0000
7	.1133	.0160	.0039	.0001	0.0000
8	.0894	.0308	.7689	.9949	.9991
9	.0661	.0420	.1543	.0037	.0006
10	.0448	.0970	.0240	.0003	.0001
11	.0467	.2109	.0161	.0002	.0001
12	.0509	.3691	.0174	.0003	.0001
13	.0552	.0951	.0045	.0001	0.0000

prior information about the nature of the change, the shift point method would most likely ignore this information and is inappropriate. For example, if one knew the average value of the dependent variable was a continuous function of the dependent variable over the domain where the shift will occur, this information should be a part of one's prior information about the model. From the Bayesian viewpoint, first Bacon and Watts (1971), then Tsurumi (1977) and later Salazar (1980), use a transition function to model both abrupt and smooth changes in linear models. Bacon and Watts work with a simple linear regression model, Tsurumi with simultaneous equations models, and Salazar (1980) with time series processes.

Now, Salazar's work will be presented for a regression model with autocorrelated errors. Suppose ψ is a non-negative function of a real variable such that

(i) $\psi(0) = 0$

Table 7.17
Posterior Distribution of m for $\rho = 1.25$ and $m^* = 3$

m	\Delta				
	.2	.3	.5	.7	.8
1	.0738	.0495	.0054	.0002	.0001
2	.0519	.0636	.0153	.0009	.0002
3	.0750	.2400	.8932	.9946	.9986
4	.1661	.1255	.0123	.0004	.0001
5	.0289	.0238	.0034	.0002	0.0000
6	.1094	.0843	.0091	.0004	.0001
7	.0346	.0312	.0048	.0002	.0001
8	.0442	.0364	.0049	.0002	.0001
9	.0528	.0398	.0049	.0002	0.0000
10	.0731	.0630	.0095	.0005	.0001
11	.1637	.1297	.0171	.0008	.0002
12	.0691	.0620	.0110	.0007	.0002
13	.0573	.0511	.0089	.0005	.0001

(ii) $\lim_{n_t \to \infty} \psi(n_t) = 1$, where $n_t = s_t/\gamma$ and (7.48)

$$s_t = \begin{cases} 0, & t \leq t^* \\ t - t^*, & t > t^* \end{cases}$$

and t^* is a fixed real number. Then ψ is called a transition function and γ the transition parameter and indicates the abruptness of the structural change which begins at time t^*. Consider a regression model with autocorrelated errors, with k independent variables $x_1, x_2, \ldots, x_k$ and k regression parameters $\beta_1, \beta_2, \ldots, \beta_k$ and suppose β_1 changes from β_{10} to $\beta_{10} + \beta_{11}$ beginning at time t^*, then at time t

$$Y_t = x_{t1}\beta_{10} + x_{t1}\psi(s_t/\gamma)\beta_{11} + x_{t2}\beta_2 + \cdots + x_{kt}\beta_k + a_t, \quad (7.49)$$

STRUCTURAL STABILITY IN OTHER MODELS

$$a_t = \rho a_{t-1} + \varepsilon_t, \quad t = 1, 2, \ldots, T,$$

where x_{ti}, $i = 1, 2, \ldots, k$, is the value of the i-th independent variable at time t, y_t the dependent variable at time t, and the errors follow a first order autoregressive process with autocorrelation parameter $\rho \in R$.

The model may be rewritten as

$$Y_t = \rho Y_{t-1} + (z_t - \rho z_{t-1})'\beta + \varepsilon_t, \quad t = 1, 2, \ldots, T, \text{ where} \quad (7.50)$$

$$z_t = [x_{t1}, x_{t1}\psi(n_t), x_{t2}, \ldots, x_{tk}]',$$

$$\beta = (\beta_{10}, \beta_{11}, \beta_2, \ldots, \beta_k)' \in R^{k+1},$$

the ε_t are i.i.d. $n(0, \tau^{-1})$, $\tau > 0$, and z_0 and Y_0 are known initial values of the independent and dependent variables, respectively.

The transition function ψ depends on unknown parameters γ and t^* and the likelihood function may be written as

$$L(\beta, t^*, \gamma, \rho, \tau | y) \propto \tau^{T/2} \exp -\frac{\tau}{2}(V - z\beta)'(V - z\beta), \quad (7.51)$$

for $\beta \in R^{k+1}$, $1 < t^* < T$, $\gamma > 0$, $\tau > 0$, $\rho \in R$ and where $V = (y_1 - \rho y_0, y_2 - \rho y_1, \ldots, y_T - \rho y_{T-1})'$ and $z = (z_1 - \rho z_0, z_2 - \rho z_1, \ldots, z_T - \rho z_{T-1})'$.

The conjugate family for this model is unknown, thus let

$$\pi_0(\beta, t^*, \gamma, \rho, \tau) \propto \tau^{a-1} e^{-\tau b}, \quad \tau > 0, \quad \beta \in R^{k+1},$$

$$1 < t^* < T, \quad \rho \in R, \quad \gamma > 0 \quad (7.52)$$

be the joint prior density for all the parameters, which implies β, t^*, γ, ρ, and τ are independent and that β, t^*, γ, and ρ have marginal constant densities and that τ is gamma with parameters a and b.

Combining the prior density with the likelihood function and integrating the product with respect to β and τ over $R^{k+1} \times (0, \infty)$, the joint posterior density of t^* and γ is

$$\pi(t^*, \gamma | y) \propto \int_R |z'z|^{-1/2} (b + v'Pv/2)^{-(T-k-1)/2 - a/2} d\rho,$$

$$1 < t^* < T, \qquad (7.53)$$

$\gamma > 0$, where $P = I - z(z'z)^{-1}z'$.

To find the marginal posterior densities of t^* and γ one must use numerical integration procedures since they cannot be expressed in analytical form. If $\rho = 0$, the model reduces to simple linear regression.

Consider the following example of a gradual change ($\gamma > 0$) using a set of $T = 30$ standardized normal random deviates. The marginal posterior density function of t^* was calculated with data generated from

$$Y_t = 3x_t + \Delta_t x_t + a_t, \qquad t = 1, 2, \ldots, 30$$
$$a_t = \rho a_{t-1} + \varepsilon_t, \qquad (7.54)$$

where $\Delta_t = 0$ for $t = 1, 2, \ldots, t_0^* - 1$, $\Delta_{t_0^*} = .1$, $\Delta_{t_0^*+1} = .3$, $\Delta_{t_0^*+2} = .5$, $\Delta_{t_0^*+3} = .8$, $\Delta_{t_0^*+4} = 1$, $\Delta_{t_0^*+5} = 1.2$, $u_0 = .5$, $x_0 = 0$ and $y_0 = .5$. The ε_t are i.i.d. $n(0,1)$ and the joint prior density of $\beta \in R$, $\tau > 0$, t^*, $1 < t^* < 30$, $\rho \in R$ and $\gamma > 0$ is

$$\pi_0(\beta, t^*, \gamma, \rho, \tau) \propto \frac{1}{\tau}.$$

Inferences about t^* are illustrated in Figures 7.7 and 7.8, where the posterior densities of t^* were computed for $\rho = 0$, .5, 1.25 and $t_0^* = 14$ and 19. It is shown that for non-explosive models, the center of the posterior densities of t^* are insensitive to changes in ρ. For example when $t_0^* = 19$ and $\rho = 0$ or .5, they are tightly concentrated around the modes of 18.5 and 19 shown in Figure 7.7. For $t_0^* = 14$, $\rho = 0$, the posterior density is spread between 1 and 17 with modal value of 10.5. The posterior density of t^* when $t_0^* = 14$ and $\rho = 1.25$ is shown in Figure 7.8 and has a mode at 12.5 and a long left tail. The two densities of this figure show the marginal posterior density of t^* is sensitive to changes in ρ. The mode changes from 9.5 to 12.5 when ρ changes from 0 to 1.25, when the change in parameter begins at 14.

STRUCTURAL STABILITY IN OTHER MODELS

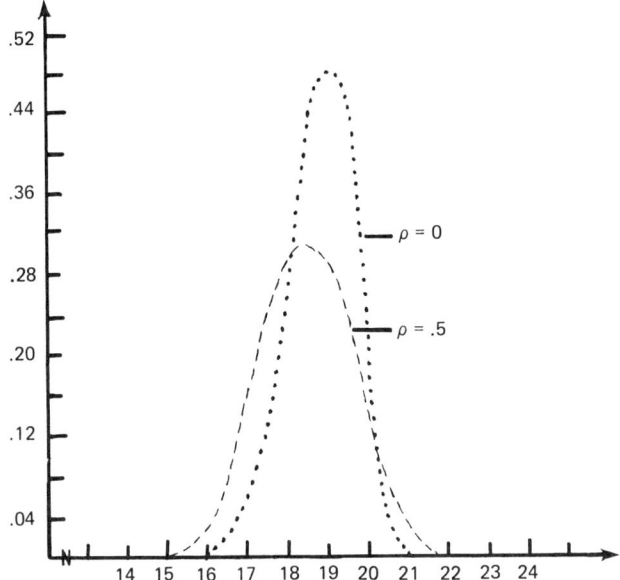

Figure 7.7. Posterior density of t* for $\rho = 0$ and .5 and $t_0^* = 19$, the true value of t* (T = 30).

Structural Change in Autoregressive Parameters

Only elementary versions of the autoregressive process will be considered; however, the posterior analysis of a general autoregressive model is easily found.

The p-th order process (5.3) was defined in Chapter 5 but for now consider a first order process with one change,

$$Y_t = \beta_{11} + \beta_{12} Y_{t-1} + \varepsilon_t, \quad t = 1, 2, \ldots, m \qquad (7.55)$$

$$Y_t = \beta_{21} + \beta_{22} Y_{t-1} + \varepsilon_t, \quad t = m+1, \ldots, T,$$

where m is the shift point, $m = 1, 2, \ldots, n-1$, $\beta_1 = (\beta_{11}, \beta_{12})' \in R^2$, $\beta_2 = (\beta_{21}, \beta_{22}) \; R^2$ are the autoregressive parameters, $\beta_1 \neq \beta_2$, Y_t is the t-th observation on the dependent variable, the ε_t are i.i.d. $n(0, \tau)$, $\tau > 0$, and y_0 is a known initial value. Thus the first m observations follow an AR(1) process with parameters β_1 and the remaining observa-

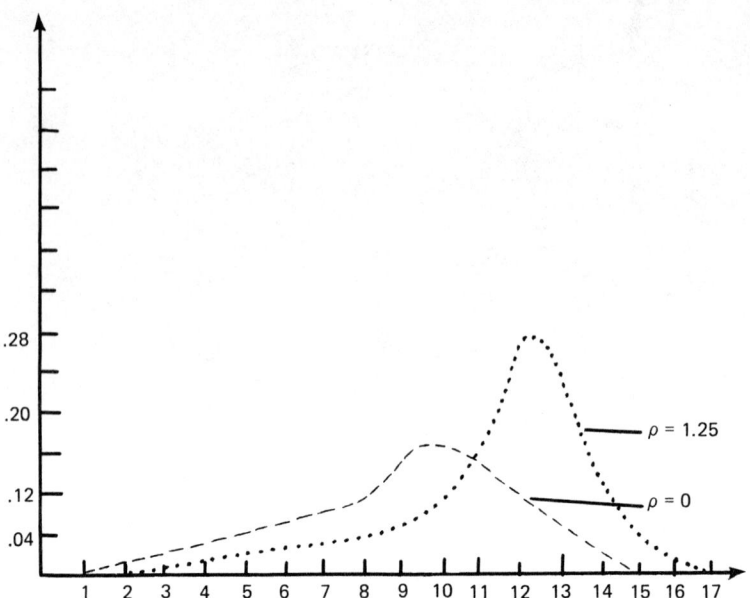

Figure 7.8. Posterior density of t* for $\rho = 0$ and 1.25 and $t_0^* = 14$, the true value of t* (T = 30).

tions follow an AR(1) process with parameters β_2, and all the parameters are unknown. The AR(p) process allows a conjugate family of normal-gamma distributions, which will be modified for an AR(1) with one change.

Suppose, a priori, that m and (β, τ) are independent where $\beta = (\beta_1', \beta_2')' \in R^4$, and that m has a uniform distribution over $1, 2, \ldots, n-1$ and that β given τ is a four variate normal with mean $\mu = (\mu_1', \mu_2')'$, $\mu_i \in R^2$, $i = 1, 2$, and precision matrix τP, where P is positive definite and

$$P = \begin{pmatrix} P_{11} & P_{12} \\ P_{21} & P_{22} \end{pmatrix},$$

where P_{11} is 2×2 and also suppose τ is a gamma with parameters a and b, then one may show:

STRUCTURAL STABILITY IN OTHER MODELS

Theorem 7.4. The marginal posterior mass function of the shift point is

$$\pi(m|y) \propto C_1^{-(2a+T)/2}(m) \, |A(m)|^{-1/2}, \quad m = 1, 2, \ldots, T-1, \tag{7.56}$$

where

$$A(m) = \begin{pmatrix} A_{11} & P_{12} \\ P_{21} & A_{22} \end{pmatrix},$$

$$A_{11} = P_{11} + G_1,$$

$$A_{22} = P_{22} + H_1,$$

$$G_1 = \begin{pmatrix} m & \sum_{1}^{m} y_{t-1} \\ \sum_{1}^{m} y_{t-1} & \sum_{1}^{m} y_{t-1}^2 \end{pmatrix},$$

$$H_1 = \begin{pmatrix} T-m & \sum_{m+1}^{T} y_{t-1} \\ \sum_{m+1}^{T} y_{t-1} & \sum_{m+1}^{T} y_{t-1}^2 \end{pmatrix},$$

$$C_1(m) = b + \frac{1}{2} [S_1(\hat{\beta}_1) + S_1(\hat{\beta}_2) + \mu'P\mu + \hat{\beta}_1'G_1\hat{\beta}_1 + \hat{\beta}_2'H_1\hat{\beta}_2]$$

$$- (\tilde{\beta}_1', \tilde{\beta}_2')A(m)\begin{pmatrix} \tilde{\beta}_1 \\ \tilde{\beta}_2 \end{pmatrix},$$

$$S_1(\hat{\beta}_1) = \sum_{1}^{m} (y_t - \hat{\beta}_{11} - \hat{\beta}_{12} y_{t-1})^2,$$

$$S_1(\hat{\beta}_2) = \sum_{m+1}^{T} (y_t - \hat{\beta}_{21} - \hat{\beta}_{22} y_{t-1})^2, \tag{7.57}$$

$$\hat{\beta}_1 = G_1^{-1} \left(\sum_{1}^{m} y_t, \sum_{1}^{m} y_t y_{t-1} \right)',$$

$$\hat{\beta}_2 = H_1^{-1} \left(\sum_{m+1}^{T} y_t, \sum_{m+1}^{T} y_t y_{t-1} \right)',$$

$$\tilde{\beta}_1 = (A_{11} - P_{12} A_{22}^{-1} P_{21})^{-1} [P_{11} \mu_1 + G_1 \hat{\beta}_1 + P_{12}$$

$$\times (\mu_1 - A_{22}^{-1} (P_{22} \mu_1 + H_1 \hat{\beta}_2)) - P_{12} A_{22}^{-1} P_{21} \mu_1],$$

$$\tilde{\beta}_2 = (A_{22} - P_{21} A_{11}^{-1} P_{12})^{-1} [P_{22} \mu_2 + H_1 \hat{\beta}_2 + P_{21}$$

$$\times (\mu_1 - A_{11}^{-1} (P_{11} \mu_1 + G_1 \hat{\beta}_1)) - P_{21} A_{11}^{-1} P_{12} \mu_2].$$

The marginal posterior density of β is easily found to be a mixture of multivariate t densities.

The marginal posterior density of τ is

$$\pi(\tau|y) = \sum_{m=1}^{T-1} \pi(m|y) g(\tau|m, y), \quad \tau > 0 \tag{7.58}$$

where $g(\tau|m, y)$ is a gamma density with parameters $a + T/2$ and $C_1(m)$.

The results of this theorem completely parallel Theorem 7.2, which is not surprising since the latter theorem gave the posterior analysis for a changing linear model. There are three results to the theorem, namely the three marginal posterior distributions of m, β, and τ, and of course more results follow. For example, the marginal densities of β_1 and β_2 easily follow from the density of β.

STRUCTURAL STABILITY IN OTHER MODELS 365

This analysis employed a shift point parameter m and will be generalized to include smoother changes by use of the transition function, (7.48).

Suppose the shift parameter m is of primary importance and the other parameters regarded as nuisance parameters, then one's inference about m will be based on (7.56).

Suppose an AR(1) model changes once according to the model

$$Y_t = 4 + \beta_{12} Y_{t-1} + \varepsilon_t, \quad t = 1, 2, \ldots, 9$$

$$Y_t = 4 + (\beta_{12} + \Delta) Y_{t-1} + \varepsilon_t, \quad t = 10, \ldots, 20$$
(7.59)

where the ε_t are i.i.d. $n(0,1)$ and $y_0 = 5$.

The prior distribution of the parameters was chosen so that m is independent of (β, τ) and uniform over $1, 2, \ldots, 19$, where $\beta = (\beta_{11}, \beta_{12}, \beta_{21}, \beta_{22})'$, and the conditional prior distribution of β given τ is normal with mean vector $(4,1,4,1)'$ and precision matrix τI_4, and the marginal distribution of τ is gamma with parameters $a = 1$ and $b = 2$.

Table 7.18 shows the results of the posterior mass function of m when the data were generated according to (7.59) with $\beta_{12} = .5$ and $\beta_{22} = \beta_{12} + \Delta$, where $\Delta = .5, 1, 1.49$. Thus there is one shift beginning with the 10th observation and this corresponds to a nonexplosive series since $|\beta_{12}| < 1$ and $|\beta_{22}| < 1$, however the process is not stationary. From the table one sees the p.m.f. of m would lead one to perhaps erroneously conclude the shift occurred at $m = 17$, when $\Delta = .5$. On the other hand, when $\Delta \geq 1$, the p.m.f. of m correctly identifies the shift.

With regard to Table 7.19 the results give the p.m.f. of m when $\beta_{12} = 1$, $\Delta = .1, .2, .3$, and $m^* = 10$, thus the data generated according to (7.59) were from an explosive model, since $|\beta_{12}| > 1$, and structural change is easier to detect in this case than when the data are generated from a nonexplosive model. For a small change of .1, it is difficult to identify the shift point, but when $\Delta \geq .2$, there is no question the change is correctly given at $m = 10$.

Now let's use a transition function to model a change in an AR(1) model, thus let

$$Y_t = \beta_1 + \beta_{20} Y_{t-1} + \beta_{21} \psi(n_t) Y_{t-1} + \varepsilon_t,$$
(7.60)

for $t = 1, 2, \ldots, T$, where ψ is the transition function given by (7.48), $n_t = S_t/\gamma$, and

Table 7.18
Posterior Distribution of m for $\beta_{12} = .5$ and $m^* = 10$

m	\Delta .5	1.	1.49
1	.000	.000	.000
2	.016	.011	.002
3	.018	.014	.004
4	.019	.016	.004
5	.015	.013	.003
6	.019	.021	.007
7	.016	.018	.006
8	.024	.043	.023
9	.021	.036	.015
10	.058	.291	.647
11	.082	.347	.278
12	.062	.093	.006
13	.044	.024	.001
14	.037	.014	.000
15	.030	.012	.000
16	.065	.014	.000
17	.469	.029	.001
18	.000	.000	.000
19	.000	.000	.000

$$S_t = \begin{cases} 0, & t \leq t^* \\ t - t^*, & t > t^*. \end{cases}$$

This equation represents an AR(1) process with a change in the autoregressive parameter from β_{20} to β_{21} beginning at time t^* and γ, the transition parameter, measures the degree of change, where $\gamma = 0$ indicates an abrupt change. Now assuming the ε_t are n.i.d. $(0, \tau^{-1})$,

STRUCTURAL STABILITY IN OTHER MODELS

Table 7.19
Posterior Distribution of m for $\beta_{12} = 1.$ and $m^* = 10$

m	Δ		
	.10	.20	.30
1	.000	.000	.000
2	.291	.003	.000
3	.116	.001	.000
4	.048	.001	.000
5	.042	.000	.000
6	.033	.000	.000
7	.031	.000	.000
8	.043	.002	.000
9	.034	.005	.000
10	.091	.980	1.000
11	.028	.004	.000
12	.015	.001	.000
13	.012	.001	.000
14	.013	.000	.000
15	.041	.000	.000
16	.041	.000	.000
17	.116	.000	.000
18	.000	.000	.000
19	.000	.000	.000

where $\tau > 0$, the joint density of the observations, given β_1, β_{20}, t^*, τ, and γ is

$$L(t^*,\gamma,\beta,\tau|s) \propto \tau^{T/2} \exp - \frac{\tau}{2} \sum_{t=1}^{T} [y_t - \beta_1 - \beta_{20} y_{t-1}$$
$$- \beta_{21} \psi(n_t) y_{t-1}]^2 \qquad (7.61)$$

where $s = (y_0, y_1, \ldots, y_n)$ is the sample of observations and $\beta' = (\beta_1, \beta_{20}, \beta_{21}) \in R^3$, $1 < t^* < T$, $\gamma \geq 0$ and $\tau > 0$.

Now suppose prior information is represented by the density

$$\pi(t^*, \gamma, \beta, \tau) \propto \tau^{a-1} \varepsilon^{-\tau b}, \quad \tau > 0, \quad \gamma > 0, \quad \beta \in R^3,$$

$$1 < t^* < T \quad (7.62)$$

i.e., the marginal prior density of τ is gamma with parameters $a > 0$ and $b > 0$, and τ is independent of the others t^*, γ, and β, which have a constant prior density.

By Bayes theorem, one may show the joint posterior density is

$$\pi(t^*, \gamma, \beta, \tau | s) \propto \tau^{(T+2a)/2-1} \exp -\frac{\tau}{2}[(\beta - \hat{\beta})'H(t^*,\gamma)(\beta - \hat{\beta})$$

$$+ s_1(\hat{\beta}) + 2b], \quad (7.63)$$

where

$$H(t^*,\gamma) = \begin{pmatrix} T & \sum_1^T y_{t-1} & \sum_1^T \psi(n_t) y_{t-1} \\ \sum_1^T y_{t-1} & \sum_1^T y_{t-1}^2 & \sum_1^T y_{t-1}^2 \psi(n_t) \\ \sum_1^T y_{t-1} \psi(n_t) & \sum_1^T y_{t-1}^2 \psi(n_t) & \sum_1^T y_{t-1}^2 \psi^2(n_t) \end{pmatrix},$$

$$\hat{\beta} = H^{-1}(t^*,\gamma) V_1,$$

$$V_1 = \left[\sum_1^T y_t, \sum_1^T y_t y_{t-1}, \sum_1^T y_{t-1} \psi(n_t) y_t \right],$$

$$s_1(\hat{\beta}) = \sum_1^T [y_t - \hat{\beta}_1 - \hat{\beta}_{20} y_{t-1} - \hat{\beta}_{21} \psi(n_t) y_{t-1}]^2.$$

The marginal posterior density of t^* and γ is

SUMMARY AND COMMENTS

$$\pi(t^*,\gamma) \propto |H(t^*,\gamma)|^{-1/2} [b + s_1(\hat{\beta})/2]^{-(T-3+a)/2},$$

$$\gamma > 0, \quad 1 < t^* < T, \quad (7.64)$$

and is found by integrating the joint posterior density (7.63) with respect to β and τ.

It does not appear possible to eliminate either parameter from (7.64) except by numerical integration procedures.

Very little work has appeared for structural change in time series processes, except for what we have seen here, namely for autoregressive models and regression models with autocorrelated errors. Clearly, the results given above for the posterior analysis of autoregressive models can be extended to autoregressive models of any order. Also, it should be relatively easy to develop Bayesian predictive techniques for autoregressive models which exhibit structural change.

Structural change in moving average models and autoregressive moving average models has yet to be derived. In Chapter 5 we saw that the posterior analysis of moving average models is more difficult than that for autoregressive processes, and we would expect the same situation when one puts structural changes into these two models, nevertheless it should be possible to build an operational posterior and predictive analysis for MA(q) and ARMA(p,q) processes.

SUMMARY AND COMMENTS

This chapter has introduced the reader to the Bayesian analysis of linear models which exhibit structural change in the parameters. Normal sequences, two-phase regression models, general linear models, regression models with autocorrelated errors and autoregressive processes have been examined.

The instability of a process was represented either by a change-point parameter or a transition function and the posterior analysis focused on the posterior mass function of the change point or the posterior density of the two parameters of the transition function.

A brief introduction to detecting instability was given on pp. 346-351, and an exhaustive numerical study of a changing linear model was presented on pp. 321-345, where the posterior mass function of the change point and the posterior densities of the other parameters were evaluated.

Although much has been written about structural change, much remains to be done. For example, time series processes need to be examined in more detail, and forecasting methods which incorporate structural change into the model need to be developed.

EXERCISES

1. Verify equation (7.8), that the marginal posterior distribution of θ is a mixture of bivariate t densities and that the mixing distribution is that of the marginal posterior distribution of the shift point. From this equation, derive the mean and variance of θ.
2. Referring to equation (7.15), show the predictive density of Y_{n+1}, one future observation, is a mixture of n − 1 univariate t densities and that the mixing distribution is of the shift point m.
3. Verify equation (7.44), of the marginal posterior mass function of m for a regression model with autocorrelated errors, (7.43).
4. For a regression model with autocorrelated errors, and one change, represented by (7.43), find the predictive distribution of one future observation Y_{n+1}, using the prior density given in Theorem 7.3.
5. One may use either a shift point or a transition function to model structural change, as for example was presented with the regression model with autocorrelated errors, (7.43). Explain when it is appropriate to use the shift point (not the transition function) and when it is appropriate to use the transition function but not the shift point. Also, present other techniques to model structural change in statistical models. Read Poirier (1976).
6. Consider the general linear model with one change (7.17), show that if one uses the improper prior density

$$\S(m,\beta,\tau) \propto \tau^{-1}, \quad \tau > 0, \quad \beta \in R^{2p}, \quad m = 1, 2, \ldots, n-1$$

that the marginal mass function of m does not exist over the integers $1, 2, \ldots, n-1$, but does exist over $p, p+1, \ldots, n-p$.
7. The parameter β of a linear model with one change (7.17) may be estimated from either the marginal posterior density function of β, (7.21), or the conditional posterior density function of β, (7.26). Explain under what conditions the latter is likely to be a "good" approximation to the former. What are the advantages of using the conditional posterior density of β (to make inferences about β) when it is a "good" approximation to the marginal posterior density of β?
8. Consider a sequence of independent Poisson random variables, say $X_1, X_2, \ldots, X_n$, where the first m have mean θ_1 and the

last a mean of θ_2, where $\theta_1 \neq \theta_2$, and $\theta_1 > 0$ and $\theta_2 > 0$. Suppose $1 \leq m \leq n - 1$ and that a priori m, θ_1, and θ_2 are independent such that m is uniform over $1, 2, \ldots, n - 1$, and θ_i is gamma with parameters $\lambda_i = i = 1, 2$. Find:

(a) the marginal posterior mass function of m.
(b) the marginal posterior density of θ_1 and θ_2.
(c) an HPD region for θ_2.

9. Consider a sequence of normal variables with one change, given by (7.10) but assuming $\theta_1 > \theta_2$, where $\theta_1 \in R$, $\theta_2 \in R$, and m = $1, 2, \ldots, n - 1$. Formulate a reasonable prior density for the parameters and derive the joint posterior density of θ_1 and θ_2.

10. Consider an autoregressive process of first order with one change at time point m, $1 \leq m \leq n - 1$, with parameters τ, β_{11}, β_{12}, β_{21}, and β_{22}. Refer to equation (7.55) for the explanation of the model, and Theorem 7.4 for the posterior analysis. In this theorem it is stated that the marginal posterior distribution of β is a mixture of multivariate t distributions, however no formula is given. Derive the formula.

REFERENCES

Bacon, D. W. and D. G. Watts (1971). "Estimating the transition between two intersecting lines," Biometrika, Vol. 58, pp. 524-534.

Bhattacharyya, G. K. and B. A. Johnson (1968). "Non parametric tests for shift at an unknown time point," Annals of Mathematical Statistics, Vol. 39, pp. 1731-1734.

Box, G. E. P. and G. C. Tiao (1965). "Multiparameter problems from a Bayesian point of view," Annals of Mathematical Statistics, Vol. 36, pp. 1468-1482.

Box, G. E. P. and G. C. Tiao (1973). *Bayesian Inference in Statistical Analysis*, Addison-Wesley, Reading, Mass.

Broemeling, L. D. (1972). "Bayesian procedures for detecting a change in a sequence of random variables," Metron, Vol. XXX-N-1-4, pp. 1-14.

Broemeling, L. D. (1974). "Bayesian inference about a changing sequence of random variables," Communications in Statistics, Vol. 3 (3), pp. 243-255.

Broemeling, L. D. (1977). "Forecasting future values of a changing sequence," Communications in Statistics, Vol. A6(1), pp. 87-102.

Broemeling, L. D. and J. H. Chin Choy (1981). "Detecting structural change in linear models," Communications in Statistics, Vol. A10(24), pp. 2551-2562.

Brown, R. L., J. Durbin, and J. M. Evans (1975). "Techniques for testing the constancy of regression relations over time" (with discussion), Journal of the Royal Statistical Society, Series B, pp. 49-92.

Chernoff, H. and S. Zacks (1964). "Estimating the current mean of a normal distribution which is subjected to changes over time," Annals of Mathematical Statistics, Vol. 35, pp. 999-1018.

Chi, Albert Yu-Ming (1979). *The Bayesian Analysis of Structural Change in Linear Models*, doctoral dissertation, Oklahoma State University, Stillwater, Oklahoma.

Chin Choy, J. H. (1977). *A Bayesian Analysis of a Changing Linear Model*, doctoral dissertation, Oklahoma State University, Stillwater, Oklahoma.

Chin Choy, J. H. and L. D. Broemeling (1980). "Some Bayesian inferences for a changing linear model," Technometrics, Vol. 22(1), pp. 71-78.

Chow, G. (1960). "Tests of the equality between two sets of coefficients in two linear regressions," Econometrica, Vol. 28, pp. 591-605.

Farley, J. U., M. J. Hinich, and T. W. Mcguire (1975). "Some comparisons of tests for a shift in the slopes of a multivariate linear time series," J. of Econometrics, Vol. 3, pp. 297-318.

Feder, P. (1975). "The log likelihood ratio in segmented regression," Annals of Statistics, Vol. 3, pp. 84-97.

Ferreira, R. E. (1975). "A Bayesian analysis of a switching regression model: known number of regimes," Journal of the American Statistical Association, Vol. 70, pp. 370-374.

Goldfield, S. M. and R. E. Quandt (1973a). "The estimation of structural shifts by switching regressions," Annals of Economic and Social Measurement, Vol. 2, pp. 475-485.

Goldfield, S. M. and R. E. Quandt (1973b). "A Markov model for switching regressions," Journal of Econometrics, Vol. 1, pp. 3-16.

Goldfield, S. M. and R. E. Quandt (1974). "Estimation in a disequilibrium model and the value of information," unpublished manuscript.

Green, C. V. and E. Fekette (1933). "Differential growth in the mouse," J. Experimental Zoology, Vol. 66, pp. 351-370.

Hartley, M. J. and P. Mallela (1975). "The asymptotic properties of a maximum likelihood estimator for a model of markets in disequilibrium," Discussion paper no. 329, State University of New York at Buffalo.

Hinkley, D. V. (1969). "Inference about the intersection in two-phase regression," Biometrika, Vol. 56, pp. 495-504.

REFERENCES

Hinkley, D. V. (1971). "Inference in two-phase regression," J. of American Statistical Association, Vol. 66, pp. 736-793.

Holbert, D. and L. D. Broemeling (1977). "Bayesian inference related to shifting sequences and two phase regression," Communications in Statistics, Vol. A6(3), pp. 265-275.

Hudson, D. (1966). "Fitting the segmented curves whose endpoints have to be estimated," Journal of the American Statistical Association, Vol. 61, pp. 1097-1109.

Kander, Z. and S. Zacks (1966). "Test procedures for possible changes in parameters of statistical distributions occurring at unknown time points," Annals of Mathematical Statistics, Vol. 37, pp. 1196-1210.

Lindley, D. V. (1965). *Introduction to Probability and Statistics from a Bayesian Viewpoint*, Cambridge University Press, Cambridge.

Madala, G. S. and F. D. Nelson (1974). "Maximum likelihood methods for models of markets in disequilibrium," Econometrica, Vol. 42, pp. 1013-1030.

Needham, A. E. (1935). "On relative growth in the jaws of certain fishes," Proceedings of Zoological Society of London, Vol. 2, pp. 773-784.

Ohki, K. (1974). "Manganese nutrition of cotton under two boron levels. II. Critical M_n level," Agronomy Journal, Vol. 66, pp. 572-575.

Page, E. S. (1954). "Continuous inspection schemes," Biometrika, Vol. 41, pp. 100-114.

Page, E. S. (1957). "On problems in which a change in parameter occurs at an unknown point," Biometrika, Vol. 44, pp. 248-252.

Poirier, D. J. (1976). *The Econometrics of Structural Change*, North-Holland Publishing Company, Amsterdam.

Pool, J. and C. F. Borchgrevinck (1964). "Comparison of rat liver response to coumarin administered in vivo versus in vitro," American Journal of Physiology, Vol. 206(1), pp. 229-238.

Quandt, R. E. (1958). "The estimation of the parameters of a linear regression system obeying two separate regimes," Journal of the American Statistical Association, Vol. 53, pp. 873-880.

Quandt, R. E. (1960). "Tests of hypotheses that a linear regression system obeys two separate regimes," Journal of the American Statistical Association, Vol. 55, pp. 324-330.

Quandt, R. E. (1972). "A new approach to estimating switching regressions," Journal of the American Statistical Association, Vol. 67, pp. 306-310.

Quandt, R. E. (1974). "A comparison of methods for testing non-nested hypotheses," Review of Economics and Statistics, Vol. 56, pp. 92-99.

Salazar, Diego (1980). *The Analysis of Structural Changes in Time Series and Multivariate Linear Models*, doctoral dissertation, Oklahoma State University, Stillwater, Oklahoma.

Salazar, Diego, L. Broemeling, and A. Chi (1981). "Parameter changes in a regression model with autocorrelated errors," Communications in Statistics, Vol. A10(17), pp. 1751-1758.

Sen, A. K. and M. S. Srivastava (1973). "On multivariate tests for detecting change in mean," Sankhya, Vol. A35, pp. 173-185.

Sen, A. K. and M. S. Srivastava (1975a). "On tests for detecting a change in mean," Annals of Statistics, Vol. 3, pp. 98-108.

Sen, A. K. and M. S. Srivastava (1975b). "On tests for detecting change in mean when variance is unknown," Annals of the Institute of Statistical Mathematics, Vol. 27, pp. 479-486.

Sen, A. K. and M. S. Srivastava (1975c). "Some one-sided tests for change in level," Technometrics, Vol. 17, pp. 61-64.

Smith, A. F. M. (1975). "A Bayesian approach to inference about a change point in a sequence of random variables," Biometrika, Vol. 62, pp. 407-416.

Smtih, A. F. M. (1977). "A Bayesian analysis of some time-varying models," in *Recent Developments in Statistics* (edited by J. R. Barra et al.), North-Holland, Amsterdam, pp. 257-267.

Smith, A. F. M. (1980). "Change-point problems: approaches and applications," in *Bayesian Statistics*, edited by J. M. Bernardo, University Press, Valencia, pp. 83-98.

Sprent, P. (1961). "Some hypotheses concerning two-phase regression lines," Biometrics, Vol. 17, pp. 634-645.

Srivastava, M. S. (1980). "On tests for detecting change in the multivariate mean," Technical Report no. 3, University of Toronto.

Tsurumi, H. (1977). "A Bayesian test of a parameter shift with an application," J. of Econometrics, pp. 371-380.

Tsurumi, H. (1978). "A Bayesian test of a parameter shift in a simultaneous equation with an application to a macro savings function," Economic Studies Quarterly, Vol. 24(3), pp. 216-230.

Zellner, A. (1971). *An Introduction to Bayesian Inference in Econometrics*, John Wiley and Sons, Inc., New York.

8
MULTIVARIATE LINEAR MODELS

INTRODUCTION

This chapter concludes the book with a study of some multivariate linear models which are useful in regression, time series, and designed experiments. Multivariate models are generalizations of some of the univariate linear models which were considered in the earlier chapters, thus the Bayesian analysis of the multivariate versions of the regression design and autoregressive models is introduced.

The Bayesian analysis of multivariate linear models is well known and is presented in the books of Box and Tiao (1973), Press (1982), and Zellner (1971) and we will review these analyses in the next three sections. The analysis of vector time series models such as vector ARMA models has not been developed, except for some special cases. For example, the univariate autoregressive model has been examined by Zellner (1971) and was presented in Chapter 5, however, very little has appeared in the multivariate case; however Litterman (1980) and Chow (1975) have done some work and the vector AR model is quite similar to the multivariate linear model. Nothing has appeared for vector moving average processes (because very little has been done with univariate MA models) therefore only the vector AR model is studied.

Structural change in linear models is rapidly being developed from a Bayesian viewpoint. For example Salazar (1981,1982) has developed an analysis for changing multivariate linear models and autoregressive processes, and Tsurumi (1978) has examined structural change in simultaneous equations of econometrics. Structural change in linear models is of course related to dynamic linear systems, which were studied in Chapter 6. Linear dynamic models are linear models with parameters which always change, while structural change in linear models deal with parameters which change only once (or only occasionally).

MULTIVARIATE REGRESSION MODELS

Suppose one has m correlated dependent variables and k independent variables and we want to investigate the relationship between the two sets, then if the average values of the dependent variables are linear functions of the independent variables,

$$Y = X\theta + e, \tag{8.1}$$

where Y is a n × m matrix representing n observations on the m dependent variables, X is a n × k matrix of n observations on k independent variables, θ is a k × m matrix of unknown regression parameters, and e is a n × m matrix of error vectors such that the n 1 × m rows of e, say $e_1', e_2', \ldots, e_n'$ are independent normal random vectors with mean vector zero and unknown precision matrix P. Thus as with the univariate model of Chapter 1, we will develop the prior, posterior, and predictive analysis.

It should be noted that the rows of the Y matrix are independent normal random vectors and that the likelihood function of the parameters is

$$L(\theta, P|Y,X) \propto |P|^{n/2} \exp -\frac{1}{2} \mathrm{Tr}(Y - X\theta)'(Y - X\theta) P, \tag{8.2}$$

where θ is a n × k real matrix and P an unknown m × m positive definite symmetric matrix. The likelihood function is the conditional density of Y given X (nonstochastic), P, and θ, and can also be written as

$$L(\theta, P|Y,X) \propto |P|^{n/2} \exp -\frac{1}{2} \mathrm{Tr}[S + (\theta - \hat{\theta})'X'X(\theta - \hat{\theta})] P, \tag{8.3}$$

where S is the m × m matrix

$$S = (Y - X\hat{\theta})'(Y - X\hat{\theta}), \tag{8.4}$$

and

MULTIVARIATE REGRESSION MODELS

$$\hat{\theta} = (X'X)^{-1}X'Y,$$

if X is n × k of full rank. Of course, $\hat{\theta}$ is the least squares estimator of θ and $(Y - X\hat{\theta})'(Y - X\hat{\theta})$ is the residual matrix. It is interesting to note that the likelihood function may also be written as

$$L(\theta, P|Y,X) \propto |P|^{n/2} \exp - \frac{\text{TrS } P}{2} \exp - \frac{1}{2}(\theta_c - \hat{\theta}_c)'P \otimes X'X(\theta_c - \hat{\theta}_c), \quad (8.5)$$

where

$$\theta_c = \begin{pmatrix} \theta_1 \\ \theta_2 \\ \cdot \\ \cdot \\ \cdot \\ \theta_m \end{pmatrix},$$

$$\hat{\theta}_c = \begin{pmatrix} \hat{\theta}_2 \\ \hat{\theta}_2 \\ \cdot \\ \cdot \\ \cdot \\ \hat{\theta}_m \end{pmatrix},$$

and θ_i is the i-th column of θ, i = 1,2, ..., m, and $\hat{\theta}_i$ is the i-th column of $\hat{\theta}$, thus $\hat{\theta}_c$ is the maximum likelihood of θ_c which is normally distributed with mean θ_c and precision matrix $P \otimes X'X$ or dispersion matrix $P^{-1} \otimes (X'X)^{-1}$. Thus the sampling distribution of $\hat{\theta}_c$ is such that the variance-covariance matrix of $\hat{\theta}_c$ places restrictions on the variances and covariances of $\hat{\theta}_c$, and the same restrictions are placed on the least squares estimators of θ. This is a consequence of the definition of the model and the principle of maximum likelihood, and we will encounter this restriction in the Bayesian analysis.

Let us first use the conjugate family for the prior distribution of θ and P, which we see from (8.3) is a normal-Wishart, namely

$$\xi(\theta, P) = \xi_1(\theta|P)\, \xi_2(P), \qquad (8.6)$$

where

$$\xi_1(\theta|P) \propto |P|^{k/2} \exp -\frac{1}{2} \text{Tr}(\theta - \mu)'A(\theta - \mu)P \qquad (8.7)$$

and

$$\xi_2(P) \propto |P|^{(v-m-1)/2} \exp -\frac{1}{2} \text{Tr} B P, \qquad (8.8)$$

where μ is a $k \times m$ known matrix, A is a known $k \times k$ matrix, $v > m - 1$, and B is a known $m \times m$ positive definite symmetric matrix, thus a priori, P has a marginal posterior density which is Wishart with v degrees of freedom, and precision matrix B, while the conditional prior density of θ given P is normal such that $E(\theta) = \mu$ and $\theta^* = (\theta'_{(1)}, \theta'_{(2)}, \ldots, \theta'_{(k)})'$, when $\theta_{(i)}$ is the i-th row of θ, has precision matrix $A^{-1} \otimes P$. Now multiplying (8.7) and (8.8) gives the joint prior density of θ and P and if P is eliminated by integration (using the properties of the Wishart distribution, see DeGroot, 1972, p. 57), one will see the marginal prior density of θ is

$$\xi(\theta) \propto |B + (\theta - \mu)'A(\theta - \mu)|^{-(v+k)/2}$$

which is a matrix t distribution. Its properties are explained in Press (1982) and Box and Tiao (1973). Box and Tiao show the mean of θ is μ and the variance-covariance matrix of θ^* is $(v - m - 1)^{-1} A^{-1} \otimes B$, and we see the parameters of the prior distribution of θ are functions of the parameters of the marginal prior distribution of P and the conditional prior density of θ given P, which is the consequence of using the conjugate prior distribution.

Combining the prior (8.6) with the likelihood function (8.2) (note X'X can be singular), the posterior density is normal-Wishart, namely

$$\xi(\theta, P|Y, X) \propto \xi_1(\theta|P, Y, X)\, \xi_2(P|Y, X), \qquad (8.9)$$

where

$$\xi_1(\theta|P, X, Y) \propto |P|^{k/2} \exp -\frac{\text{Tr}}{2}(\theta - \tilde{\theta})'(X'X + A)(\theta - \tilde{\theta})P,$$

$$(8.10)$$

$$\xi_2(P|X,Y) \propto |P|^{(n+v-m-1)/2} \exp -\frac{\text{Tr}}{2} [B + \mu'A\mu + Y'Y$$

$$- (X'Y + A\mu)'(X'X + A)^{-1}(X'Y + A\mu)]P, \qquad (8.11)$$

where

$$\tilde{\theta} = (X'X + A)^{-1}(X'Y + A\mu). \qquad (8.12)$$

Now (8.10) is the conditional density of θ given P, which is normal, and ξ_2 the marginal density of P, which is Wishart. The marginal density of θ is a matrix t with density

$$\xi(\theta|X,Y) \propto |C + (\theta - \tilde{\theta})'(X'X + A)(\theta - \tilde{\theta})|^{-(n+v+k)/2}, \qquad (8.13)$$

where

$$C = B + \mu'A\mu + Y'Y - (X'Y + A\mu)'(X'X + A)^{-1}(X'Y + A\mu),$$

and the posterior mean of θ is $E(\theta|X,Y) = \tilde{\theta}$. Also, the posterior dispersion matrix of θ^* is

$$D(\theta^*|X,Y) = (n + v - m - 1)^{-1}(X'X + A)^{-1} \otimes C. \qquad (8.14)$$

The last equation shows certain restrictions are placed on the posterior variances and covariances of the regression matrix θ, which was noted by Rothenberg (1963). We have seen such restrictions occur if one uses the method of maximum likelihood, that is the same restrictions are placed on the variance-covariance matrix (with $A = 0$) of the sampling distribution of the maximum likelihood estimator $\hat{\theta} = (X'X)^{-1}X'Y$ of θ.

Can these restrictions be avoided? Suppose one uses Jeffreys' improper prior density

$$\xi(\theta, P) \propto |P|^{-(m+1)/2}, \qquad (8.15)$$

then the marginal posterior density of θ is

$$\xi(\theta|X,Y) \propto \frac{1}{|S + (\theta - \hat{\theta})'X'X(\theta - \hat{\theta})|^{n/2}} \qquad (8.16)$$

and the same restrictions on the variances and covariances of θ^* occur as with the maximum likelihood estimator of θ. See Zellner (1971, p. 229).

Can one develop a Bayesian analysis which avoids these restrictions? Press (1982) and Zellner (1971) develop a posterior analysis which somewhat avoids placing restrictions on the variances and covariances of the posterior distribution of θ. They use a generalized natural conjugate prior density (so called by Press) where θ and P are independent and θ has a normal density and P a Wishart, however the joint and marginal posterior densities of θ and P are not standard and one must use approximations to obtain a standard form. If this is done (see Press, 1982), the large sample approximation to the marginal posterior distribution is a normal, but the precision matrix of θ is $D + G \otimes (X'X)$, where D and G are known matrices and D^{-1} is the dispersion matrix of the marginal prior density of θ_c, which is normal with some mean μ. The restrictions are avoided to some extent by adding D to the Kronecker product $G \otimes (X'X)$, the part of the sum which places restrictions on the marginal prior dispersion matrix of θ. The problem with this approach is that one's prior knowledge of the dispersion of θ may not be of such a nature to remove undesirable restrictions on θ. Of course another problem is that the posterior distributions of θ are large sample approximations and their accuracy in small samples needs to be studied. The Bayesian analysis of the multivariate linear model based on the generalized conjugate prior will not be pursued here but the reader should read Press (1982) or Zellner (1971).

Now consider the problem of forecasting or predicting future observations with the multivariate linear model. Suppose one plans a future experiment

$$W = Z\theta + \varepsilon, \qquad (8.17)$$

where Z is the known "future" n × k design matrix, W the n × m matrix of n future observations on the m dependent variables, and ε the future error matrix with the same properties as the error vector e of (8.1), then (8.17) represents a future replication of the experiment given by the linear model (8.1).

Predicting W is to be based on the Bayesian predictive density of W, which is the conditional density of W given Y, which is

$$g(W|Y,X,Z) = \int_{\Omega^*} \xi(\theta, P|Y,X) \, h(W|Z, \theta, P) \, d\theta \, dP, \qquad (8.18)$$

where $\xi(\theta, P|Y,X)$ is the joint posterior density of the parameters and h is the conditional density of W given Z, θ, and P, namely

$$h(W|Z, \theta, P) \propto |P|^{n/2} \exp - \frac{\mathrm{Tr}}{2} (W - Z\theta)'(W - Z\theta)P, \qquad (8.19)$$

MULTIVARIATE REGRESSION MODELS   381

and Ω^* is the parameter space where θ is a $k \times m$ real matrix and P a $m \times m$ positive definite symmetric matrix of order m. It is assumed that e and ε are independent.

If one uses the conjugate prior density (8.6), the corresponding posterior density of the parameters is given by (8.9) and if this is used in the predictive density g of (8.18), the forecasting procedure will be based on conjugate prior information. Doing this, one may show the predictive density of (8.18) reduces to

$$g(W|X,Y,Z) \propto |(W - W^*)'D(W - W^*) + E|^{-(2n+v)/2} \qquad (8.20)$$

where

$$W^* = [I - Z(Z'Z + X'X + A)^{-1}Z']^{-1} Z(Z'Z + X'X + A)^{-1}(X'Y + A\mu), \qquad (8.21)$$

$$D = I - Z(Z'Z + X'X + A)^{-1}Z', \qquad (8.22)$$

and

$$E = Y'Y + \mu'A\mu + B + (X'Y + A\mu)'(Z'Z + X'X + A)^{-1}(X'Y + A\mu)$$
$$- W^{*\prime}Z(Z'Z + X'X + A)^{-1}(X'Y + A\mu). \qquad (8.23)$$

We see the predictive distribution of W is a matrix t with mean W^* and the covariance matrix of $W = [W_{(1)}, W_{(2)}, \ldots, W_{(n)}]'$ is $(n + v - m - 1)^{-1} D^{-1} \otimes E$, $W'_{(i)}$ is the i-th row of W. This predictive distribution was derived from (8.18) by completing the square on θ, eliminating θ using the properties of the normal density, eliminating P from the properties of the Wishart distribution, and finally completing the square on W to identify its distribution as a matrix t.

We now consider the problem of making inferences for the parameters θ and P of the multivariate linear model. If one only has vague prior knowledge about the parameters, one would not use the conjugate prior density (8.6), which was used in the derivation of the posterior analysis; instead we could consider Jeffreys' vague prior

$$\xi(\theta,P) \propto \frac{1}{|P|^{(m+1)/2}} \qquad (8.24)$$

which is the density employed by Box and Tiao (1973), Zellner (1971) and Press (1982), in which case the marginal posterior density of θ is given by (8.16), which is obtained from the marginal posterior density (8.13) by letting $A \to 0(k \times k)$, $B \to 0(m \times m)$, and $v = -k$, and assuming X'X is nonsingular. Inferences for θ are based on this distribution and one would estimate θ by

$$E(\theta|X,Y) = (X'X)^{-1}X'Y,$$

the usual least squares estimator. More sophisticated inferences require the properties of the matrix t distribution which is explained by Box and Tiao (1973, pp. 441-452), and Zellner (1971, pp. 396-399), where the latter author refers to the matrix t distribution as a generalized Student t denoted by GS_t. For example, in order to find HPD regions for subsets of the elements of θ (say the first column or row), Box and Tiao (p. 448) explain the procedure. If a row or column of the design matrix is of primary interest, one may show the row or column has a multivariate t distribution which we have used many times in the previous chapters.

Let θ_1 be the first column of the regression matrix θ, then Box and Tiao (1973, p. 443) show θ_1 has a multivariate t distribution, thus it is quite easy using the material of Chapter 1 to construct HPD regions for θ_1.

In most applications of the multivariate linear model, the regression parameters θ are of primary concern and the precision matrix P is a nuisance parameter. If P is also of interest, one's inferences about the elements of P would be based on its marginal density (8.11) which is a Wishart distribution with $n + v$ degrees of freedom and precision matrix (see DeGroot, 1970)

$$P(P|X_1Y) = B + Y'Y + \mu'A\mu - (X'Y + A\mu)'(X'X + A)^{-1}(X'Y + A\mu). \quad (8.25)$$

This distribution was derived assuming a conjugate prior normal Wishart distribution (8.6), so if one uses a vague prior density (8.24) instead, one would modify the parameters of (8.11) by letting $v = -k$, $A = 0$, and $B = 0$, thus a posteriori, P has a Wishart distribution with $n - k$ degrees of freedom and precision matrix

$$P(P|X,Y) = Y'Y - Y'X(X'X)^{-1}X'Y, \quad (8.26)$$

the matrix of the sum of squares and cross products of the residuals. Thus by assuming vague prior knowledge, one perhaps would estimate P

$$E(P|X,Y) = (n - k)[Y'Y - Y'X(X'X)^{-1}X'Y]^{-1}, \qquad (8.27)$$

the marginal posterior mean matrix of P. The properties of the Wishart distribution are given by Press (1982) as well as by Box and Tiao (1973) and Zellner (1971), however, Press gives a very detailed analysis of the Wishart and inverse Wishart distributions. For example (p. 107), he gives the variance-covariance matrix of the elements of the precision matrix. HPD regions for some of the elements of P can be found. For example, the diagonal elements of P have gamma distributions, thus for them such regions are easily constructed numerically, however for arbitrary subsets of the elements of P the density is unknown, and one would have to use multidimensional integration techniques of the Wishart density.

An examination of Box and Tiao (1973), Press (1982), and Zellner (1971) shows that the first authors give an interesting and thorough Bayesian analysis of a chemical process with two dependent variables, yield of product and yield of by-product, and one independent variable, temperature, thus the model is

$$Y_{i1} = \theta_{11} X_{i1} + \theta_{21} X_{i2} + e_{i1}, \qquad (8.28)$$

$$Y_{i2} = \theta_{12} X_{i1} + \theta_{22} X_{i2} + e_{i2} \qquad (8.29)$$

where $X_{i1} = 1$, and X_{i2} is the i-th setting of the temperature corresponding to product yield Y_{i1} and by-product yield Y_{i2}, for $i = 1, 2, \ldots, n$. The regression matrix is

$$\theta = \begin{pmatrix} \theta_{11} & \theta_{12} \\ \theta_{21} & \theta_{22} \end{pmatrix},$$

and the n independent bivariate error vectors $e_i' = (e_{i1}, e_{i2})$ are normal with mean (0, 0) and unknown 2 × 2 precision matrix P. Assuming a vague prior density for the parameters θ and P, Box and Tiao (1973, pp. 453-459) determine the marginal posterior distributions (bivariate-t distributions) of θ_1 and θ_2 where $\theta = (\theta_1, \theta_2)$ and the elements of $\theta_{(1)}$ are the regression parameters of the product yield equation (8.28). They plot some of the contours of the posterior distribution of θ_1 and θ_2 and from these test hypotheses about the parameters. This example is quite interesting and demonstrates the advantages of the Bayesian approach to inference. Zellner also gives an example of using the model to analyze multivariate data (p. 244, 1971), but it is not as detailed as the one given by Box and Tiao.

MULTIVARIATE DESIGN MODELS

The Bayesian analysis of multivariate linear models for designed experiments is now being developed and most of the work has been done by Press (1980) and Press and Shigemasu (1982). Box and Tiao (1973) give a detailed treatment of multivariate regression analysis but devote very little to the analyses of designed experiments.

The analysis presented here is similar to Press and Shigemasu (1982) with one important difference. Prior distributions for the parameters of the model are expressed with a natural conjugate prior distribution instead of assuming the precision matrix P and the other parameters θ (treatment and block means, etc.) to be independent.

The Bayesian analysis of univariate design models for one-way and two-way layouts was given in Chapter 3 and, as mentioned there, was done earlier by Lindley and Smith (1972) where they used exchangeable prior distributions. Press and Shigemasu (1982) also employ the idea of exchangeability for the prior distribution in their treatment of the multivariate case.

Press in his 1980 paper uses a vague constant prior for the design parameters and a Jeffreys' vague prior for the precision matrix of the observations, therefore the analysis presented now is similar to his in that the posterior distribution of the parameters is normal-Wishart which is the conjugate family to the multivariate normal linear model.

Let us first consider a one-way layout consisting of k treatments assigned at random to n experimental units such that n_i, $i = 1, 2, \ldots,$ k receive treatment i. On each unit one observes m correlated responses, then if Y_{ij} is the $m \times 1$ vector of responses obtained from the j-th unit which received treatment i, the normal theory model is

$$Y = X\theta + \varepsilon \qquad (8.30)$$

where

$$Y' = [Y_{11}, \ldots, Y_{1n_1}; \ldots; Y_{k1}, Y_{k2}, \ldots, Y_{kn_k}],$$

$$\varepsilon' = [\varepsilon_{11}, \ldots, \varepsilon_{1n_1}; \ldots; \varepsilon_{k1}, \varepsilon_{k2}, \ldots, \varepsilon_{kn_k}],$$

$$\theta' = [\theta_{(1)}, \theta_{(2)}, \ldots, \theta_{(k)}],$$

and

$$X = \text{DIAG}(J_{n_1}, \ldots, J_{n_m}),$$

a block diagonal matrix where J_{n_i} is a $n_i \times 1$ vector of ones, and this is the model (8.30), namely

$$Y_{ij} = \theta_{(i)} + \varepsilon_{ij} \qquad (8.31)$$

where $\theta_{(i)}$ is a real unknown $m \times 1$ vector representing the effect of treatment i (the mean vector of population i), and the ε_{ij}'s are such that the rows of e are independent normal random vectors with mean zero and $m \times m$ unknown precision matrix P.

Of course the linear model for designed experiments has the same form as the multivariate regression model (8.16). The main parameters of interest are the treatment effects $\theta_{(1)}, \theta_{(2)}, \ldots, \theta_{(k)}$ of $m \times 1$ vectors which represent the means of the k normal populations, and one must specify prior information about them as well as the precision matrix P.

We have seen that if one uses the Jeffreys' improper prior density

$$\xi(\theta, P) \propto \frac{1}{|P|^{(m+1)/2}}$$

that the joint posterior density of θ and P is normal-Wishart and that the marginal density of θ is

$$\xi(\theta|X,Y) \propto |S + (\theta - \hat{\theta})'X'X(\theta - \hat{\theta})|^{-n/2} \qquad (8.32)$$

where

$$\theta = \begin{pmatrix} \theta_{(1)}' \\ \theta_{(2)}' \\ \vdots \\ \theta_{(k)}' \end{pmatrix}$$

is a real $k \times m$ matrix, $\hat{\theta} = (X'X)^{-1}X'Y$ is the least squares estimator of θ and $S = Y'Y - Y'X(X'X)^{-1}$ is the matrix of residual sum of squares and cross products. This density of θ is called a matrix t density and θ is said to have a matrix t distribution with parameters $\hat{\theta}$, $(X'X)^{-1}$, S, and $d = n - k - m + 1$. Box and Tiao (1973) use the notation

$$\theta \sim t_{km}[\hat{\theta}, (X'X)^{-1}, S, d] \tag{8.33}$$

to denote the distribution of θ, and we have seen that $E(\theta|X,Y) = \hat{\theta}$ and the posterior dispersion matrix of $\theta^* = [\theta'_{(1)}, \theta'_{(2)}, \ldots, \theta'_{(k)}]'$ is $D(\theta^*|X,Y) = (d-2)^{-1}(X'X)^{-1} \otimes S$.

If on the other hand one uses a conjugate prior density for θ and P, how does one choose the hyperparameters? Suppose the prior density of θ and P is

$$\xi(\theta, P) \propto \xi_1(\theta|P) \xi_2(P) \tag{8.34}$$

and

$$\xi_1(\theta|P) \propto |P|^{k/2} \exp -\frac{\text{Tr}}{2}(\theta - \mu)'A(\theta - \mu)P \tag{8.35}$$

and

$$\xi_2(P) \propto |P|^{(v-m-1)/2} \exp -\frac{\text{Tr}}{2} BP, \tag{8.36}$$

where μ is a real $k \times m$ matrix, A is $k \times k$, B is $m \times m$, and $v > m - 1$, then we know the marginal prior density of θ is

$$\xi(\theta) \propto |B + (\theta - \mu)'A(\theta - \mu)|^{-(k+v)/2} \tag{8.37}$$

Thus, if one uses a conjugate prior density for the parameters, one is assuming that $E(\theta) = \mu$ and that

$$D(\theta^*) = (v - m - 1)^{-1}A^{-1} \otimes B.$$

Now A is $k \times k$ and B is $m \times m$ and the Kronecker product is the $km \times km$ matrix

$$(v - m - 1)^{-1} \begin{pmatrix} a_{11}^{-1}B & a_{12}^{-1}B & \cdots & a_{1k}^{-1}B \\ & a_{22}^{-1}B & \cdots & a_{2k}^{-1}B \\ & & \ddots & \vdots \\ & & & a_{kk}^{-1}B \end{pmatrix},$$

a_{ij}^{-1} is the ij-th element of A^{-1}. If A is chosen to be a diagonal matrix and the diagonal elements of A are equal, then the treatment effects $\theta_{(1)}, \theta_{(2)}, \ldots, \theta_{(k)}$ are such that they are uncorrelated and have the same precision matrix. If one chooses the rows of μ to be the same, then one can represent prior information about the treatment effects so that they all have the same mean vector as well as the same covariance matrix.

One still has the problem of choosing the hyperparameters, but ignoring that for now let us consider the posterior analysis.

The marginal posterior density of θ is given by (8.13), that is

$$\theta | X, Y \sim t_{km} [\tilde{\theta}, (X'X + A)^{-1}, C, n + v - m + 1], \quad (8.38)$$

where

$$\tilde{\theta} = (X'X + A)^{-1}(X'Y + A\mu),$$

and

$$C = B + \mu'A\mu + Y'Y - (X'Y + A\mu)'(X'X + A)^{-1}(X'Y + A\mu).$$

(8.39)

The matrix θ of treatment effects has posterior mean $\tilde{\theta}$ and the covariance matrix of θ^* is

$$D(\theta | X, Y) = (n + v - m - 1)^{-1}(X'X + A)^{-1} \otimes C.$$

How does one make inferences for the treatment effects? In problems like these, one is usually interested in the treatment effects $\theta_1, \theta_2, \ldots, \theta_k$, each of which is a $m \times 1$ vector, and especially in the equality of treatment effects. In the classical approach to statistics one tests for equality of treatment effects by the analysis of variance and one should consult Press (1982), Anderson (1958), Morrison (1967), and Srivastava and Khatri (1979).

When one studies a univariate one-way layout by Bayesian techniques (see Chapter 3), one first finds the marginal posterior distribution of θ, which is a multivariate t distribution. Then one transforms to the F-distribution and finds an HPD region to test for equality of treatment effects. When one uses a Jeffreys' improper prior distribution, the HPD region test is equivalent to the analysis of variance, but what does one do in the multivariate case?

Consider the hypothesis

$$H_0: \theta_1 = \theta_2 = \ldots = \theta_k \qquad (8.40)$$

then we know the marginal posterior distribution of θ is the matrix t (8.38), assuming a conjugate prior density. The null hypothesis is equivalent to

$$H_0: \eta = 0,$$

where $\eta = Q\theta$ and

$$Q = \begin{pmatrix} 1, & -1, & 0, & \ldots & 0 \\ 0, & 1, & -1, 0, & \ldots & 0 \\ 0, & 0, & 0, & \ldots & 1, & -1 \end{pmatrix} \qquad (8.41)$$

which is a $(k-1) \times k$ matrix. It can be shown, see Box and Tiao (1973, p. 447), that η is distributed as $t_{(k-1)m}[Q\tilde{\theta}, Q(X'X + A)^{-1}Q', C, v + n - m + 1]$, and one may test for equality of treatment effects by finding an HPD region for η. A $(1 - \alpha)$ HPD region for η is given by

$$C_\eta(\alpha) = \{\eta: U(\eta) \leqslant k(\alpha)\} \qquad (8.42)$$

where $k(\alpha)$ depends on α and the distribution of $U(\eta)$, which can be determined by the methods of Box and Tiao (1973). They find the exact distribution of $U(\eta)$ when $m = 1$ and 2 and give an approximation to the distribution of $U(\eta)$ for $m \geqslant 3$, and in fact they show $-\phi d \cdot \log U(\eta)$ has an approximate chi-square distribution with $m(k - 1)$ degrees of freedom, where $\phi = 1 + (2d)^{-1}(m + k - 1 - 3)$ and $d = n + v - m + 1$. Thus equality of treatment effects is easily tested by use of the chi-square tables and is a straightforward extension of the testing procedure in the univariate case.

THE VECTOR AUTOREGRESSIVE PROCESS

Press and Shigemasu (1982) show how one may analyze a two-way layout with and without interaction; however they do not use a conjugate prior density, but if one does, it appears possible to test hypotheses about the row and/or column effects by finding an HPD region for these parameters. Press and Shigemasu (1982) also extend their analyses to models which include covariables, thus giving a Bayesian alternative to the analyses of covariance procedures of classical statistics.

THE VECTOR AUTOREGRESSIVE PROCESS

Autoregressive processes are the most used of parametric time series models and the univariate case was considered in Chapter 5. The Bayesian analysis of the vector autoregressive process is being developed by Litterman (1980), however much remains to be done. The Bayesian identification of vector AR processes has yet to be developed and the posterior and predictive analysis has yet to be completed, although Chow (1975) has given us the moments of the predictive distribution.

By using a conjugate prior density for the parameters, the prior and predictive analysis will be developed.

First let us consider a m vector AR(p) process, namely

$$Y(t) = \sum_{i=1}^{p} \theta_i Y(t - i) + \varepsilon(t),$$

where $t = 1, 2, \ldots, n$, $Y(t)$ is a $m \times 1$ real observation at time t, θ_i is a $m \times m$ real matrix, θ_p is a nonzero matrix, $p = 1, 2, \ldots$ and the $\varepsilon(t)$ are n.i.d.$(0, P^{-1})$, where P is a $m \times m$ positive definite symmetric precision matrix.

The model is a special case of the multivariate linear model (8.1), where

$$Y = X\theta + \varepsilon \qquad (8.43)$$

and: y is a $n \times m$ matrix with t-th row $Y'(t)$, while the t-th row of the $n \times pm$ matrix X is $Y'(t-1), Y'(t-2), \ldots, Y'(t-p)$, the $pm \times m$ regression matrix θ of autoregressive parameters is

$$\theta = \begin{pmatrix} \theta'_1 \\ \theta'_2 \\ \vdots \\ \theta'_p \end{pmatrix} \qquad (8.44)$$

and ε is $n \times m$ with t-th row $\varepsilon'(t)$, $t = 1, 2, \ldots, n$. It is assumed that $Y(0), Y(-1), \ldots, Y(1-p)$ are known "initial" values.

Conditions of stationarity will not be imposed on the parameters. The likelihood function is

$$L(\theta, P | X, Y) \propto |P|^{n/2} \exp - \frac{\text{Tr}}{2} (Y - X\theta)'(Y - X\theta) P$$

and we will use a conjugate prior density for the parameters, namely

$$\xi(\theta | P) \propto |P|^{mp/2} \exp - \frac{\text{Tr}}{2} (\theta - \mu)' A (\theta - \mu) P$$

as the conditional prior density of θ given P and

$$\xi(P) \propto |P|^{(v-m-1)/2} \exp - \frac{\text{Tr}}{2} BP$$

is the marginal prior density of P and is Wishart with v degrees of freedom and precision matrix B. The marginal prior density of θ is the matrix t density

$$\xi(\theta) \propto |B + (\theta - \mu)' A (\theta - \mu)|^{-(mp+v)/2}, \qquad (8.45)$$

where B is a positive definite $m \times m$ matrix and A is $mp \times mp$. The prior dispersion matrix of θ^* is

$$D(\theta^*) = (v - m - 1)^{-1} A^{-1} \otimes B, \qquad (8.46)$$

where θ^* is the $m^2 p \times 1$ vector

$$\theta^* = (\theta_{11}, \ldots, \theta_{1m}; \theta_{21}, \ldots, \theta_{2m}; \ldots, \theta_{p1}, \ldots, \theta_{pm})', \qquad (8.47)$$

where θ_{ij} is the j-th row of θ'_i. Thus the hyperparameters of the prior distribution of the parameter may be chosen so that the autoregressive matrices are uncorrelated, each having the same mean matrix and dispersion matrix. If $m = p = 2$, the 4×4 matrix in the upper left hand portion of the 8×8 matrix $D(\theta^*)$ is

$$(v - m - 1)^{-1} \begin{pmatrix} a_{11}^{-1} B & a_{12}^{-1} B \\ a_{21}^{-1} B & a_{22}^{-1} B \end{pmatrix} \qquad (8.48)$$

where a_{ij}^{-1} is the ij-th element of A^{-1}, and is the variance-covariance matrix of the first autoregressive matrix θ_1'. This would be a second order model and the variance-covariance matrix of the second autoregressive matrix would be the 4×4 matrix in the lower right hand corner of $D(\theta^*)$. The covariance matrix between θ_1' and θ_2' is the 4×4 matrix

$$\begin{pmatrix} a_{13}^{-1} B & a_{14}^{-1} B \\ a_{23}^{-1} B & a_{24}^{-1} B \end{pmatrix} (v - 1)^{-1},$$

and by choosing a_{13}, a_{14}, a_{23}, and a_{24} to be zero, one may put zero correlation between the two autoregressive matrices of the model.

In general $A^{-1} \otimes B$ is given by the partitioned matrix

$$\begin{pmatrix} A_{11}^{-1} \otimes B & A_{12}^{-1} \otimes B & \cdots & A_{1p}^{-1} \otimes B \\ A_{21}^{-1} \otimes B & A_{22}^{-1} \otimes B & \cdots & A_{2p}^{-1} \otimes B \\ & & \cdot & \\ & & \cdot & \\ & & \cdot & \\ & & & A_{pp}^{-1} \otimes B \end{pmatrix} = A^{-1} \otimes B,$$

(8.49)

where the $mp \times mp$ matrix A^{-1} is partitioned as

$$A^{-1} = (A_{ij}^{-1}), \ i = 1, 2, \ldots, p; \ j = 1, 2, \ldots, p.$$

Therefore $A_{ij}^{-1} \otimes B(v-1)^{-1}$ is the covariance matrix between θ_i' and θ_j' if $i \ne j$, and the dispersion matrix of θ_i', if $i = j$. If one chooses $A_{ij} = 0(m \times m)$, $i \ne j$, and $A_{11} = A_{22} = \ldots = A_{pp}$, the regression matrices $\theta_1', \theta_2', \ldots, \theta_p'$ are uncorrelated and each have the same dispersion matrix. We also see the prior variance-covariance matrix places certain restrictions, a priori, on the variances and covariances of the elements of the regression matrices of the model, and if these do not conform to one's prior information about the parameters, the prior density used here is not appropriate. This problem was encountered with the multivariate linear model.

The posterior analysis of the vector autoregressive model is the same as that for the multivariate linear model. In particular the marginal posterior density of θ is

$$\theta \sim t_{(pm)m}[\tilde{\theta}, (X'X + A)^{-1}, C, n + v - m + 1] \qquad (8.50)$$

where

and

$$\tilde{\theta} = (X'X + A)^{-1}(X'Y + A\mu)$$

$$C = B + \mu'A\mu + Y'Y - \tilde{\theta}'(X'Y + A\mu),$$

Thus the posterior mean of θ is $\tilde{\theta}$ and the dispersion matrix of θ^* is

$$D(\theta^*|X, Y) = A^* \otimes C(n + v - m - 1)^{-1}, \qquad (8.51)$$

where $A^* = (X'X + A)^{-1}$.

Recall that θ^* is $m^2 p \times 1$ and given by $\theta^* = [\theta_{11}, \theta_{12}, \ldots, \theta_{1m}; \theta_{21}, \theta_{22}, \ldots, \theta_{2m}; \ldots; \theta_{p1}, \theta_{p2}, \ldots, \theta_{pm}]'$, where θ_{12} is the second row of θ_1', and θ_{12} is $1 \times m$.

Let

$$\theta^* = \begin{pmatrix} \theta_1^* \\ \theta_2^* \\ \cdot \\ \cdot \\ \cdot \\ \theta_p^* \end{pmatrix},$$

where the j-th row of θ_i^* is θ_{ij} and

$$\theta_i^* = \begin{pmatrix} \theta_{i1}' \\ \theta_{i2}' \\ \cdot \\ \cdot \\ \cdot \\ \theta_{im}' \end{pmatrix}, \quad i = 1, 2, \ldots, p, \qquad (8.53)$$

which is a $m^2 \times 1$ vector.

If A^* is partitioned into p^2 submatrices A_{ij}^* of order m, then the dispersion matrix of θ_i^* is

$$D(\theta_i^* | X, Y) = A_{ii}^* \otimes C (n + v - m - 1)^{-1} \qquad (8.54)$$

and the covariance matrix between θ_i^* and θ_j^* is

$$\text{Cov}(\theta_i^*, \theta_j^* | X, Y) = (n + v - m - 1)^{-1} A_{ij}^* \otimes C. \qquad (8.55)$$

In making inferences for the regression parameters of this model, one bases them on the matrix t distribution (8.50), and we saw how to use this distribution for the multivariate design model with a one-way layout.

Often the autoregressive model is used to predict future observations and this will be done, but sometimes one is more interested in knowing the values of the parameters θ and P as in some time series which occur in the analysis of designed experiments of psychology as in Glass, Willson, and Gottman (1975).

One's knowledge of the θ parameters is gained from a posterior analysis of the model which is based on the matrix t distribution $t_{(pm)m}[\tilde{\theta}, (X'X + A)^{-1}, C, n + v - m + 1]$. We have seen that one may test hypotheses about the regression parameters by constructing HPD regions.

Consider the bivariate AR(1) model

$$Y_1(t) = \theta_{11} Y_1(t-1) + \theta_{12} Y_2(t-1) + \varepsilon_1(t)$$

$$Y_2(t) = \theta_{21} Y_1(t-1) + \theta_{22} Y_2(t-1) + \varepsilon_2(t)$$

where $t = 1, 2, \ldots, n$, which is a model for relating two univariate time series $\{Y_1(t): t = 1, 2, \ldots\}$ and $\{Y_2(t): t = 1, 2, \ldots\}$. Thus each component of the bivariate time series is a linear function of the previous observations on both series plus an error term $\varepsilon'(t) = [\varepsilon_1(t), \varepsilon_2(t)]$ which has a bivariate normal distribution with mean $(0,0)$ and 2×2 precision matrix P. In terms of the general model $m = 2$, $p = 1$ and $\theta = \theta_1'$, where θ_1 is the 2×2 regression matrix

$$\theta_1 = \begin{pmatrix} \theta_{11} & \theta_{12} \\ \theta_{21} & \theta_{22} \end{pmatrix}.$$

If a normal-Wishart distribution of θ and P is used, we have seen the marginal posterior density of θ is the matrix t $t_{22}[\tilde{\theta}, (X'X + A)^{-1}, C, n + v - 1]$ and one would estimate θ by $\tilde{\theta}$, while the dispersion matrix of

$$\theta^* = \begin{pmatrix} \theta_{11} \\ \theta_{21} \\ \theta_{12} \\ \theta_{22} \end{pmatrix}$$

is

$$D(\theta^* | X, Y) = (n + v - 3)^{-1} A^* \otimes C,$$

where $A^* = (X'X + A)^{-1}$ and

$$A^* \otimes C = \begin{pmatrix} a^*_{11} C & a^*_{12} C \\ a^*_{21} C & a^*_{22} C \end{pmatrix},$$

where a^*_{ij} is the ij-th element of A^* and C is 2×2. The posterior dispersion matrix of $(\theta_{11}, \theta_{21})'$ is $(n + v - 3)^{-1} a^*_{11} C$, the dispersion matrix of $(\theta_{12}, \theta_{22})'$ is $(n + v - 3)^{-1} a^*_{22} C$, and the covariance matrix between $(\theta_{11}, \theta_{21})'$ and $(\theta_{12}, \theta_{22})'$ is $a^*_{12} C = a^*_{21} C$.

Now,

$$\theta = \begin{pmatrix} \theta_{11} & \theta_{21} \\ \theta_{12} & \theta_{22} \end{pmatrix}$$

and one can make inferences about this matrix by the techniques of the earlier sections. For example, the hypothesis $\theta_{12} = 0$ may be tested as follows. We know the first column of θ has a bivariate t distribution and that θ_{12} has a univariate t distribution, then an HPD region for θ_{12} can be constructed and H_0 rejected if $\theta_{12} = 0$ is not in the interval. Also H_0: $\theta_{21} = 0$ can be tested by a similar procedure. Other inferences about the θ matrix can be developed by the methods given by Box and Tiao (1973), Press (1982), and Zellner (1971).

We now consider predicting one step ahead for $Y(n + 1)$ with a vector autoregressive process. The general formula for this is given by (8.20), namely

$$g[Y(n + 1) | X, Y, X^*] \propto |E + [Y(n + 1) - Y^*]' D^*$$

$$\times [Y(n + 1) - Y^*]|^{-(2n+v)/2}, \qquad (8.56)$$

where

$$Y^* = [I - X^*(X^{*'}X^* + X'X + A)^{-1} X^{*'}]^{-1} X^*(X^{*'}X^* + X'X + A)^{-1}$$

$$\times (X'Y + A\mu),$$

$$E = Y'Y + \mu'A\mu + B + (X'Y + A\mu)'(X^{*'}X^* + X'X + A)^{-1}(X'Y + A\mu)$$

$$- Y^{*'} X^*(X^{*'}X^* + X'X + A)^{-1}(X'Y + A\mu),$$

and

$$D^* = I - X^*(X^{*\prime}X^* + X'X + A)^{-1}X^{*\prime}.$$

The other term that needs to be defined is X^* which is the X matrix of the future observation $Y(n + 1)$, where $Y(n + 1) = X^*\theta + e(n + 1)$ or $Y'(n + 1) = Y'(n)\theta'_1 + Y'(n - 1)\theta'_2 + \ldots + Y'(n - p - 1)\theta'_p + e'(n + 1)$. The $(n + 1)$-st error term is $n_m(0, p^{-1})$ and independent of the error term ε of (8.43), thus X^* is the $1 \times mp$ matrix

$$X^* = [Y'(n), Y'(n - 1), \ldots, Y'(n - p - 1)]. \qquad (8.57)$$

The predictive density (8.56) is based on the conjugate prior density which is given by (8.45). Since $Y(n + 1)$ is a $m \times 1$ vector the predictive distribution of $Y(n + 1)$ is a multivariate t distribution with mean vector

$$E[Y(n + 1) | X, Y, X^*] = Y^*,$$

precision matrix

$$P[Y(n + 1) | X, Y, X^*] = (2n + v - m - 2)^{-1} D^{*-1} \otimes E,$$

and $2n + v - m - 2$ degrees of freedom.

If one uses Y^* as a point forecast of $Y(n + 1)$, one is simultaneously forecasting the m future components of $Y(n + 1)$, but suppose one is interested only in the first component of the future observation, then how would one construct its point forcast? Of course one would most likely use the first component of Y^* and the precision of the forecast would be the precision of the first component of a random vector which has a multivariate t distribution and the formulas are well-known and given in Chapter 1 or in DeGroot (1970).

An HPD region for $Y(n + 1)$ is based on the random variable

$$F[Y(n + 1)] = [Y(n + 1) - Y^*]'D^*[Y(n + 1) - Y^*]m^{-1}, \qquad (8.58)$$

which has an F distribution with m and $2n + v - m$ degrees of freedom and a $(1 - \alpha)$ region is given by the ellipse

$$\{Y(n + 1): F[Y(n + 1)] \leq F_{2, m, 2n+v-m}\}.$$

If one wants to predict two or more steps ahead one very easily can derive the appropriate Bayesian predictive density but it is not a standard distribution.

THE VECTOR AUTOREGRESSIVE PROCESS

The posterior and predictive analyses of the vector autoregressive process were derived with a conjugate prior density but if one believes the Jeffreys' prior

$$\xi(\theta, P) \propto |P|^{-(m+1)/2}$$

is appropriate, the posterior and predictive densities are obtained by letting $v = -k(=m)$, $A \to 0(k \times k)$ and $B \to 0(m \times m)$ in the appropriate formulas.

Model identification is an important matter when one uses an autoregressive process to model a time series and this problem was discussed in Chapter 5 on time series analysis. In that chapter a Bayesian identification method of Diaz and Farah (1981) was explained for a univariate autoregressive process. This method is easily extended to a vector process as follows. Consider a m dimensional vector AR process of order p,

$$Y'(t) = \sum_{j=1}^{p} Y'(t-j)\theta_{pj} + \varepsilon'(t), \quad t = 1, 2, \ldots, n, \tag{8.59}$$

where $p = 1, 2, \ldots, L$, and L is the maximum known order of the model with $L = 1, 2, \ldots$. The observation at time t is $Y(t)$, which is a $m \times 1$ vector and the θ_{pj}, $j = 1, 2 \ldots, p$, $p = 1, 2 \ldots, L$ are $m \times m$ real matrices, $Y(0)$, $Y(-1)$, ..., $Y(1-p)$ are known initial values, and $\varepsilon(1)$, ..., $\varepsilon(n)$ are independent $n_m(0, P^{-1})$ random $m \times 1$ vectors.

Then the model is

$$Y = X\theta_p + E, \tag{8.60}$$

where Y is $n \times m$ and

$$Y = \begin{pmatrix} Y'(1) \\ Y'(2) \\ \cdot \\ \cdot \\ \cdot \\ Y'(n) \end{pmatrix},$$

and X is $n \times mp$ and

$$X = \begin{pmatrix} Y'(0) & Y'(-1), & \ldots, & Y'(1-p) \\ Y'(1) & Y'(0), & \ldots, & Y'(2-p) \\ \cdot & \cdot & & \cdot \\ \cdot & \cdot & & \cdot \\ \cdot & \cdot & & \cdot \\ Y'(n-1) & Y'(n-2), & \ldots, & Y'(n-p) \end{pmatrix}.$$

Also

$$\theta_p = \begin{pmatrix} \theta_{p1} \\ \theta_{p2} \\ \cdot \\ \cdot \\ \cdot \\ \theta_{pp} \end{pmatrix}$$

and

$$E = \begin{pmatrix} \varepsilon'(1) \\ \varepsilon'(2) \\ \cdot \\ \cdot \\ \cdot \\ \varepsilon'(n) \end{pmatrix}.$$

The prior distribution for p, θ_p, and P are chosen as follows: First the marginal mass function of p is $\xi(p)$, $p = 1, 2, \ldots, L$, and given p, θ_p and P have a normal-Wishart distribution with density

$$\xi(\theta_p, P|p) = \xi_1(\theta_p|P,p)\, \xi_2(P), \qquad (8.61)$$

where

$$\xi_1(\theta_p|P,p) = (2\pi)^{-mp/2} |P|^{mp/2} |A_p|^{m/2}$$

$$\times \exp -\frac{\text{Tr}}{2} (\theta_p - \mu_p)' A_p (\theta_p - \mu_p) P$$

where μ_p is a known $mp \times 1$ vector and A_p is a known $mp \times mp$ matrix, and

… THE VECTOR AUTOREGRESSIVE PROCESS

$$\xi_2(P) \propto |P|^{(v-m-1)/2} \exp - \frac{\text{Tr}}{2} BP,$$

where B is a known PDS m × m matrix. Note, the conditional prior distribution of θ_p depends on p and P but the marginal distribution of P which is Wishart does not depend on p, but only on $v > m - 1$ and B.
Combining the likelihood function

$$L(\theta_p, P, p | X, Y) \propto |P|^{n/2} \exp - \frac{\text{Tr}}{2} (Y - X\theta_p)'(Y - X\theta_p) P \tag{8.62}$$

with the prior density (8.61) gives

$$\xi(\theta_p, p, P | X, Y) \propto \xi(p)(2\pi)^{-mp/2} |A_p|^{m/2} |P|^{(n+v+mp-m-1)/2}$$

$$\times \exp - \frac{1}{2} \text{Tr}[B + (\theta_p - \mu_p)' A_p (\theta - \mu_p)$$

$$+ (Y - X\theta_p)'(Y - X\theta_p)] P, \tag{8.63}$$

and we see the density is in the form of a normal-Wishart with respect to θ_p and P given p, that is given p, θ and P have a normal-Wishart distribution, which we knew from the introduction of this section. The marginal posterior mass function of p is found from (8.63) by completing the square in θ_p and integrating with respect to θ_p, then eliminating P using the properties of the Wishart distribution.

If this is done one may show the posterior mass function of p is

$$\xi(p|X,Y) \propto$$

$$\frac{\xi(p)|A_p|^{m/2}}{|A_p + X'X|^{m/2}|Y'Y + B - (X'Y + A_p\mu_p)'(A_p + X'X)(X'Y + A_p\mu_p) + \mu_p'A_p\mu_p|^{(n+v)/2}} \tag{8.64}$$

where p = 1, 2, ..., L.

By examining the posterior distribution of p, one may identify the order of the autoregressive process, but one must be careful in choosing the order of the model. If the mode of p occurs at p* with a posterior probability of .99, one would most likely choose p* as the order of the model but if the mass function of p is more or less uniform over the mass points 1, 2, ..., L, then the choice of the order is not so easy. Another problem with using the mass function (8.64) is that the hyperparameters μ_p and A_p, p = 1, 2, ..., L must be specified.

One can avoid the specification of the hyperparameters by using the prior density

$$\xi(p, \theta_p, P) \propto \xi(p) |P|^{-(m+1)/2}, \quad p = 1, 2, \ldots, L \qquad (8.65)$$

where $\xi(p)$ is the marginal prior mass function of p, θ is the mp × m matrix of regression matrices, and P is the precision matrix of the m variables of each observation. Now applying Bayes theorem and combining the prior density (8.65) with the likelihood function (8.62), the marginal posterior mass function of p is

$$\xi(p|X,Y) \propto \frac{\xi(p)(2\pi)^{mp/2}}{|X'X|^{m/2} |Y'Y - Y'X(X'X)^{-1}X'Y|^{(n-mp)/2}},$$

$$p = 1, 2 \ldots, L \qquad (8.66)$$

These Bayesian procedures for identification have been employed by Al Mahmeed (1982) to identify the order of univariate autoregressive processes, and he has compared them to several non-Bayesian methods of Akaike (1969, 1974, 1979).

This section on time series analysis provides only a short introduction to Bayesian inferences for multivariate processes and it would be nice if one could extend the analyses to vector moving average models and to vector autoregressive moving average models. It seems possible to give a complete Bayesian approach for univariate ARMA processes but until this is accomplished the posterior and predictive analysis of vector models will be delayed. Therefore instead of examining ARMA models, we will look at other models which are used for the analysis of time series.

OTHER TIME SERIES MODELS

This section will consider the regression model with autocorrelated error terms and multiple regression variables and the autoregressive

OTHER TIME SERIES MODELS

transfer function model, which is a special case of the vector autoregressive model.

The univariate (one regressor) regression model with autocorrelated errors is examined by Zellner (1971) and Chapter 5 of this book, where Zellner's development was reviewed and where certain extensions of the analysis were made. For example, Bayesian modal estimates of the parameters were constructed by adopting the method of Lindley and Smith (1972) and the same estimation technique will be employed here with the multiple linear regression model.

The transfer function autoregressive model is a special case of the vector autoregressive model and is also a special case of the transfer function model which was analyzed by Newbold (1973). Newbold's analysis developed the posterior distribution of the parameters of the general transfer function model, however because of the generality of the process approximations to the distributions had to be made.

Multiple Regression with Autocorrelated Errors

Consider a series of observations $Y(t)$, $t = 1, 2, \ldots$ and a corresponding series of k scalar nonstochastic regressor variables $X_1(t), X_2(t), \ldots, X_k(t)$, then the regression model with autocorrelated errors is

$$Y(t) = \sum_{j=1}^{k} \theta_j X_j(t) + \varepsilon(t), \quad t = 1, 2, \ldots, n \qquad (8.67)$$

where the θ_j are real unknown regression parameters and the $\varepsilon(t)$ satisfy a p-th order autoregressive process

$$\varepsilon(t) = \sum_{j=1}^{p} \rho_j \varepsilon(t-j) + e(t), \quad t = 1, 2, \ldots, n \qquad (8.68)$$

where the ρ_j are real unknown scalar parameters and $e(1), e(2), \ldots, e(n)$ are n.i.d. $(0, \tau)$ with $\tau > 0$ and unknown. If the ρ_j are chosen so that (8.68) is stationary, the regression model is said to be nonexplosive, otherwise it is explosive. Thus if $p = 1$ and $|\rho_1| < 1$, the model is nonexplosive. Only the first order model is considered, however the posterior and predictive analysis is easily accomplished for any p, $p = 1, 2, \ldots$.

One may represent this model in matrix notation (see Zellner, 1971, p. 93) as

$$Y = X\theta + \varepsilon$$

$$\varepsilon = \rho\varepsilon_1 + e \qquad (8.69)$$

where

$$\varepsilon' = [\varepsilon(1), \varepsilon(2), \ldots, \varepsilon(n)], \quad \varepsilon'_1 = [\varepsilon(0), \ldots, \varepsilon(n-1)],$$

$$\theta' = [\theta_1, \theta_2, \ldots, \theta_k], \quad Y' = [Y(1), Y(2), \ldots, Y(n)],$$

$$X = \begin{pmatrix} X_1(1), & X_2(1), & \ldots, & X_k(1) \\ X_1(2), & X_2(2), & \ldots, & X_k(2) \\ \vdots & \vdots & & \vdots \\ X_1(n), & X_2(n), & \ldots, & X_k(n) \end{pmatrix}.$$

If there are "initial" known values for the dependent and independent variables the model may be written

$$Y = \rho Y_{-1} + (X - \rho X_{-1})\theta + e,$$

where

$$X_{-1} = \begin{pmatrix} X_1(0), & \ldots, & X_k(0) \\ X_1(1), & \ldots, & X_k(1) \\ \vdots & & \vdots \\ X_1(n-1), & \ldots, & X_k(n-1) \end{pmatrix}$$

and

$$Y'_{-1} = [Y(0), \ldots, Y(n-1)].$$

The initial values of the k independent variables are given by the first row of X_{-1}, while the initial value of the dependent variable is the first component of Y_{-1}. There are other ways to handle the initial value problem and they are discussed by Zellner.

OTHER TIME SERIES MODELS 403

How does one choose a prior distribution for the parameters? For the univariate case Zellner uses an improper prior and in Chapter 5 a modified conjugate prior distribution was employed. Let us use an improper prior density

$$\xi(\theta, \tau, \rho) \propto \frac{1}{\tau}, \quad \theta \in R^p, \quad \tau > 0, \quad \rho > 0 \qquad (8.70)$$

then the joint posterior density of the parameters is

$$\xi(\theta,\rho,\tau|X,Y) \propto \tau^{n/2-1} \exp -\frac{\tau}{2} [(Y - \rho Y_{-1}) - (X - \rho X_{-1})\theta]'$$

$$\times [(Y - \rho Y_{-1}) - (X - \rho X)\theta], \qquad (8.71)$$

where $\rho \in R$, $\tau > 0$, and $\theta \in R^p$. It is seen that the conditional posterior distribution of τ given θ and ρ is gamma, the conditional distribution of θ given ρ is a multivariate t, and the conditional distribution of ρ given θ is a univariate t. Furthermore, it is seen the joint posterior distribution of ρ and θ is not a $(p + 1)$-dimensional multivariate t, and that the marginal distributions of θ and ρ may be isolated but they are not standard well-known distributions.

The joint density of θ and ρ is

$$\xi(\theta,\rho|X,Y) \propto \{[(Y - \rho Y_{-1}) - (X \rho X_{-1})\theta]'$$

$$\times [(Y - \rho Y_{-1}) - (X - \rho X_{-1})\theta]\}^{-n/2},$$

where $\theta \in R^p$ and $\rho \in R$, and we see that given ρ, θ has a multivariate t density and that given θ, ρ has a univariate t distribution, but that θ and ρ jointly do not have a multivariate t distribution.

The marginal density of ρ is obtained by first integrating (8.71) with respect to θ then τ and the result is

$$\xi(\rho|X,Y) \propto \xi_1(\rho,X)\, \xi_2(\rho,X,Y), \quad \rho \in R, \qquad (8.72)$$

where

$$\xi_1(\rho,X) = \frac{1}{|(X - \rho X_{-1})'(X - \rho X_{-1})|^{1/2}}$$

and

$$\xi_2(\rho,X,Y) =$$

$$\frac{1}{\{(Y - \rho Y_{-1})'(Y - \rho Y_{-1}) - (Y - \rho Y_{-1})'(X - \rho X_{-1})[(X - \rho X_{-1})'}$$

$$\times (X - \rho X_{-1})]^{-1}(X - \rho X_{-1})'(Y - \rho Y_{-1})\}^{(n-1)/2}.$$

This density is not of a standard form but is easily graphed since ρ is a scalar. Also the posterior moments of ρ are found by numerical integration.

In the same way, the marginal posterior density of θ can be derived, however since θ is p-dimensional it may be difficult to find the marginal moments of the components of θ. Another approach is to find the conditional distribution of θ given ρ, which is a multivariate t distribution with n degrees of freedom, location vector

$$E(\theta|X,Y,\rho) = [(X - \rho X_{-1})'(X - \rho X_{-1})]^{-1}(X - \rho X_{-1})'(Y - \rho Y_{-1}) \quad (8.73)$$

and precision matrix

$$P(\theta|X,Y,\rho) = \frac{n(X - \rho X_{-1})'(X - \rho X_{-1})}{S(\rho,X,Y)}, \quad (8.74)$$

where

$$S(\rho,X,Y) = (Y - \rho Y_{-1})'[I_n - (X - \rho X_{-1})[(X - \rho X_{-1})'$$

$$\times (X - \rho X_{-1})]^{-1}(X - \rho X_{-1})'](Y - \rho Y_{-1}).$$

The marginal mean vector of θ is

$$E(\theta|X,Y) = \underset{\rho}{E}\, E(\theta|X,Y,\rho), \quad (8.75)$$

where $\underset{\rho}{E}$ denotes expectation with respect to ρ.

The marginal posterior mean of θ is computed by the numerical integration of $\xi(\rho|X,Y)\, E(\theta|X,Y,\rho)$ with respect to ρ, thus only p one-dimensional integrations are involved in order to determine this moment. If one first found the marginal density of θ, the means of

OTHER TIME SERIES MODELS 405

each of the p components of θ would involve $(p - 1)$ dimensional integrations, thus if $p \geq 3$, one has more computing efficiency from formula (8.75). In the same way one may find the dispersion matrix of θ by averaging the conditional dispersion matrix $(n - 2)^{-1} P^{-1}(\theta|X,Y,\rho)$ of θ with respect to the marginal density of ρ.

To make inferences for θ and ρ, one may base them on the approximate marginal posterior distribution, but how does one estimate and test hypotheses about the precision τ? It can be shown that the marginal density of τ cannot be obtained in closed form, however it can be shown the conditional density of τ given ρ is

$$\xi(\tau|X,Y,\rho) \propto \tau^{(n-p)/2 - 1} \exp - \frac{\tau}{2} S(\rho,X,Y), \quad \tau > 0 \quad (8.76)$$

where $S(\rho,X,Y)$ is given by (8.74), that is the conditional distribution of τ given ρ is a gamma with parameters $\alpha = (n - p)/2$ and $\beta = S(\rho,X,Y)/2$. The conditional posterior mean of τ^{-1} given ρ is

$$E(\tau^{-1}|\rho,X,Y) = S(\rho,X,Y)(n - p - 2)^{-1}. \quad (8.77)$$

One may estimate τ either from the marginal distribution of τ or from the conditional distribution of τ given ρ by conditioning on an estimate of ρ, say $E(\rho|X,Y)$, which is computed via (8.72), or one may find the marginal posterior mean of τ^{-1} by averaging $E(\tau^{-1}|\rho,X,Y)$ with respect to the marginal density of ρ, that is

$$E(\tau^{-1}|X,Y) = \int_R E(\tau^{-1}|X,Y,\rho) \xi(\rho|X,Y) d\rho$$

may be found numerically. Of course in the same way one may compute the precision or variance of the marginal posterior distribution of τ.

An alternative method of estimation of the parameters θ, τ, and ρ of the multiple regression model with autocorrelated errors is to determine the mode of the joint posterior density (8.71), as was done in the univariate case of the model in Chapter 5. Lindley and Smith (1972) show the joint mode of the distribution may be found by solving simultaneously the modal equations, which are found from the conditional distributions of θ given ρ and τ, ρ given θ and τ, and τ given θ and ρ. From an examination of the joint density (8.71), it can be shown that the conditional posterior density of τ given θ and ρ is gamma with parameters $\alpha = n/2$ and β, where $2\beta = [(Y - \rho Y_{-1}) - (X - \rho X_{-1})\theta]'$ $\times [(Y - \rho Y_{-1}) - (X - \rho X_{-1})\theta]$, that the conditional posterior distribution of θ given ρ and τ is normal with mode

$$E(\theta|\rho,\tau) = [(X - \rho X_{-1})'(X - \rho X_{-1})]^{-1}(X - \rho X_{-1})'(Y - \rho Y_{-1}),$$

(8.78)

and that ρ given θ and τ is also normal with mode

$$E(\theta|\rho,\tau) = [(Y_{-1} - X_{-1}\theta)'(Y_{-1} - X_{-1}\theta)]^{-1}(Y_{-1} - X_{-1}\theta)'(Y - X\theta).$$

(8.79)

The mode of conditional distribution of τ given θ and ρ is

$$m(\tau|\theta,\rho) = \frac{n-2}{\beta}.$$

(8.80)

The three equations

$$\theta = E(\theta|\rho,\tau),$$
$$\rho = E(\rho|\theta,\tau),$$
$$\tau = m(\tau|\theta,\rho)$$

(8.81)

are solved simultaneously by an iterative scheme of beginning with initial guesses for ρ, which allows one to solve for θ and τ and the cycle is repeated until the solutions stabilize. Of course, convergence is not guaranteed and if convergence occurs the solution perhaps is the mode of the joint density. Obviously this method of estimation requires investigation and should be used with caution.

Thus, there are two ways to estimate the parameters of the joint density, the first is to find the means of the marginal distributions along with their variances and the second is with joint mode. The first way will give the marginal Bayesian estimators of the parameters with respect to a squared error loss function and the second way will give additional information about the joint posterior distribution.

One may easily repeat the above posterior analysis when one uses a proper prior density for the parameters. For example consider

$$\xi(\theta,\rho,\tau) = \xi_1(\tau)\xi_2(\theta,\rho|\tau), \quad \tau > 0, \quad \theta \in R^p, \quad \rho \in R, \quad (8.82)$$

where

$$\xi_1(\tau) \propto \tau^{a-1} e^{-\tau b}, \quad \tau > 0,$$

$$\xi_2(\theta,\rho|\tau) \propto \tau^{1/2} \exp{-\frac{\tau}{2}[(Y^* - \rho Y^*_{-1}) - (X^* - \rho X^*_{-1})\theta]'}$$

$$\times [(Y^* - \rho Y^*_{-1}) - (X^* - \rho X^*_{-1})\theta], \quad \theta \in R^p, \quad \rho \in R,$$

where Y^*, Y^*_{-1}, X^*, and X^*_{-1} are $n \times 1$, $n \times 1$, $n \times p$, and $n \times p$ matrix hyperparameters. Indeed these quantities may be chosen from an hypothetical future "experiment" (or with past data), where

$$Y^* = X^*\theta + \varepsilon^*. \tag{8.83}$$

If this is done the posterior and predictive distributions are of the same type that were derived with the improper prior density (8.70). The predictive density of a future observation was not found but was obtained for the simple linear regression model of Chapter 5.

A Transfer Function Model

A transfer function model relates two time series $\{Y(t): t = 1, 2, \ldots\}$ and $\{X(t): t = 1, 2, \ldots\}$ in such a way that

$$Y(t) = \sum_{i=1}^{p} \theta_i Y(t-i) + \sum_{i=1}^{q} \phi_i X(t-i) + \varepsilon(t)$$

$$X(t) = \sum_{j=1}^{r} \eta_j X(t-j) + \varepsilon(t), \tag{8.84}$$

where $t = 1, 2, \ldots, n$. The first equation is an autoregressive model of the Y variable plus a linear function of the values of another variable X, and the second equation is an autoregressive process for the X variable. This model can also be thought of as an autoregressive process with a concomitant stochastic time variable X, which satisfies an autoregressive process. This transfer function model is a special case of one studied by Newbold (1973), who gave the complete Bayesian posterior analysis. It is assumed that $\theta = (\theta_1, \theta_2, \ldots, \theta_p)' \in R^p$, $\phi = (\phi_1, \phi_2, \ldots, \phi_q) \in R^q$, and $\eta = (\eta_1, \eta_2, \ldots, \eta_r) \in R^r$ are unknown vectors, that $\varepsilon(1), \ldots, \varepsilon(n)$ are n.i.d. $(0, \tau_1^{-1})$, that $e(1), e(2), \ldots, e(n)$ are n.i.d. $(0, \tau_2^{-1})$, $\tau_1 > 0$, $\tau_2 > 0$, that the last two sequences are independent, that $Y(0), Y(1), \ldots, Y(1-p)$; $X(0), \ldots, X[\max(r,q)]$ are known scalars and that τ_1 and τ_2 are unknown.

Under these conditions, how does one estimate the unknown parameters? How does one perform a Bayesian analysis of such a model? It is obvious that if the X values are fixed, the model can be analyzed as one would analyze any multivariate linear model. The density of $Y(1), Y(2), \ldots, Y(n)$ given θ, ϕ, τ_1, and $X(1), X(2), \ldots, X(n)$ is

$$g_1[Y|\theta,\phi,\tau_1,X] \propto \tau_1^{n/2} \exp - \frac{\tau_1}{2} \sum_{t=1}^{n} [\theta(B)Y(t) - \phi(B)X(t)]^2,$$

$$Y \in R^n \quad (8.85)$$

where

$$Y = [Y(1), Y(2), \ldots, Y(n)]',$$

$$\theta(B) = 1 - \theta_1 B - \theta_2 B^2 - \ldots - \theta_p B^p,$$

$$\phi(B) = 1 - \phi_1 B - \phi_2 B^2 - \ldots - \phi_q B^q,$$

$$B^s Y(t) = Y(t-s),$$

and the density of $X = [X(1), X(2), \ldots, X(n)]'$ is

$$g_2[X|\eta,\tau_2] \propto \tau_2^{n/2} \exp - \frac{\tau_2}{2} \sum_{t=1}^{n} [\eta(B)X(t)]^2, \quad X \in R^n,$$

$$(8.86)$$

where

$$\eta(B) = 1 - \eta_1 B - \eta_2 B^2 - \ldots - \eta_r B^r.$$

It is obvious that if one assumes $\xi(\eta,\tau_2) \propto 1/\tau_2$, $\eta \in R^r$, $\tau_2 > 0$ is the prior density of η and τ_2, that the marginal posterior density of η and τ_2 is a normal-gamma, because (8.86) is the density of an AR(r) process. The marginal posterior density of η is an r-th dimensional multivariate t and the marginal posterior distribution of τ_2 is a gamma.

OTHER TIME SERIES MODELS   409

Given X and assuming (η, τ_2) are independent, a priori, of (θ, ϕ, τ_1), where the marginal prior density of the latter is $\xi(\theta, \phi, \tau_1) \propto 1/\tau_1$, $\theta \in R^p$, $\phi \in R^q$, $\tau_1 > 0$ it can be shown that the marginal posterior density of these parameters is also a normal-gamma, that is the marginal posterior distribution of τ_1 is gamma and that of θ and ϕ is a $(p + q)$-dimensional t distribution.

The joint posterior density of θ, ϕ, and η is

$$\xi(\theta, \phi, \eta | X, Y) = \xi_1(\theta, \phi | X, Y) \xi_2(\eta | X), \tag{8.87}$$

where

$$\xi_1(\theta, \phi | X, Y) \propto \left\{ \sum_{t=1}^{n} [\theta(B)Y(t) - \phi(B)X(t)]^2 \right\}^{-n/2},$$

$$\theta \in R^p, \quad \phi \in R^q, \tag{8.88}$$

and

$$\xi_2(\eta | X) \propto \left\{ \sum_{t=1}^{n} [\eta(B)X(t)]^2 \right\}^{-n/2}, \qquad \eta \in R^r. \tag{8.89}$$

It is interesting to observe that one may estimate θ and ϕ from (8.88) only, that η may be estimated from (8.89) only, and that θ, ϕ, and η jointly do not have a multivariate t distribution. Thus whether X is "fixed" or stochastic, one's inferences for θ and ϕ would be the same. This is because there was no cross correlation between the two series. Since the $\varepsilon(t)$ and $e(t)$, $t = 1, 2, \ldots, n$ were independent, the fact that the $X(t)$ were values from an AR(r) process did not add any information to the posterior distribution of θ and ϕ. If $U(t) = [\varepsilon(t), e(t)]'$ had had a bivariate normal distribution with nonzero correlation $\rho = \rho[\varepsilon(t), e(t)]$ and if $U(1), U(2), \ldots, U(n)$ were independent, the above analysis would not hold. What would be the proper analysis? See Granger and Newbold (1977) and Newbold (1973) and the section above (pp. 389-401) on vector autoregressive models.

One interesting case of this type of transfer function model (8.94) is the distributed lag model which is used in economics (see Zellner, 1971, p. 200). Consider

$$Y(t) = \alpha \sum_{i=0}^{\infty} \lambda^i X(t - i) + \sum_{i=0}^{\infty} \lambda^i U(t - i), \tag{8.90}$$

where Y(t) is the random response at time t, U(t) the t-th error term, and X(t) a stimulus at time t. It is assumed $0 \leq \lambda < 1$, $\alpha \in R$, and U(t)'s are independent normal random variables with mean zero and unknown precision $\tau > 0$. The stimulus variable influences the main response but its influence decreases geometrically with time. This model is just one of many versions of the distributed lag model and Zellner gives a full account of the subject. It can be seen this version of the model is equivalent to

$$Y(t) = \lambda Y(t-1) + \alpha X(t) + U(t), \quad (8.91)$$

where Y(0) is an "initial" value of the main response. We see the model is an autoregressive AR(1) process with a concomitant variable and is a special case of the transfer function model, with a nonstochastic regressor.

If we use an improper prior density

$$\xi(\alpha, \lambda, \tau) \propto \frac{1}{\tau}, \quad \alpha \in R, \ 0 \leq \lambda < 1, \ \tau > 0$$

the joint posterior density for λ and α is

$$\xi(\lambda, \alpha | X, Y) \propto [(Y - \lambda Y_{-1} - \alpha X)'(Y - \lambda Y_{-1} - \alpha X)]^{-n/2},$$

$$(8.92)$$

therefore since $0 \leq \lambda < 1$, the joint posterior distribution of λ and α is not a bivariate t, however one may determine the marginal posterior density of λ and the conditional density of α given λ (which is a univariate t) and compute the marginal mean of α. Also the conditional distribution of τ given λ is gamma and the posterior marginal mean of τ may be computed by averaging the conditional mean of τ given λ over the marginal distribution of λ.

The above should give the reader a good introduction to the Bayesian analysis of regression and design models and vector autoregression models. More information about these and other models, especially those used in econometrics can be found in Zellner (1971), where the simultaneous equations process is given a full Bayesian treatment. Also examined by Zellner are distributed lag and errors in variables models. For more information about advanced topics in multivariate analyses the reader should consult Press (1982).

MULTIVARIATE MODELS WITH STRUCTURAL CHANGE

Chapter 7 of this book was about the Bayesian approach to linear models which have parameters which change over the period of observation. The models considered were univariate regression and time series models, and now we will consider structural change in the multivariate regression model. Structural change problems in the multivariate linear model are more complex because the models contain more parameters and there are more possibilities for change.

Very little has appeared in the literature about change in the multivariate normal linear model, however Sen and Srivastava (1973, 1975a, 1975b, 1975c) developed tests for detecting change and examined the exact and asymptotic properties of the test statistics. Salazar (1980, 1982) considered changes in the multivariate linear model using improper and proper prior densities and a change point parameter. Moen (1982) developed a detailed analysis of the multivariate linear model and his work will be reported here. Tsurumi (1978, 1980) has been devising ways to estimate structural change in the simultaneous equations model of economics and has been using transition functions to model the change in the relationships of economic variables.

Some of the recent trends in the Bayesian analysis of structural change can be found in Broemeling (1982). The approach taken here is to model a change in the regression matrix of the multivariate linear model by a change point which indexes the observation at which a change will occur. A proper prior density for the parameters will be used.

Consider the multivariate linear model and suppose there is a change in the matrix of regression parameters at the m-th observation where $1 \leq m \leq n - 1$, then the model may be written as

$$Y_1 = X_1 \theta_1 + E_1$$
$$Y_2 = X_2 \theta_2 + E_2, \qquad (8.93)$$

where θ_1 and θ_2 are distinct $k \times p$ real matrices, Y_1 is a $m \times p$ matrix of m observations on p variables, X_1 is a $m \times k$ matrix of m observations on k independent variables, and E_1 is a $m \times p$ matrix such that the rows of E_1 are independent normal random p vectors with a zero mean vector and an unknown $p \times p$ precision matrix P. Y_2 is an $(n - m) \times p$ observation matrix, X_2 is a $(n - m) \times k$ known design matrix, θ_2 a $k \times p$ regression matrix, and E_2 a $(n - m) \times p$ error matrix with independent rows, each with a $N_p(0, P^{-1})$ distribution.

The matrix of regression parameters θ_i is $\theta_i = (\theta_{i1}, \theta_{i2}, \ldots, \theta_{ip})$, where $i = 1, 2$ and θ_{i1} is the first column of θ_i. Let $\theta_i^* = (\theta_{i1}', \theta_{i2}', \ldots, \theta_{ip}')$, thus θ_i^* is a $pk \times 1$ vector where the columns of θ_i are "stacked."

Our main interest will focus on the shift point parameter m, however the other parameters, namely θ_1, θ_2, and P will also be studied and we will find their posterior distributions where the regression matrix changes from θ_1 to θ_2. All one knows is that at least one of the elements of the matrix θ_1 changes.

Given m, $1 \leq m \leq n - 1$, θ_1, θ_2, and P the density of the n observations is

$$g(Y|\theta_1, \theta_2, P, m) \propto |P|^{n/2} \exp - \frac{Tr}{2}[(Y_1 - X_1\theta_1)'(Y_1 - X_1\theta_1)$$
$$+ (Y_2 - X_2\theta_2)'(Y_2 - X_2\theta_2)]P \quad (8.94)$$

where $Y = (Y_1' | Y_2')'$.

Note if m = 1, the first p-variate observation has regression matrix θ_1 and the remaining n - 1 have regression matrix θ_2. On the other hand if m = n - 1 only the n-th (or last) observation has regression matrix θ_2, while the first n - 1 observations have regression matrix θ_1.

We will assume a priori that a change is possible at any one of the n - 1 observations and that $\xi(m)$ is the prior mass function of m. m = 1, 2, ..., n - 1. Also, that m is independent of θ_1, θ_2, and P and that given P, θ_1 and θ_2 are independent with densities

$$\xi(\theta_i|P) \propto |P|^{k/2} \exp - \frac{Tr}{2}(\theta_i - \mu_i)'A_i(\theta_i - \mu_i)P \quad (8.95)$$

and that P has marginal prior density

$$\xi(P) \propto |P|^{(\alpha-p-1)/2} \exp - \frac{Tr}{2} BP, \quad (8.96)$$

where B is p × p and positive definite, μ_i is a k × p real matrix, and A_i is k × k positive definite, where i = 1, 2.

Thus, a priori, θ_1 and θ_2 given P have normal distributions such that θ_i^* has dispersion matrix $(P \otimes A_i)^{-1} = P^{-1} \otimes A_i^{-1}$ and by choosing A_i as diagonal the rows of θ_i are conditionally independent. This prior distribution on the parameters implies the marginal prior density of θ is

$$\xi(\theta_1, \theta_2) \propto \frac{1}{\left| \sum_{i=1}^{2} (\theta_i - \mu_i)'A_i(\theta_i - \mu_i) + B \right|^{(k+\alpha)/2}}, \quad (8.97)$$

which is a matrix t density and the dispersion matrix of

$$\theta^* = \begin{pmatrix} \theta_1^* \\ \theta_2^* \end{pmatrix}$$

is

$$D(\theta^*) = (\alpha - p - 1)^{-1} \begin{pmatrix} B \otimes A_1^{-1} & \phi \\ \phi & B \otimes A_2^{-1} \end{pmatrix},$$

where ϕ denotes a matrix of zeros.

Note θ_1 and θ_2 are uncorrelated but not independent and that the rows of θ can be made to be uncorrelated by choosing A_i, $i = 1, 2$ as diagonal.

The posterior analysis for the model will be done in the following order: First the marginal posterior mass function of m will be derived, then the marginal posterior density of θ, and finally the marginal posterior density of P.

The marginal posterior mass function of m is found by integrating the joint posterior density

$$\xi(\theta, P, m | Y) \propto |P|^{(n+2k+\alpha-p-1)/2} \exp - \frac{\text{Tr}}{2} \left\{ \sum_{i=1}^{i=2} [(Y_i - X_i\theta_i)' \times (Y_i - X_i\theta_i) + (\theta_i - \mu_i)'A_i(\theta_i - \mu_i)] + B \right\} P$$
(8.98)

with respect to θ_1, θ_2, and P.

First complete the square in θ_1 and θ_2 then integrate with respect to these parameters, then

$$\xi(\mathcal{P}, m | Y) \propto |\mathcal{P}|^{(n+\alpha-p-1)/2} |D_1|^{-P/2} |D_2|^{-P/2} \exp - \frac{\text{Tr}}{2}$$

$$\times \left\{ B + \sum_{i=1}^{2} [Y_i' Y_i + \mu_i' A_i \mu_i - (X_i' Y_i + A_i \mu_i)' \right.$$

$$\left. \times (X_i' X_i + A_i)^{-1} (X_i' Y_i + A_i \mu)] \right\} \mathcal{P} \qquad (8.99)$$

is the joint posterior density of $\mathcal{P}$ and m and

$$\xi(m|Y) \propto |D_1 D_2|^{-P/2} |B + Y'Y + \mu' A \mu - (X'Y + A\mu)'(X'X + A)^{-1}$$

$$\times (X'Y + A\mu)|^{-(n+\alpha)/2}, \qquad (8.100)$$

where

$$m = 1, 2, \ldots, n - 1,$$

$$D_i = X_i' X_i + A_i,$$

$$A = \begin{pmatrix} A_1 & \phi \\ \phi & A_2 \end{pmatrix},$$

$$\mu = (\mu_1', \mu_2')',$$

$$X = \begin{pmatrix} X_1 & \phi \\ \phi & X_2 \end{pmatrix},$$

and

$$Y = (Y_1', Y_2').$$

Now $\xi(m|Y)$ is the marginal posterior mass function of m and it is the function from which inferences about the change are to be made. One simply plots $\xi(m|Y)$ versus m and upon inspection of the graph decides the location of the change point.

MULTIVARIATE MODELS WITH STRUCTURAL CHANGE 415

Knowing where the change point is likely to be and knowing what parameter or parameters account for the change are two different aspects of structural change. We know a priori that the regression parameters change and we know their prior means, covariances and variances. Now we must check to see a posteriori which ones account for the change. In the previous chapter, we saw an example of this with a univariate linear model. One needs to compute the posterior density of θ_1 and θ_2 and see what parameters are changing.

The marginal posterior density of θ_1 and θ_2 is derived from (8.98) and is

$$\xi(\theta|Y) \propto \sum_{m=1}^{n-1} \xi(m|Y) \xi(\theta|m,Y), \qquad (8.101)$$

where $\xi(m|Y)$ is the marginal posterior mass function (8.100) and $\xi(\theta|m,Y)$ is the conditional posterior density of θ_1 and θ_2 given m and is

$$\xi(\theta|m,Y) \propto \frac{1}{|(\theta - \tilde{\theta})'(X'X + A)(\theta - \tilde{\theta}) + S(m,X,Y)|^{(n+2k+\alpha)/2}}, \qquad (8.102)$$

where

$$S(m,X,Y) = Y'Y - (X'Y + A\mu)'(X'X + A)^{-1}(X'Y + A\mu) + \mu'A\mu + B.$$

This is a matrix t density $t_{2kp}[\tilde{\theta}, (X'X + A), S(m,X,Y), d]$, where $d = n + \alpha - p + 1$ and $E(\theta|m,Y) = \tilde{\theta}$, where $\tilde{\theta} = (X'X + A)^{-1} \cdot (X'Y + A\mu)$ and the conditional dispersion matrix of θ^* given m is

$$D(\theta^*|m,Y) \propto (d - 2)^{-1} S(m,X,Y) \otimes (X'X + A)^{-1},$$

where θ^* is the $2pk \times 1$ vector of the columns of θ_1^* and θ_2^*.

We see the marginal posterior density of $\theta = (\theta_1', \theta_2')'$ is a mixture of matrix t densities and the mixing distribution is the marginal mass function of the change point, thus the posterior analysis of the changing linear multivariate model is quite similar to the posterior analysis of the changing linear univariate model of the previous chapter.

To investigate what parameters account for the change, we must compare the elements of the θ_1 matrix with the corresponding elements of θ_2, thus $k \times p$ matrix $\gamma = \theta_1 - \theta_2$ is the parameter of primary interest.

We know θ_1 and θ_2 jointly have a distribution with density (8.101), which is a mixture of matrix t densities $\xi(\theta|m,Y)$, $m = 1, 2, \ldots, n-1$; what is the density of γ?

Using formula (8.4.44) from Box and Tiao (1973), one may show the distribution of

$$\gamma = P\theta, \qquad (8.103)$$

where

$$P = [I_k, -I_k]$$

is also a mixture of matrix t densities, namely

$$\xi(\gamma|Y) \propto \sum_{m=1}^{n-1} \xi(m|Y)\, \xi(\gamma|m,Y), \qquad (8.104)$$

where $\xi(\gamma|m,Y)$ is a matrix t density and is the conditional density of γ given m and given by

$$\xi(\gamma|m,Y) \propto |(\gamma - \hat{\gamma})'[P(X'X + A)^{-1}P']^{-1}(\gamma - \hat{\gamma})$$

$$+ S(m,X,Y)|^{-(n+2k+\alpha)/2}, \qquad (8.105)$$

where $S(m,X,Y)$ is given by (8.102), $\hat{\gamma} = P\tilde{\theta}$, $\tilde{\theta} = (X'X + A)^{-1}(X'Y + A\mu)$, P is given by (8.103), i.e.,

$$\gamma \sim t_{(2k)p}[P\tilde{\theta},\ P(X'X + A)^{-1}P',\ S(m,X,Y),\ d],$$

where $d = n + \alpha - p + 1$.

Now how should one use (8.104) in order to make inferences for γ? Suppose one wants to see if the first column of the regression matrix θ_1 has changed, then one would see if the first column of γ was different than the zero vector and this can be done by examining the posterior distribution of the first column of γ, say γ_1, where

$$\gamma = (\gamma_1, \gamma_2, \ldots, \gamma_p)$$

and γ_i is the i-th column (k × 1) of γ.

Since $\gamma_1 = \gamma Q$, where Q is $p \times 1$ and $Q = (1,0,0, \ldots, 0)'$, it can be shown, again from Box and Tiao (1973, p. 447), that γ_1 has density

$$\xi(\gamma_1|Y) \propto \sum_{m=1}^{n-1} \xi(m|Y) \xi(\gamma_1|m,Y), \qquad (8.106)$$

where

$$\xi(\gamma_1|m,Y) \propto$$

$$\frac{1}{[(\gamma_1 - \hat{\gamma}_1)'[P(X'X + A)^{-1}P']^{-1}(\gamma_1 - \hat{\gamma}_1) + Q'S(m,X,Y)Q]^{(n+2k+\alpha)/2}} \qquad (8.107)$$

Now $\hat{\gamma}_1 = \hat{\gamma}Q$ and P is given by (8.103) and we see the marginal posterior distribution of γ_1 is a mixture of k dimensional t densities. Unfortunately, it is quite difficult to find an HPD region for γ_1, since its posterior distribution is a mixture, however one may test $H_0: \gamma_1 = 0 (k \times 1)$ by a conditional test from the conditional posterior distribution of γ_1 given $m = m^*$, where m^* is an estimate of m obtained from the marginal posterior distribution of m, given by (8.100).

We know the conditional distribution of γ_1 has density (8.107) and represents a t distribution with $n + k + \alpha$ degrees of freedom, location vector $\hat{\gamma}_1$, and precision matrix $(n + k + \alpha)[P(X'X + A)^{-1}P']^{-1} \times [Q'S(m,X,Y)Q]^{-1}$, thus

$$F(\gamma_1|m) = k^{-1}(\gamma_1 - \hat{\gamma}_1)'[P(X'X + A)^{-1}P']^{-1}(\gamma_1 - \hat{\gamma}_1) \qquad (8.108)$$

has a conditional distribution (given m) which is an F with k and $n + k + \alpha$ degrees of freedom and one would reject $H_0: \gamma_1 = 0$ if

$$F(0|m) > F(\beta;k, n + k + \alpha) \qquad (8.109)$$

with s significance level of β, where $F(\beta;k,n + k + \alpha)$ is the upper $100\beta\%$ point of the F distribution with k and $n + k + \alpha$ degrees of freedom. One should use the conditional test with caution because the appropriate test should be based on the marginal posterior density of γ_1 not its conditional counterpart. If the conditioning value of m, m* has a "large" posterior probability, then the conditional distribution of γ_1 given m = m* is a "good" approximation to the marginal distribution of m, in which case one may use the conditional F test (8.109).

If γ_1 is scalar, one may easily find the HPD region for γ_1 by numerical integration of the marginal distribution of γ_1 and this was done in an example of the previous chapter, but if γ_1 is not scalar it is quite difficult to numerically integrate the mixture of t densities (8.106).

Up to this point we know how to make inferences for the shift point parameter m and the matrices of regression parameters, and the only other parameters of interest are the elements of the precision matrix P, which give the correlations between the p responses, thus we need the marginal posterior distribution of P.

The joint posterior density of m and P is given by (8.99), thus the marginal density of P is

$$\xi(P|Y) = \sum_{m=1}^{n-1} \xi(m|Y) \, \xi(P|m,Y), \qquad (8.110)$$

where $\xi(m|Y)$ is the marginal posterior mass function of m and $\xi(P|m,Y)$ is the conditional posterior density function of P and is also given by (8.99) which is a Wishart distribution with $(n + \alpha)$ degrees of freedom and precision matrix

$$\hat{P}(P|m,Y) = B + Y'Y + \mu'A\mu - (X'Y + A\mu)'(X'X + A)^{-1}(X'Y + A\mu).$$

$$(8.111)$$

A point estimate of P is given by the posterior mean of P which is

$$E(P|Y) = \sum_{m=1}^{n-1} \xi(m|Y)(n + \alpha)\hat{P}^{-1}(P|m,Y), \qquad (8.112)$$

however HPD regions for P are difficult to obtain because its distribution is a mixture. Conditional HPD regions for the components of P are more easily determined from the conditional posterior distribution of P given m = m*, where m* is say the posterior mode of m. For example, one perhaps is interested in testing for zero correlation between the p components of the dependent variable.

Some questions about the posterior analysis have been left unanswered but for the most part, if one is interested in estimating or testing hypotheses about the parameters, one would be able to do these things from the information given in this section.

Often one wants to forecast future observations $W = (W_1, W_2, \ldots, W_\ell)'$ where W is a $\ell \times p$ matrix, where

$$W = V\theta_2 + E^*$$

with V a $\ell \times k$ known matrix, θ_2 the regression matrix of the changing model (8.93), and E^* an $\ell \times p$ matrix with rows which are independent and distributed as $N(0, P^{-1})$. Note V is a known matrix of ℓ observations on the k independent variables and E^* is independent of E_1 and E_2.

Changing multivariate models should be very useful in forecasting future economic activity because the models used now do not include the possibility of changing parameters, therefore it seems if the models are revised to include changing parameters the forecasts should improve.

There are many ways to represent parameters which change in a linear process other than the way we did here in this section, which was to change the regression matrix only once during the observation period. Another way is given by Chapter 6 with the linear dynamic model which has parameters which always change.

The Bayesian predictive density will be used to forecast the future observations W and is determined by finding the conditional density of W given the past observations Y.

It can be shown that the future distribution of W is a mixture of matrix t distributions and that the mixing distribution is the posterior mass function of m, the shift point.

This section is concluded with a numerical study of the robustness of the distribution of m to changes in the values of the hyperparameters μ, A, and α of the prior distribution.

In this study, the natural conjugate prior distribution of the rows of θ_i, $i = 1, 2$, namely θ'_{ij}, $j = 1, 2, \ldots, k$, given P is a multivariate normal with mean vector $\mu'_{ij} \in R^p$ and precision matrix $r_{ij}P$, $r_{ij} > 0$, and the marginal distribution of P is Wishart with α degrees of freedom and precision matrix B. One may show the posterior distribution of the change point is

$$\xi(m|Y) \propto |D_1 D_2|^{-P/2} Q(m)^{-(n+\alpha)/2}, \quad 1 \leq m \leq n - 1 \quad (8.113)$$

$$= 0 \text{ otherwise},$$

where

$$D_i = X_i'X_i + R_i, \quad i = 1, 2$$

$$R_i = \text{Diagonal}(r_{ij}), \quad i = 1, 2$$

$$Q(m) = B + \sum_{i=1}^{2} [Y_i'Y_i + \mu_i'R_i\mu_i - (X_i'Y_i + R_i\mu_i)'D_i^{-1}(X_i'Y_i + R_i\mu_i)],$$

and

$$\mu_i = (\mu_{i1}, \mu_{i2}, \ldots, \mu_{ik})', \quad i = 1, 2.$$

For the numerical study the following choices were made for the parameters of the model:

$$\theta_1 = \begin{pmatrix} 1 & 3 \\ 2 & 4 \end{pmatrix}, \quad \theta_2 = \theta_1 + \begin{pmatrix} \Delta_2 & 0 \\ \Delta_2 & 0 \end{pmatrix},$$

$\Delta_2 = .04, .05, .06,$ and $.07$,

$$\mu_1 = \begin{pmatrix} 1 & 3 \\ 2 & 4 \end{pmatrix}, \quad \mu_2 = \mu_1 + \begin{pmatrix} \Delta_1 & 0 \\ \Delta_1 & 0 \end{pmatrix},$$

$\Delta_1 = 0, .2, .4,$ and $.6$, and

$$B = \begin{pmatrix} 2 & 2\rho \\ 2\rho & 2 \end{pmatrix},$$

where $\rho = 0, \pm.2, \pm.5,$ and $\pm.7$.

The bivariate observations were generated with the 2×2 precision matrix P, where

$$P^{-1} = \begin{pmatrix} \sigma_1^2 & \rho\sigma_1\sigma_2 \\ \rho\sigma_1\sigma_2 & \sigma_2^2 \end{pmatrix},$$

and $\sigma_1^2 = \sigma_2^2 = 1$. The example is a bivariate regression model with two independent variables and the first m bivariate observations are

$$Y_{i1} = \theta_{11}X_{i1} + \theta_{21}X_{i2} + e_{i1}$$

$$Y_{i2} = \theta_{12}X_{i1} + \theta_{22}X_{i2} + e_{i2}$$
(8.114)

for $i = 1, 2, \ldots, m$, $1 \leqslant m \leqslant n - 1$. The regression parameter matrix is

$$\theta_1 = \begin{pmatrix} \theta_{11} & \theta_{12} \\ \theta_{21} & \theta_{21} \end{pmatrix}$$

and since

$$\theta_2 = \theta_1 + \begin{pmatrix} \Delta_2 & 0 \\ \Delta_2 & 0 \end{pmatrix},$$

only the first column of θ_1 changes, thus only the average value of the first component of (Y_{i1}, Y_{i2}) changes and not the average value of Y_{i2}. The n × 2 design matrix consisted of ones for the first column and the values for the second column were two-digit numbers selected at random from a random number table. Now n = 20 bivariate observations were generated and three different cases were considered for the actual change, namely m = 3, 10, and 17.

Moen (1982), using an IMSL (International Mathematical and Statistical Library) subroutine, generated the bivariate normal error term for a specified ρ, and SAS (Statistical Analysis System) programs were written for finding the posterior distribution of m, (8.113).

Tables 8.1-8.3 present the results for a sample of size n = 20 when using a natural conjugate prior when the actual point of change is at 3, 10, and 17.

For fixed values of ρ and Δ_2, changes in Δ_1 do not have much of an effect on the posterior probability of the actual point of change; however in all cases, as Δ_2 increases for fixed values of ρ and Δ_1, the posterior probability also increases. It is most often the case that for fixed Δ_1 and Δ_2, the posterior probability of the true change point

Table 8.1
Posterior Probability That m = 3, When the Actual Point of Change Is Three, Using a Natural Conjugate Prior Distribution;

$$n = 20, \quad \mu_2 = \mu_1 + \begin{pmatrix} \Delta_1 & 0 \\ \Delta_1 & 0 \end{pmatrix}, \quad \beta_2 = \beta_1 + \begin{pmatrix} \Delta_2 & 0 \\ \Delta_2 & 0 \end{pmatrix}$$

Δ_1	Δ_2	ρ					
		-.7	-.5	-.2	0	.2	.5	.7
0.	.04	.60709	.18253*	.07398*	.07374*	.10962*	.41512*	.91379
	.05	.91138	.52496	.24787*	.25171*	.37789*	.85267	.99335
	.06	.98507	.83394	.58358	.60240	.76023	.97817	.99941
	.07	.99764	.95520	.84746	.86660	.93980	.99669	.99993
.2	.04	.61913	.18900*	.07608*	.07516*	.11055*	.41126*	.90975
	.05	.91615	.53609	.25404*	.25632*	.38127*	.85168	.99311
	.06	.98612	.84069	.59226	.60917	.76384	.97820	.99939
	.07	.99784	.95760	.85253	.87036	.94125	.99672	.99993
.4	.04	.61606	.18916*	.07609*	.07451*	.10795*	.39160*	.89395
	.05	.91579	.53726	.25416*	.25460*	.37512*	.84118	.99180
	.06	.98610	.84217	.59311	.60765	.75967	.97654	.99928
	.07	.99783	.95827	.85358	.87016	.94027	.99650	.99992
.6	.04	.59747	.18299*	.07403*	.07184*	.10210*	.35773*	.86235
	.05	.91020	.52835	.24819*	.24666*	.35974*	.81991	.98892
	.06	.98500	.83838	.58607	.59775	.74746	.97285	.99903
	.07	.99763	.95728	.85060	.86594	.93673	.99595	.99989

*The largest probability occurs at the 19th data point.

MULTIVARIATE MODELS WITH STRUCTURAL CHANGE 423

Table 8.2
Posterior Probability That m = 10, When the Actual Point of Change Is Ten,
Using a Natural Conjugate Prior Distribution;

$$n = 20, \quad \mu_2 = \mu_1 + \begin{pmatrix} \Delta_1 & 0 \\ \Delta_1 & 0 \end{pmatrix}, \quad \beta_2 = \beta_1 + \begin{pmatrix} \Delta_2 & 0 \\ \Delta_2 & 0 \end{pmatrix}$$

Δ_1	Δ_2	$-.7$	$-.5$	$-.2$	0	$.2$	$.5$	$.7$
0.	.04	.53581	.20797	.08031*	.10368*	.23619*	.59613	.79931
	.05	.75976	.51787	.36097	.42065	.54068	.74536	.91593
	.06	.90374	.70831	.58655	.61133	.68529	.86086	.97162
	.07	.96846	.84290	.72901	.73916	.79954	.93511	.99162
.2	.04	.55664	.21213	.08487*	.11160*	.25076*	.61090	.81380
	.05	.78028	.53525	.37299	.43463	.55492	.75969	.92442
	.06	.91514	.72680	.60236	.62617	.69965	.87180	.97517
	.07	.97288	.85662	.74430	.75340	.81222	.94157	.99282
.4	.04	.57246	.21269	.08727*	.11653*	.25963*	.62255	.82337
	.05	.79476	.54896	.38127	.44509	.56662	.77093	.92921
	.06	.92213	.74148	.61559	.63885	.71182	.87971	.97690
	.07	.97529	.86678	.75696	.76529	.82247	.94590	.99334
.6	.04	.58229	.20952	.08738*	.11809*	.26232*	.63075	.82823
	.05	.80338	.55841	.38542	.45173	.57558	.77902	.93076
	.06	.92537	.75225	.62597	.64914	.72165	.88478	.97712
	.07	.97608	.87363	.76691	.77477	.83032	.94835	.99331

*The largest probability occurs at the 19th data point.

Table 8.3
Posterior Probability That m = 17, When the Actual Point of Change Is Seventeen, Using a Natural Conjugate Prior Distribution;

$$n = 20, \quad \mu_2 = \mu_1 + \begin{pmatrix} \Delta_1 & 0 \\ \Delta_1 & 0 \end{pmatrix}, \quad \beta_2 = \beta_1 + \begin{pmatrix} \Delta_2 & 0 \\ \Delta_2 & 0 \end{pmatrix}$$

					ρ			
Δ_1	Δ_2	-.7	-.5	-.2	0	.2	.5	.7
0.	.04	.99903	.96745	.62674	.36332*	.25127*	.39714*	.90392
	.05	.99994	.99679	.92265	.79327	.72147	.91054	.99715
	.06	1.00000	.99966	.98862	.96538	.95687	.99377	.99990
	.07	1.00000	.99996	.99836	.99505	.99473	.99955	1.00000
.2	.04	.99912	.96993	.63778	.36921*	.25168*	.38634*	.89395
	.05	.99995	.99705	.92619	.79793	.72258	.90703	.99683
	.06	1.00000	.99969	.98921	.96644	.95724	.99352	.99989
	.07	1.00000	.99996	.99845	.99522	.99479	.99953	1.00000
.4	.04	.99910	.97069	.64235	.36969*	.24728*	.36419*	.87132
	.05	.99995	.99711	.92737	.79783	.71701	.89777	.99597
	.06	1.00000	.99969	.98937	.96635	.95597	.99276	.99986
	.07	1.00000	.99996	.99847	.99520	.99462	.99948	.99999
.6	.04	.99897	.96986	.64047	.36473*	.23832*	.33235*	.83168
	.05	.99994	.99699	.92627	.79294	.70465	.88170	.99426
	.06	1.00000	.99968	.98914	.96513	.95295	.99133	.99979
	.07	1.00000	.99996	.99843	.99500	.99420	.99936	.99999

*The largest probability occurs at the 19th data point.

COMMENTS AND CONCLUSIONS

(m = 3, 10, 17) is smallest when $\rho = 0$ and increases as ρ becomes large in absolute value.

These results are what one would expect in the behavior of the posterior distribution of m. As the prior mean of θ_2 changes, the posterior probability of m does not change very much when the other parameters are constant.

Of course this is a very limited study and we see additional work is required before definite conclusions can be drawn. One interesting feature of the study is that when β_2 changes as little as .04 from β_1, it is often detected with a high probability except when ρ is close to zero in absolute value, and for values of ρ close to zero, the change is difficult to detect unless β_2 changes by as much as .07 from β_1.

Another interesting question is why changes are more difficult to detect when $\rho = 0$.

COMMENTS AND CONCLUSIONS

This chapter introduces the reader to some of the basic problems one encounters in the analysis of multivariate data. If the data can be represented by a multivariate linear model, then this chapter shows one how Bayesian inference can be made.

The first model to be considered was the multivariate regression model where prior information is given either by the normal-Wishart conjugate prior or Jeffreys' improper density. The posterior analysis consisted of finding the marginal posterior distribution of the regression matrix and from that showing how one may estimate these parameters either with an HPD region or with point estimators. Forecasting future observations was based on the predictive distribution which was a matrix t.

The multivariate linear model for a one-way layout was also examined and the results are similar to those obtained for the regression model, namely the marginal posterior distribution of the "treatment" mean vectors is a matrix t and inferences about them are obtained from the usual HPD regions.

Two vector time series models were studied, the autoregressive and the regression model with autocorrelated errors, and it was demonstrated that the posterior analysis of the former is a simple generalization of the univariate case which was examined in Chapter 5. Of particular importance with the autoregressive model is the forecasting methodology where inferences about future observations (one-step-ahead) are based on a multivariate t distribution.

The last section of the chapter was on structural change in the multivariate regression model, where one change in the regression matrix is allowed. The main emphasis here was on the change point parameter, however the marginal posterior distributions of the other parameters were obtained. Finally this section is concluded with a numerical study of the sensitivity of the posterior distribution of the change point to the parameters of the prior distribution.

One can conclude that this chapter is for the most part a review of the standard methodology given by Box and Tiao (1973), Press (1982) and Zellner (1971). Some "new" developments were also introduced, namely the Bayesian identification procedure for vector autoregressive processes and the part on structural change in multivariate regression models.

Of course, there are many interesting areas in which additional research can be done. For example, structural change problems in the vector autoregressive model need to be examined, especially the forecasting procedures. The autoregressive model is widely used in economic forecasting and it is often thought that the regression parameters are indeed changing. If this is so, forecasting methods via the Bayesian predictive density should give improved predictions over those which are based on autoregressions with parameters which do not change.

Another promising area of research is the development of a complete Bayesian analysis of vector ARMA models, but as was shown in Chapter 5, one must first solve the univariate case of the ARMA process.

EXERCISES

1. Show the likelihood function may be expressed by equations (8.2), (8.3), and (8.5).
2. Verify the matrix t density. Describe the marginal posterior distribution of θ. See equation (8.13).
3. From equation (8.19) show the predictive distribution of W is a matrix t with mean W^* and covariance matrix $(n + v - m - 1)^{-1}$. $D^{-1} \otimes E$ and explain how you would construct point and region forecasts of W.
4. Describe a way to set the values of the hyperparameters of the conjugate prior distribution of the parameters of a multivariate linear model using past data from a related experiment. See Chapter 3 and the method of the prior predictive distribution for assessing prior information.
5. Consider equation (8.30), which gives a multivariate linear model for a one-way layout k of treatments. A test for the equality of treatment effects is given by the HPD region (8.12). Does this

EXERCISES

test reduce to the one given by (3.103) of Chapter 3, which is the approximate test for the equality of treatment effects when measures a single response on each experimental unit?

6. From equation (8.71), show that:
   (a) the conditional posterior distribution of τ given θ and ρ is gamma,
   (b) the conditional distribution of θ given ρ is a multivariate t, and
   (c) the conditional distribution of ρ given θ is a univariate t.

7. Equation (8.92) is the joint posterior density of λ and α, two parameters of a distributed lag model (8.91). Explain why the joint distribution is not a multivariate t distribution.

8. Consider a multivariate regression model (8.93) with one change, then using the prior density (8.95) and (8.96), show the marginal posterior mass function of m is given by (8.100). Describe how you would forecast a future observation with this model.

9. Equation (8.30) of this chapter describes a linear model with a one-way layout where m responses are measured on the experimental units. Define a multivariate linear model (of m responses) which would be appropriate to analyze data emanating from a randomized block design. That is, generalize the univariate model (3.124) of Chapter 3 to the multivariate case. Also, develop a test for the equality of treatment effects.

10. Consider a moving average process

$$Y(t) = \varepsilon(t) - \phi_1 \varepsilon(t-1), \ t = 1, 2, \ldots, m$$

$$= \varepsilon(t) - \phi_2 \varepsilon(t-1), \ t = m+1, \ldots, n$$

where $1 \leq m \leq n-1$, $\phi_1 \in R$, $\phi_2 \in R$ and the $\varepsilon(t)$, $t = 0, 1, 2, \ldots, n$ are n.i.d. $(0, \tau^{-1})$, where τ is a positive precision and $Y(t)$ is the t-observation. Assuming $\phi_1 \neq \phi_2$, the parameter of the moving average process has changed at some unknown point m from ϕ_1 to ϕ_2. Letting $\varepsilon(0) = 0$, the residuals may be calculated recursively by

$$\varepsilon(t) = Y(t) + \phi_1 \varepsilon(t-1) \ \text{if} \ t = 1, 2, \ldots, m$$

$$= Y(t) + \phi_2 \varepsilon(t-1) \ \text{if} \ t = m+1, \ldots, n$$

where $1 \leq m \leq n-1$. Assuming m can be any one of its possible values with equal probability, describe a way to estimate ϕ_1 and ϕ_2. Hint: See problem 10 of Chapter 5.

REFERENCES

Al Mahmeed, M. (1982). *The Analysis of Autoregressive Processes: The Identification and the Prior, Posterior, and Predictive Analysis*, Ph.D. Dissertation, Oklahoma State University, Stillwater, Oklahoma.

Anderson, T. W. (1951). *An Introduction to Multivariate Statistical Analysis*, John Wiley and Sons, Inc., New York.

Box, G. E. P. and G. C. Tiao (1973). *Bayesian Inferences in Statistical Analysis*, Addison Wesley, Reading, Massachusetts.

Broemeling, L. D. (1982). *The Econometrics of Structural Change*, Special issue of *Journal of Econometrics*, July 1982.

Chow, Gregory C. (1975). "Multiperiod predictions from stochastic difference equations by Bayesian methods," in *Studies in Bayesian Econometrics and Statistics, In Honor of Leonard J. Savage*, edited by Fienberg and Zellner.

DeGroot, Morris H. (1970). *Optimal Statistical Decisions*, McGraw-Hill, New York.

Glass, Gene V., Victor L. Willson, and John M. Gottman (1975). *Design and Analysis of Time Series Experiments*, Colorado Associated University Press.

Granger, C. W. J. and P. Newbold (1977). *Forecasting Economic Time Series*, Academic Press, New York.

Litterman, Robert B. (1980). "A Bayesian procedure for forecasting with vector autoregressions," Massachusetts Institure of Technology.

Moen, David H. (1982). "The Bayesian analysis of structural change in multivariate linear models," Dissertation proposal, Department of Statistics, Oklahoma State University, Stillwater, Oklahoma.

Morrison, Donald F. (1967). *Multivariate Statistical Methods*, McGraw-Hill, New York.

Newbold, Paul (1973). "Bayesian estimation of Box-Jenkins transfer function-noise models," Journal of the Royal Statistical Society, Series B, pp. 323-336.

Press, James S. (1980). "Bayesian inference in MANOVA," in *Handbook of Statistics*, Vol. I, edited by P. B. Krishnaiah, North-Holland Publishing Company, New York.

Press, James S. (1982). *Applied Multivariate Analysis: Using Bayesian and Frequentist Methods of Inference*, Krieger Publishing Co., Melbourne, Florida.

Press, James S. and Kazuo Shigemasu (1982). "Bayesian MANOVA and MANOCOVA under exchangeability," Technical report no. 97, August 1982.

Rothenberg, T. J. (1963). "A Bayesian analysis of simultaneous equations systems," Report 6315, Econometric Institute, Netherlands School of Economics, Rotterdam.

REFERENCES

Salazar, Diego (1980). *The Analysis of Structural Changes in Time Series and Multivariate Linear Models*, Ph.D. dissertation, Oklahoma State University, Stillwater, Oklahoma.

Salazar, Diego, L. D. Broemeling, and A. Chi (1981). "Parameter changes in a regression model with autocorrelated errors," Communications in Statistics, Vol. A10, pp. 1751-1758.

Salazar, Diego (1982). "Structural changes in time series models," in *The Econometrics of Structural Change*, a special issue of the *Journal of Econometrics*, edited by L. D. Broemeling, July 1982.

Sen, A. K. and M. S. Srivastava (1973). "On multivariate tests for detecting change in mean," Sankhya A, Vol. 35, pp. 173-185.

Sen, A. K. and M. S. Srivastava (1975a). "On tests for detecting change in mean," The Annals of Statistics, Vol. 3, pp. 98-108.

Sen, A. K. and M. S. Srivastava (1975b). "On tests for detecting change in mean when variance is unknown," Annals of the Institute of Statistical Mathematics, Vol. 27, pp. 479-486.

Sen, A. K. and M. S. Srivastava (1975c). "Some one-sided tests for change in level," Technometrics, Vol. 17, pp. 61-64.

Tsurumi, H. (1978). "A Bayesian test of a parameter shift in a simultaneous equation with an application to a macro savings function," Economic Studies Quarterly, Vol. 29, pp. 216-230.

Tsurumi, H. (1980). "A Bayesian estimation of structural shifts by gradual switching regressions with an application to the U.S. gasoline market," in *Bayesian Analysis in Econometrics and Statistics: Essays in Honor of Harold Jeffreys*, edited by A. Zellner, North-Holland Publishing Company, New York.

Zellner, Arnold (1971). *An Introduction to Bayesian Inference in Econometrics*, John Wiley and Sons Inc., New York.

9
LOOKING AHEAD

INTRODUCTION

This chapter concludes the book with some material on what is in store for the future of the Bayesian analysis of linear models. But first I will give a brief review of what the book presented. Then, some specific problems will be identified and possible solutions to them will be proposed.

A REVIEW OF THE BAYESIAN ANALYSIS OF LINEAR MODELS

What was accomplished by writing this book? First of all the book contains material which is well known in the literature. The books of Box and Tiao (1973), Press (1982), and Zellner (1971) have laid the foundation for the way a Bayesian would analyze data which were generated by such linear models as those for designed experiments, mixed and random models, some time series models, and some econometric models. This book discusses most of the models which are covered by these three books; however some of the econometric models, as given in Zellner (1971) and some of the multivariate models presented by Press (1982) are not examined.

Second, I have introduced material not found in the standard references. For example, in Chapter 4 on mixed models, the approach is completely different than that given in Box and Tiao (1973). A complete theory of making inferences for the variance components and fixed effects was developed in such a way that all the marginal posterior distributions of these parameters are specified by standard densities for any mixed model. Then in Chapter 5 on time series, the Bayesian analysis of the autoregressive process including identification procedures goes beyond what has appeared in Zellner (1971). A new approach to the analysis of moving average models proved to be only partially successful because only the exact analysis of a first order model was completed.

The material of Chapter 6 over linear dynamic models is not usually presented in statistics textbooks; thus this chapter is essentially "new" material for the statistics graduate student. The Bayesian approach to the study of linear dynamic systems yielded some interesting results beginning with the Kalman filter. Filtering, smoothing, and prediction of the states of the system were explained by finding their joint posterior distribution and by using a normal approximation to the poly-t distribution; Bayesian adaptive filtering and control procedures were devised and illustrated with numerical examples.

A relatively new topic in the statistical literature is that of structural change, the material of Chapter 7. Beginning with sequences of independent normal variables and ending with autoregressive processes, changes in the parameters of these models were studied from the Bayesian viewpoint. The main parameter of interest is the shift point parameter which indexes where the unknown change will occur and its marginal posterior distribution is found. It was always possible to explicitly determine this distribution without using approximations, thus exact small sample inferences were found. Also studied was the distribution of the other parameters of models which have changing parameters.

Multivariate design and regression models were introduced in Chapter 8 and given the same analysis as that of Box and Tiao (1973) and Press (1982). Also studied were multivariate autoregressive models and multivariate regression models (with changing parameters) and the Bayesian analysis of these two types of processes introduced new material to the reader. For example, future observations were forecasted using the Bayesian predictive density and a numerical example computed the marginal posterior distribution of the shift point of a changing bivariate regression model.

The latter part of the book, Chapters 4-8, contains material, most of it new developments in the posterior and predictive analysis of linear models, after the Preface, Introduction, and the first three chapters laid the foundation for Bayesian inferences.

Many of the older books on the theory of linear models begin with a study of the so-called general linear model. For example, Chapter Six of Graybill (1961) develops the classical approach to estimation (least squares and maximum likelihood) and hypothesis testing (the likelihood ratio test) of the linear hypothesis of full rank models. The counterpart in this book is Chapter 1, which introduces the reader to the standard Bayesian approach of the prior, posterior and predictive analyses. Prior information is expressed with a vague improper density or with a proper conjugate density, and the posterior analysis consists of showing that the marginal posterior density of the regression parameters is a multivariate t and that the precision parameter has a gamma posterior density. The predictive density of future observations was shown to be a multivariate t and inferences about the parameters and future observations were made on the basis of the moments or HPD intervals of these distributions. Chapter 1 thus presents those principles which are used throughout the book.

Those models appearing in Chapters 3-9 are introduced in Chapter 2, where each one is discussed in detail. Thus by reading Chapter 2 one is able to see what to expect in the rest of the book. For the most part, this book is mainly concerned with the technical details of the prior, posterior, and predictive analyses of a linear model, but in Chapter 2 some history and philosophy of Bayesian inferences were given. The two controversial aspects of a Bayesian analysis were explained, namely the ideas of subjective probability and prior information. It is hoped the reader will appreciate the idea that the subjective interpretation of a probability distribution of a parameter θ is no more subjective than the frequency interpretation of probability of an observable random variable X. In my view there are no problems with accepting a subjective interpretation of probability, however problems do remain with implementing prior information.

Prior information was implemented with either a vague improper density or with a proper conjugate density, in the former case if one has little or no information and in the latter case if one has a better idea of the values of the parameters. The problem with using proper prior densities is that one must know the values of the hyperparameters, and how they should be chosen. Chapters 2 and 3 advocated the use of the prior predictive density to choose the value of the hyperparameters, where past or future values of the observations are used to fit the hyperparameters. This seems to me to be very reasonable because these methods are based on the sampling model $f(x|\theta)$, $x \in s$, $\theta \in \Omega$, of the experiment and if one believes the model f is true for the data x of the experiment, one should believe what the data (past or future) imply for the prior values of the parameter.

The traditional models of statistical practice, the regression models and the models of designed experiments are special cases of the general linear model, but because they are so important they were extensively studied in Chapter 3. Again most of this material is found in Box and Tiao (1973) and Zellner (1971) but the Bayesian analysis of the models corresponding to completely randomized and randomized block designs is given a different presentation from that usually made.

FUTURE RESEARCH

A review of the book gives one ideas of some interesting problems that remain to be solved. Each chapter was an introduction to a particular class of linear models thus challenging problems are not difficult to identify. With the exception of the general linear model of Chapter 1 much work remains to be done with the other classes of models.

We begin our search for future research problems with the mixed linear model of Chapter 4. The marginal posterior distribution of each variance component was found by averaging the conditional distribution of the component given the vector of random effects with respect to the marginal distribution of the random effects, which is a poly-t distribution. Since the poly-t distribution is quite cumbersome to work with, it was approximated by a multivariate normal distribution and in this way, one could develop computing formulas for the posterior mean and variance of each variance component in the model. The accuracy of the approximation was tested in a limited numerical study of the one-way random model. It was concluded that if one had "precise" prior knowledge of the components and fixed effects the approximation was adequate.

An interesting problem is to devise a better approximation to the poly-t distribution than that provided by the normal distribution, or to find the exact moments of a poly-t random variable.

An obvious challenging problem in regard to mixed models is to extend the analysis of Chapter 4 involving univariate responses to the multivariate case.

Lastly, the conditional posterior distribution of the fixed effects given the random effects is a multivariate t; now what is the exact marginal posterior distribution of the fixed effects? Can the marginal mean of the fixed effects be easily computed without resorting to an approximation?

As I mentioned in Chapter 5, the Bayesian analysis of autoregressive moving average models is in its infancy and as was seen, only a complete Bayesian posterior analysis of the autoregressive model is complete. The exact posterior and predictive analysis of the MA(1) model is possible, but what about the higher order moving average models? Can anything be done?

The likelihood function of moving average processes is very difficult to evaluate over the parameter space because one must numerically invert matrices at each point of the parameter space. Thus the obvious solution to this vexing problem is to find an adequate approximation to the likelihood functions which avoids numerical inversions of matrices and evaluation of determinants.

Often the main purpose of a time series analysis is forecasting future observations and the one-step-ahead predictions were based on the Bayesian predictive density. An open problem for autoregres-

FUTURE RESEARCH											435

sive processes is to characterize exactly the Bayesian predictive density of k future observations.

The Diaz-Farah (1981) method is the Bayesian method of identifying the order of an autoregressive process and it was explained in Chapter 5. It should be compared to the Akaike (1979) way of identification. One problem with comparing a Bayesian method with a non-Bayesian technique is specifying the prior distribution of the Diaz-Farah identification procedure.

One may conclude that once one solves the posterior analysis of moving average models, one will be able to do a Bayesian analysis of the ARMA class of models.

Very little research has been done by statisticians with linear dynamic systems but as was noted in Chapter 6, there exists a large literature on the subject, mostly written by electrical engineers. If one takes a strictly Bayesian approach to filtering, smoothing, prediction, and control one is continuing in the tradition of Kalman (1960) who gave us the mean of the posterior distribution of the successive states of the system, and of course this is now known as the Kalman filter.

Suppose one is interested in filtering the states of the system where the observation and system error precision matrices are unknown but one has prior information about these matrices in the form of Wishart distributions, then it was shown that the marginal distribution of the successive states was a 2/0 poly-t distribution. This implies difficulty in computing estimates of the successive states, thus the poly-t was approximated by the normal distribution and an approximate filter (the mean of the normal distribution) was found, however one should investigate the adequacy of the approximation. The numerical example indicated few difficulties with the approximation but it was such a limited study further work should be done.

Chapter 7 gave a very thorough analysis of linear models with changing parameters but most of the examination was focused on the location parameters of the model. That is only location parameters (means of normal sequences, regression parameters, etc.) were changing. One perhaps should study changing precision parameters of linear models, thus suppose the variance of a normal independent sequence changes once over the period of observation, then what is the posterior distribution of the change point and the parameters of the model? Not very much has appeared in the literature concerning such problems, however an interesting study of switching precision in the errors of autoregressive processes was done by Abraham (1979).

There is a large overlap between the topic of structural change and the topics of slippage and outliers in the statistical and econometric literature. By reading Broemeling (1982) (a special issue on structural change in econometrics) and Barnet and Lewis (1978) one would be able to identify the similarities between structural change problems and problems with outliers. The statistical models are often the same in both areas, thus perhaps some of the solutions to slippage and outlier problems can be found in the literature on structural change and vice versa.

Perhaps the most challenging question to the Bayesian is how does one implement prior information? I have tried to provide a solution to the question by advocating the use of the prior predictive distribution where the hyperparameters of the proper prior density (often a conjugate density) are fit to either past data or to hypothetical data given by the experimenter. The reader should refer to Chapter 2 or Chapter 3 where this idea is explained. I am not happy with this method because it does not provide unique values for the hyperparameters, but I believe it is better than always using a vague improper prior.

One other related approach is the following: Suppose $f(x|\theta)$, $x \in s$, $\theta \in \Omega$ is one's statistical model for the observations X, that is $f(x|\theta)$ is the conditional density of the observations given the parameter θ. If one believes this is a satisfactory model for one's observations and if one has very little prior information (prior to observing X) about θ how should one implement prior knowledge of θ? My answer is as follows: If one believes f is an adequate model for X, then one should know with some confidence what some of the future values of X will be. (How else could you build a model for X?)

Suppose the experimenter provides one with some future values (prior values of X) of X, say $x_1^*, x_2^*, \ldots, x_m^*$, then what do these values imply about the value of the parameter θ? If one believes f is a satisfactory model, if one believes $x_1^*, x_2^*, \ldots, x_m^*$ are reasonable prior values of X, and lastly if one believes the prior information about θ is contained in the likelihood function based on $x_1^*, x_2^*, \ldots, x_m^*$, then to me the function

$$L^*(\theta|x_1^*, x_2^*, \ldots, x_m^*) = \prod_{i=1}^{m} f(x_i^*|\theta), \quad \theta \in \Omega \quad (9.1)$$

contains one's prior information of θ.

If L^* is integrable on Ω, then it is reasonable to use it as one's proper prior density $\S(\theta)$ for a Bayesian analysis. Hence the posterior density of θ is

$$\S(\theta|x_1^*, x_2^*, \ldots, x_n^*) \propto \prod_{1}^{n} f(x_i|\theta) \, L^*(\theta|x_1^*, x_2^*, \ldots, x_m^*),$$

$$\theta \in \Omega \quad (9.2)$$

where $x_1, x_2, \ldots, x_n$ are the observations of the experiment.

Consider the example of sampling from a $n(\theta, \tau^{-1})$ population, then one's model is

$$f(x|\theta,\tau) = \frac{\tau^{1/2} e^{-(\tau/2)(x-\theta)^2}}{\sqrt{2\pi}}, \qquad x \in R, \qquad \theta \in R, \qquad \tau < 0$$

and if one believes this is the model of the observation X and is willing to guess prior values of X as $x_1^*, x_2^*, \ldots, x_m^*$, then one's prior knowledge of θ and τ is given by

$$L^*(\theta,\tau|x_1^*,x_2^*,\ldots,x_m^*) \propto \tau^{m/2} e^{-\frac{\tau}{2}\left[\sum_1^m (x_i^* - \bar{x})^2 + m(\theta - \bar{x})^2\right]}, \quad \theta \in R, \qquad \tau > 0$$

where $\bar{x} = \Sigma_1^m x_i^*/m$. Of course as a function of θ and τ, L^* is a normal gamma density, thus one's prior knowledge of θ and τ is expressed with a conjugate prior density. One may modify the procedure by multiplying L^* by $\S_0(\theta, \tau)$ some function of θ and τ which represents one's prior knowledge of the parameters before guessing prior values of X, perhaps by using Jeffreys' prior density $\S_0(\theta, \tau) = 1/\tau$, $\theta \in R$, $\tau > 0$.

There is one controversial point about using L^* as a prior density, namely why should one treat L^* as a density function in the first place? My answer is that if one believes the prior information about θ is contained in L^* and if one believes prior information about θ can be expressed probabilistically L^* is a reasonable choice.

This method has many advantages because it is based on the sampling model and is justified somewhat by classical principles. Also it gives one a prior density which easily combines with the likelihood function and will be a member of the conjugate class when the class exists. This method does not suffer the disadvantage of the prior predictive method, which does not give a unique prior density, and as is seen, one does not have to fit hyperparameters to the prior values of X with the proposed technique.

One last problem which needs to be addressed is the technical problem of handling the poly-t distribution. See Chapters 3, 4, and 6 and the appendix to the book. As we have seen this distribution frequently occurs in a Bayesian analysis of linear models and Dreze (1977) gives many more examples. Virtually nothing is known about the distribution although Press (1982) and Box and Tiao (1970) have developed approximations. For example, the normalizing constant and thus the moments are unknown, and one must work with it using numerical integration techniques.

If one can either find the exact moments or an adequate approximation to the poly-t distribution, one would be making an important contribution to the Bayesian analysis of linear models.

CONCLUSIONS

We have seen that this book introduces the reader to the way a Bayesian perhaps would analyze data generated by a linear model and that the analysis consists of a prior, a posterior, and a predictive analysis.

In addition to the usual variety of linear models which are introduced in the introductory linear models course, the reader will learn about mixed and random models, ARMA processes, linear dynamic systems, models which incorporate structural change, and some econometric models.

The prior analysis consists of expressing prior information by either a vague improper density or a conjugate density and that if one uses a conjugate density, one may fit the hyperparameters to past data or the prior guesses of future data.

Bayes theorem gives the posterior density of the parameters from which inferences about the parameters are made. If one is interested in testing hypotheses then one may construct a region of highest posterior density for the parameter of interest and if one is interested in estimation, certain moments or other characteristics of the posterior distribution can be computed.

In some problems such as those involving time series one is primarily interested in forecasting and we saw that the Bayesian predictive distribution will provide one with the necessary methodology.

This concluding chapter gave a brief review of the book and identified some interesting problems for future research.

REFERENCES

Akaike, H. (1979). "A Bayesian extension of the minimum AIC procedure of autoregressive model building," Biometrika, Vol. 66, No. 2, pp. 237-242.

Abraham, Bovas and William W. S. Wei (1979). "Inferences in a switching time series," American Statistical Association Proceedings of the Business and Economic Statistics Section, pp. 354-358.

REFERENCES

Barnet, Vic and Toby Lewis (1978). *Outliers in Statistical Data*, John Wiley and Sons, Inc., New York.

Box, G. E. P. and G. C. Tiao (1973). *Bayesian Inference in Statistical Analysis*, Addison-Wesley, Reading, Mass.

Broemeling, Lyle (1982). "Introduction," in *Structural Changes in Econometrics*, pp. 1-5, edited by Lyle Broemeling, Annals of Applied Econometrics 1982-2, A Supplement to the Journal of Econometrics, Vol. 19, No. 1.

Diaz, Joaquin and Jose Farah (1981). "Bayesian identification of autoregressive processes," 22nd NBER-NSF Seminar on Bayesian Inference in Econometrics.

Dreze, Jacque H. (1977). "Bayesian regression analysis using poly-t distributions," in *New Developments in the Application of Bayesian Methods*, edited by Aykac and Brumat, North-Holland Publishing Co., Amsterdam.

Graybill, F. A. (1961). *An Introduction to Linear Statistical Models*, McGraw-Hill, New York.

Kalman, R. E. (1960). "A new approach to linear filtering and prediction problems," Transactions, ASME, Journal of Basic Engineering, Vol. 82-D, pp. 34-45.

Press, James S. (1982). *Applied Multivariate Analysis: Using Bayesian and Frequentist Methods of Inference*, Kreiger Publishing Co, Melbourne, Florida.

Zellner, Arnold (1971). *An Introduction to Bayesian Inference in Econometrics*, John Wiley and Sons Inc., New York.

APPENDIX

INTRODUCTION

The material in the appendix is essential for an understanding of the book. It contains the distributions which are used throughout the text and if one learns them one will be able to easily master the Bayesian analysis of linear models.

Only a few well-known distributions (with the exception of the poly-t distribution) are required for the prior, posterior, and predictive analysis and they are: the normal, the gamma, the normal-gamma, the univariate t, the multivariate normal, the Wishart, the normal-Wishart, the multivariate t, the poly-t distribution and the matrix t.

Each of the distributions is defined by a density function, some of their properties are explained, and then each is related to the relevant topic in the book.

THE UNIVARIATE NORMAL

Let X be a real random variable with probability density freedom

$$f(x|\mu,\tau) = \frac{\tau^{1/2}}{\sqrt{2\pi}} \exp -\frac{\tau}{2}(x-\mu)^2, \quad x \in R$$

thus X is said to have a normal distribution with mean $\mu \in R$ and precision $\tau > 0$ (variance τ^{-1}) and this relationship is denoted by $X \sim N(\mu, \tau^{-1})$.

THE GAMMA

Let Y be a positive random variable with probability density function

$$f(y|\alpha,\beta) = \frac{\Gamma(\alpha)}{\beta^\alpha} y^{\alpha-1} e^{-y\beta}, \quad y > 0,$$

where $\alpha > 0$ and $\beta > 0$, thus Y is said to have a gamma distribution with parameters α and β. This is denoted by $Y \sim G(\alpha,\beta)$ and it can be shown that

$E(Y) = \alpha/\beta$,

$Var(Y) = \alpha/\beta^2$,

$E(Y^{-1}) = \beta/(\alpha - 1), \quad \alpha > 1$

and

$$Var(Y^{-1}) = \beta^2/(\alpha - 1)^2 (\alpha - 2), \quad \alpha > 2.$$

THE NORMAL-GAMMA

Let X be a real random variable and Y a positive random variable, then X and Y are said to have a normal-gamma distribution if the density of X and Y is

$$f(x,y|\mu,\tau,\alpha,\beta) \propto y^{1/2} \exp\left[-\frac{\tau y}{2}(x-\mu)^2\right] y^{\alpha-1} e^{-y\beta},$$

$$x \in R, \quad y > 0,$$

where $\mu \in R$, $\tau > 0$, $\alpha > 0$, and $\beta > 0$.

This is a four-parameter density and as is shown in Chapter 1 is a member of the class which is conjugate to the two-parameter normal family. Thus, the prior and predictive analysis of a two-parameter normal population depends on the normal-gamma distribution.

The marginal distribution of X is a univariate t distribution.

THE MULTIVARIATE NORMAL

THE UNIVARIATE t DISTRIBUTION

A real random variable X is said to have a t distribution with parameters d, μ, and p if the density of X is

$$f(x|d,\mu,p) \propto [1 + p(x-\mu)^2/d]^{-(d+1)/2}, \quad x \in R$$

where $\mu \in R$, $d > 0$, and $p > 0$. It can be shown that

$$E(X) = \mu,$$

and

$$Var(X) = d/(d-2), \quad d > 2.$$

In the prior and posterior analysis of a two-parameter normal population inferences about the population mean are made via the t distribution and predictive inferences about future observations are based on this distribution. See Chapter Two, Chapter Four and Chapter Nine of DeGroot (1970).

THE MULTIVARIATE NORMAL

Let X be a real m dimensional random vector then X has a m-dimensional normal distribution with mean vector

$$E(X) = \mu \in R^p$$

and m × m precision matrix P, which is symmetric and positive definite if the density of X is

$$f(x|\mu,P) \propto |P|^{1/2} \exp -\frac{1}{2}(x-\mu)'P(x-\mu), \quad x \in R.$$

It can be shown the variance-covariance matrix of X is

$$D(X) = P^{-1}.$$

We denote this relationship by $X \sim N_m[\mu, P^{-1}]$.

What is the marginal distribution of $X^{(1)}$ where $X^{(1)}$ is the vector consisting of the first m_1 components, $1 \leq m_1 \leq m$ of X? It can be shown that

$$X^{(1)} \sim N_{m_1}\{\mu^{(1)}, [P_{11} - P_{12}P_{22}^{-1}P_{21}]^{-1}\}$$

where $\mu^{(1)}$ is the vector of the first m_1 components of μ and P is partitioned as

$$P = \begin{pmatrix} P_{11} & P_{12} \\ P_{21} & P_{22} \end{pmatrix},$$

where P_{11} is $m_1 \times m_1$.

It can also be shown that the conditional distribution of $X^{(1)}$ given $X^{(2)}$ is normal, but for the details see DeGroot (1970), Chaptere Five.

THE WISHART DISTRIBUTION

Let V be a random $m \times m$ symmetric matrix with density $f(v|d,T) \propto |T|^{d/2} |v|^{(d-m-1)/2} \exp - (1/2) \text{Tr}(Tv)$ if v is positive definite, otherwise zero, where T is a $m \times m$ positive definite symmetric matrix and $d > m - 1$, then V is said to have a Wishart distribution with d degrees of freedom and precision matrix T. It can be shown the $E(V) = nT^{-1}$, but for additional properties of the Wishart see Press (1982) and DeGroot (1970).

THE NORMAL-WISHART

Let X be a $m \times 1$ real random vector and V a $m \times m$ symmetric random matrix; then X and V have a normal-Wishart distribution if the density of X and V is such that the conditional distribution of X given, V = v is normal with mean $\mu \in R^m$ and precision matrix av, where $a > 0$, and v is $m \times m$ positive definite and the marginal distribution of V is Wishart with d degrees of freedom and precision matrix T where T is symmetric positive definite and $d > m - 1$. The density of X and V is

$$f(x,v|\mu,a,d,T) = f_1(x|v,\mu,a) f_2(v|d,T),$$

where

$$f_1(x|v,\mu,a) \propto |v|^{1/2} \exp -\frac{1}{2}(x-\mu)'v(x-\mu), \quad x \in R^m$$

and

$$f_2(v|d,T) \propto |T|^{d/2}|v|^{(d-m-1)/2} \exp -\frac{\text{Tr}}{2}(Tv),$$

v is positive definite.

We see f_1 is the conditional density of X given V = v and f_2 the marginal density of V.

When one analyzes a multivariate normal population with both parameters unknown, the normal Wishart is the conjugate class. See Chapter 8.

THE MULTIVARIATE t DISTRIBUTION

There are many ways to define multivariate t distributions, see Johnson and Kotz (1972) for example, but for the purposes of this book, only the following t distribution occurs in the Bayesian analysis of linear models. Let X be a m × 1 real random vector, then X has a t distribution with d degrees of freedom, location $\mu \in R^m$ and precision matrix T if the density of X is

$$f(x|d,\mu,T) \propto \left[1 + \frac{(x-\mu)'T(x-\mu)}{d}\right]^{-(d+m)/2}, \quad x \in R^m$$

where d > 0 and T is m × m symmetric positive definite matrix.

First, it can be shown that when m = 1, X has a univariate t distribution as defined earlier and that $(x-\mu)T^{1/2}$ has Student's t distribution with d degrees of freedom. An interesting property of the multivariate t distribution is that if $m \geq 2$, the components of X cannot be independent but they can be uncorrelated since the dispersion matrix of X is

$$D(X) = \frac{d}{d-2} T^{-1}, \quad d > 2.$$

The mean vector of X is $E(X) = \mu$ and the marginal distribution of $X^{(1)}$ is also a t with d degrees of freedom, location $\mu^{(1)}$ and precision matrix $T_{11} - T_{12} T_{22}^{-1} T_{21}$, where

$$T = \begin{pmatrix} T_{11} & T_{12} \\ T_{21} & T_{22} \end{pmatrix}$$

and T_{11} is of order m_1. Here $X^{(1)}$ consists of the first m_1 components of X and μ is partitioned as

$$\mu = \begin{pmatrix} \mu^{(1)} \\ \mu^{(2)} \end{pmatrix}$$

where $\mu^{(1)}$ is $m_1 \times 1$.

The conditional distribution of $X^{(1)}$ given $X^{(2)}$ is also a t where the conditional mean vector of $X^{(1)}$ is

$$E[X^{(1)}|X^{(2)} = x_2] = \mu^{(1)} - T_{11}^{-1} T_{12} [x_2 - \mu^{(2)}],$$

and the conditional precision matrix is

$$\frac{(d + m_2) T_{11}}{d + [x_2 - \mu^{(2)}]'[T_{22} - T_{21} T_{11}^{-1} T_{12}][x_2 - \mu^{(2)}]}.$$

Unlike the normal distribution, the precision matrix of the conditional distribution of $X^{(1)}$ given $X^{(2)}$ depends on the value of the conditioning variable.

The multivariate t distribution is encountered many times in the Bayesian analysis of linear models; see Chapters 1, 2, and 3 in particular.

HPD regions for the regression parameters θ of a general linear model are constructed on the basis of the F distribution, which is a particular function of a t random vector X. That is to say if X has a multivariate t distribution with parameters d, μ, and T then

$$Q(X) = m^{-1} (X - \mu)' T(X - \mu)$$

has an F distribution with m and d degrees of freedom. See Chapter 1.

THE POLY-t DISTRIBUTION

The poly-t distribution is the name of the distribution of a real random vector X of m dimensions which has density

$$f(x|d,\mu,T) \propto \prod_{j=1}^{k} f_j(x|d_j, \mu_j, T_j), \quad x \in R^m$$

where

$$f_j(x|d_j, \mu_j, T_j) = [1 + (x - \mu_j)' T_j (x - \mu_j)]^{-d_j/2}, \quad x \in R^m.$$

The parameters of the distribution are $d = (d_1, d_2, \ldots, d_k)'$

$$\mu = \begin{pmatrix} \mu_1 \\ \mu_2 \\ \cdot \\ \cdot \\ \cdot \\ \mu_k \end{pmatrix}$$

and

$$T = [T_1, T_2, \ldots, T_k],$$

where $d_j > 0$, μ_j is $m \times 1$, and T_j is symmetric and semi-positive definite and $\sum_1^k T_j$ is positive definite. We see if $k = 1$, X has a multivariate t distribution but when $k \geq 2$, very little is known about it. For example, the normalizing constant and other moments are unknown and it is very difficult to work with, but unfortunately it occurs quite often in a Bayesian analysis of linear models. Actually there are two basic forms of the distribution and the one above is called the product form, the other the quotient form. Only the product form appears in this book.

The poly-t distribution first occurs in Chapter 3 when one analyzes several normal populations with a common mean, where the poly-t is the marginal posterior distribution of the common mean. Next it occurs in the Bayesian analysis of mixed linear models, where it is the marginal posterior distribution of the random effects of the model.

Lastly, it is found as the posterior distribution of the states of a dynamic system in Chapter 6. For other examples of this vexing distribution see Dreze (1977), who gives us an excellent account of the problems of working with the poly-t distribution.

If X is a scalar or of dimension less than three one may use numerical integration techniques, but in general one will probably have to develop an adequate approximation.

A normal approximation to the poly-t was used in Chapters 4 and 6 where each factor f_j of the density was approximated by a normal density with the same mean vector and dispersion matrix as f_j. Since the poly-t distribution is in general multimodal and asymmetric, we know the normal distribution is an inadequate approximation, and further research is required.

THE MATRIX t DISTRIBUTION

The matrix t distribution is a generalization of the multivariate t distribution, of which the univariate t is a special case. We have seen that the matrix t distribution occurs in the analysis of multivariate linear models, namely as the posterior distribution of the matrix of regression coefficients and as the one-step-ahead predictive distribution of a future observation.

Press (1982) and Box and Tiao (1973) define the matrix t distribution as follows. The pq elements of the random p × q matrix T follow a matrix t distribution if the density of T is

$$f(t|A,B) = \frac{c(m,p,q)}{|A|^{(m-q)/2} |B|^{p/2}} |A^{-1} + t B^{-1} t'|^{m/2},$$

where t is a real p × q matrix, A is of order p symmetric and positive definite, B is of order q, symmetric and positive definite, and $m > p + q - 1$.

The constant is given by

$$c^{-1}(m,p,q) = \frac{\pi^{pq/2} \Gamma_p\left(\frac{m-p}{2}\right)}{\Gamma_q(m/2)},$$

where

$$\Gamma_p(\lambda) = \pi^{p(p-1)/4} \Gamma(\lambda) \Gamma\left(\lambda - \frac{1}{2}\right) \cdots \Gamma\left(\lambda - \frac{p}{2} + \frac{1}{2}\right),$$

and Γ is the gamma function.

Box and Tiao and Press both give additional properties of the distribution and in particular they show that any row or any column of T has a multivariate t distribution.

COMMENTS

The normal-gamma distribution is a special case of the normal-Wishart and the univariate t a special case of the multivariate t which in turn is a special case of the matrix t distribution. The multivariate t is also a special case of the poly-t (when $k = 1$), but for the most part the poly-t is in a class by itself.

We have seen that with the exception of the poly-t distribution, all are easy to work with in the sense that the moments and other characteristics are well known.

For additional information about these distributions see Box and Tiao (1973), DeGroot (1970), Dreze (1977), Press (1982), and Zellner (1971).

REFERENCES

Box, G. E. P. and G. C. Tiao (1973). *Bayesian Inference in Statistical Analysis,* Addison-Wesley, Reading, Mass.

DeGroot, M. H. (1970). *Optimal Statistical Decisions,* McGraw-Hill Book Company, New York.

Dreze, Jacque H. (1977). "Bayesian regression analysis using poly-t distributions," in *New Developments in the Application of Bayesian Methods,* edited by Aykac and Brumat, North-Holland Publishing Co., Amsterdam.

Johnson, Normal L. and Samuel Kotz (1972). *Distributions In Statistics; Continuous Multivariate Distributions,* John Wiley and Sons, Inc., New York.

Press, James S. (1982). *Applied Multivariate Analysis: Using Bayesian and Frequentist Methods of Inference,* Kreiger Publishing Co., Melbourne, Florida.

Zellner, Arnold (1971). *An Introduction to Bayesian Inference in Econometrics,* John Wiley and Sons, Inc., New York.

INDEX

A

Adaptive control, 278-284, 289-293
Adaptive estimation, 273-278 (*see also* Dynamic models; Estimation)
Adaptive filter, 285-289
Analysis:
  of covariance, 134
  with interaction, 131
  posterior, 53 (*see also* Mixed model)
  predictive, 54
  prior, 47
  of two factors, 123
  of variance, 159
Approximations:
  examples of, 161-165
  to random effects, 149
  to variance components, 153
Autoregressive structural change of, 360-368
Autoregressive univariate process, 182-186
Autoregressive vector process, 388-399

B

Bayes theorem, 2-3

C

Confidence intervals and regions:
  HPD posterior, 54
  HPD predictive, 55
Control problems, 35
Control strategies, 273-293

D

Designs:
  completely randomized, 118
  randomized block, 129
Distributed lag models:
  multivariate, 408-409
  univariate, 213-215
Dynamic models:
  adaptive estimation, 273-278 (*see also* Adaptive estimation)
  control, 243-250
  estimation, 238

[Dynamic models]
  filtering, 238
  Kalman filter, 239, 250-254
  nonlinear, 266-273
  observation equation, 237
  prediction, 242-243
  system equation, 237
  transition matrix, 237

E

Eigenvalues, 188, 196
Eigenvectors, 188, 196
Estimation (*see also* Adaptive estimation; Dynamic models):
  adaptive, 273-278
  least squares, 87
  maximum likelihood, 50, 143
  modal, 170-172

F

Filtering:
  adaptive, 273-278
  Kalman, 239

G

Gamma distribution, 441
General linear model:
  point estimation, 8-11
  posterior estimation, 5-14
  predictive analysis, 14-18
  prior analysis, 3-5
  region estimation, 11-12
  testing hypotheses, 12-14

H

Hypothesis testing:
  by HPD regions, 54
  by posterior odds, 54

I

Identification:
  by Diaz-Farah method, 396-399, 202-206
  of multivariate AR, 396-399
  of univariate AR, 202-206
Interval estimation, 54
Iterative technique, 77, 165, 174-177

M

Marginal posterior density, 53
Mixed model:
  between component, 161-169
  fixed factors, 142
  maximum likelihood, 143
  modal estimation, 165-176
  nested models, 144, 171
  one-way layout,
  posterior analysis, 146-158
  random factors, 142
  variance components, 143
  within component, 161-169
Models:
  additive, 125
  autocorrelated errors, 38
  autoregressive, 25, 36
  autoregressive moving average, 25, 31, 37
  dynamic, 25, 36
  econometric, 25
  general linear, 1
  moving average, 25, 30 (*see also* Moving average)
  multivariate, 374-427
  time series, 25, 30
  traditional, 65-141
Moving average, 186-195 (*see also* Models)
Multiple regression:
  univariate case, 94-100
Multivariate case, 375-383
Multivariate distributions:

INDEX 453

[Multivariate distributions]
   matrix-t, 444
   normal, 442 (*see also* Normal multivariate)
   poly-t, 446
   Wishart, 443
Multivariate models:
   autocorrelated errors, 388-400
   autoregressive, 400-406
   for designed experiments, 383-388
   regression, 375-383
   time series, 388-608
   transfer function, 406-409
Natural conjugate distributions:
   normal-gamma, 441 (*see also* Normal distribution)
   normal-Wishart, 443 (*see also* Normal distribution)

N

Normal distribution, 440
Normal-gamma, 441 (*see also* Natural conjugate distributions)
Normal linear model, 1
Normal multivariate, 442 (*see also* Multivariate distribution)
Normal populations, 70, 81
Normal-Wishart, 443 (*see also* Natural conjugate distributions)
Nonlinear dynamic models, 104-117
Nonlinear regression models, 266-273
Nuisance parameters, 53 (*see also* Parameters)

P

Parameters:
   hyperparameters, 83
   nuisance, 82 (*see also* Nuisance parameters)

[Parameters]
   precision, 1
Posterior joint distribution, 53
Posterior marginal distribution, 53-54
Posterior odds ratio, 54
Precision matrix, 1
Predictive analysis, 43-44
Predictive distribution, 56
Predictive region, 56
Prior distributions:
   conjugate, 47-48, 50
   exchangeable, 52
   Jeffreys, 47
   noninformative, 47
   vague, 47
Principle:
   of imaginary results, 49
   of insufficient reason, 40
   of precise measurement, 48
Probability:
   frequency interpretation, 46
   models of, 45
   subjective interpretation, 42

R

Regression:
   multivariate linear, 375-383
   univariate linear, 84-90
   univariate multiple, 91-103
   univariate nonlinear, 104
Research problems, 433-438

S

Statistical analysis system (SAS), 159
Structural change:
   of autocorrelated errors, 351-360
   detection of, 345-350
   of linear models, 314-345
   of normal sequences, 307-314
   sensitivity to, 322-325
   of time series, 360-368

T

t distribution (*see also* Univariate distributions):
  multivariate, 444
  univariate, 442
Testing hypotheses:
  by HPD regions, 54
  by posterior odds, 54
Time series:
  autocorrelated errors, 206-212
  autoregressive moving average, 195-202
  autoregressive processes, 182-186, 215-231
  distributed lag, 212-215
  identification, 202-206
  invertible, 195
  multivariate, 388-399
  prediction with, 184-185, 191, 201, 210
  stationary, 183-199
  vector processes, 388-399

Transition:
  function, 355-358
  matrix, 237

U

Univariate distributions:
  gamma, 441
  normal, 440
  t, 442 (*see also* t distribution)

V

Vague prior distribution, 47
Variance components, 29, 143, 148, 155

W

Wishart distribution, 443

12003500